Recursion: Complexity in Cognition

STUDIES IN THEORETICAL PSYCHOLINGUISTICS

For further volumes:
http://www.springer.com/series/6555

Tom Roeper • Margaret Speas
Editors

Recursion: Complexity in Cognition

Springer

Editors
Tom Roeper
Margaret Speas
Department of Linguistics
University of Massachusetts
Amherst, MA, USA

ISSN 1873-0043
ISBN 978-3-319-05085-0 (Hardcover) ISBN 978-3-319-05086-7 (eBook)
ISBN 978-3-319-15574-6 (Softcover)
DOI 10.1007/978-3-319-05086-7
Springer Cham Heidelberg New York Dordrecht London

Library of Congress Control Number: 2014939309

Printed on acid-free paper

Springer is part of Springer Science+Business Media (www.springer.com)

Contents

Contents

Contributors

Noam Chomsky Department of Linguistics, MIT, Cambridge, MA, USA

Jill de Villiers Department of Psychology, Smith College, Northampton, MA, USA

Koji Fujita Graduate School of Human and Environmental Studies, Kyoto University, Kyoto, Japan

Wolfram Hinzen Catalan Institute for Advanced Studies and Research (ICREA), Barcelona, Spain

Department of Linguistics, Universitat de Barcelona, Barcelona, Spain

Kathryn Hobbs Department of Psychology, Harvard University, Cambridge, MA, USA

Bart Hollebrandse Dutch Department, University of Groningen, Groningen, The Netherlands

Aravind K. Joshi Department of Computer and Information Science and Institute for Research in Cognitive Science, University of Pennsylvania, Philadelphia, PA, USA

Eva Juarros-Daussà Department of Romance Languages and Literatures, University at Buffalo, Buffalo, NY, USA

Hilda Koopman Department of Linguistics, UCLA, Los Angeles, CA, USA

Manfred Krifka Center for General Linguistics (ZAS), Humbolt University, Berlin, Germany

Peter Ludlow Department of Philosophy, Northwestern University, Evanston, IL, USA

Tom Roeper Department of Linguistics, University of Massachusetts, Amherst, MA, USA

Margaret Speas Department of Linguistics, University of Massachusetts, Amherst, MA, USA

Edward P. Stabler Department of Linguistics, UCLA, Los Angeles, CA, USA

Recursion: Complexity in Cognition

Margaret Speas

Introduction

The hypothesis put forward just over 10 years ago by Hauser, Chomsky and Fitch (HCF) (2002) that the narrow faculty of language "only includes recursion and is the only uniquely human component of the faculty of language" (p. 1569) has engendered a rich research program, challenging the strongest forms of this hypothesis, seeking to discover the formal properties of the narrow faculty of language and exploring the ways in which the narrow language faculty interfaces with the systems governing language use and interpretation. This hypothesis came in the wake of the progression of generative grammar over 40 years, from the theory of the 1960s that posited a complex and domain-specific faculty for language to a theory that explains complex constructions in terms of a bare minimum of computational machinery. The proposal of HCF was made in the context of a comparative overview of the cognitive and communicative systems of humans, other primates and other species.

There have been interesting challenges to both the claim that the language faculty includes nothing but recursion (Everett 2005, 2009, Pinker and Jackendoff 2005, a.o.) and the claim that recursion is uniquely human (Gentner et al. 2006, a.o.) and books exploring the consequences of the hypothesis for linguistic theory (Gärtner and Sauerland 2007) and challenging the hypothesis on empirical and formal grounds (van der Hulst 2010[1]). The papers in the present volume arose from

[1]The papers in van der Hulst 2010 were presented at a conference at the University of Illinois that took place in 2007, and early versions of some of the papers had been circulated before the UMass conference.

M. Speas (✉)
Department of Linguistics, University of Massachusetts, Amherst, MA, USA
e-mail: pspeas@linguist.umass.edu

a conference at the University of Massachusetts, Amherst, held in May 2009, whose goal was to move beyond controversies about HCF's specific claim and foster an interdisciplinary approach to solving the new questions that arose from the claim. To that end, the conference brought together specialists in computational linguistics, syntax, language acquisition, psychology, semantics, biology and philosophy, in a format that allowed for extensive discussion. Three central themes emerged from that discussion, and the papers in this volume reflect these themes. Below we begin with a bit of background. We then elaborate on the themes that emerged from the conference, and conclude by describing more specifically how each paper in this volume contributes to our understanding of the issues.

Background

To set the stage, it is important to understand what HCF meant by the "narrow faculty of language," and what role recursion is claimed to have in language. Human language requires a constellation of diverse cognitive and sensory-motor abilities. The motor control needed to articulate sounds or form manual signs, the ability to comprehend false beliefs, the understanding of symbol-referent relations, the ability to parse strings of acoustic or visual signals into words, the ability to recognize and repeat patterns, the composition of phrase meanings from word meanings, the ability to construct and parse recursive phrase structures: these are distinct abilities which must converge for human language to arise. HCF's review of the literature on studies of related abilities in non-humans reveals that some form of most of these abilities can be found in other species. What other species seem to lack is a specific type of ability to use a recursive procedure to construct phrases from words and sentences from phrases. HCF therefore distinguish the ability to construct natural language phrases and sentences from the other abilities involved in human language, calling the former the "narrow faculty of language (FLN)" and the latter the "broad faculty of language (FLB)." This distinction is sometimes misconstrued in the literature as implying that linguists should study only the narrow faculty. On the contrary, HCF emphasize the importance of studying all of the various abilities contributing to language. Their claim is just that FLN is a property that is not found in the communicative or cognitive systems of any other species.

That property, they claim, crucially involves recursion. Much of the debate surrounding this claim has been hampered by a certain lack of clarity in the definition of recursion. A number of recent papers, including van der Hulst (2010) and several of the papers therein, have brought clarity to this issue, and the contributions to the present volume show convergence on several basic points. van der Hulst's introduction includes a very thoughtful overview of the issues, so here I will just summarize the problem and the points of convergence.

A recursive procedure is one that can apply to its own output. In linguistic discussions this is often illustrated with the structures that result from such a procedure, in which one can see a certain category dominated by another instance of the same category. For example, in (i) the output of the procedure that combines the NP *rain* with the P *of* constitutes part of the input of the procedure that creates the whole NP *buckets of rain*. The resulting structure is an NP that contains another NP.

(i) $[_{\mathbf{NP}}$ buckets $[_{PP}$ of $[_{\mathbf{NP}}$ rain]]]

But as van der Hulst (2010) points out, this XP-in-XP property, which he calls "specific recursion," is not a necessary result of a recursive procedure, nor does the existence of such structures entail a recursive procedure. First of all, in a language where each lexical item had a distinct part of speech, sentences formed through a recursive procedure might never have a particular category dominating another instance of the same category. Second, one might learn a finite set of constructions that happen to have a category X contained within another instance of category X without having learned a recursive procedure for generating them (as HCF suggest is the case for primates that have learned to "count" to a certain number). Finally, in Chomsky's current minimalist framework, the fact that the objects constructed by MERGE are sent off to the interface levels at several points in the derivation (known as "phases") means that the syntactic structure that would manifest specific recursion may have been erased or rendered invisible by the time a higher instance of the same category is merged. For example, within certain interpretations of phase theory, a CP will never contain another CP within it, since the lower CP structure was erased when the vP in which it is embedded was sent off to spellout:

(i) derivation of "... that Pat thinks that Mary has a cat"
a. $[_{CP}$ that Mary has a cat]
b. MERGE *thinks* : $[_{VP}$ thinks $[_{CP}$ that Mary has a cat]]
c.: spellout: $[_{VP}$ thinks that Mary has a cat]
d. MERGE *Pat*: $[_{IP}$ Pat $[_{VP}$ thinks that Mary has a cat]]
e. MERGE *that*: $[_{CP}$ that $[_{IP}$ Pat $[_{VP}$ thinks that Mary has a cat]]

So, in such a theory sentences are built by means of a recursive procedure, and the resulting representations are hierarchical, but in many cases[2] the representation does not include a phrasal category containing another instance of the same category. It is also important to point out that the existence of phases should not need to be stipulated as an additional aspect of language. Cedric Boeckx discussed this issue in a presentation at the UMass conference, and explained how the existence of phases can arise from general principles. He chose not to include his paper in this volume, but much of the material can be found in Boeckx (2012).

In short, the central claim of HCF is not that the language faculty involves embedding some specific XP within another XP. Rather, the claim is simply that

[2]Possibly in all cases, except for instances of adjunction, and perhaps in DPs and/or PPs, if these are not phases.

the only computational procedure for language is MERGE, which combines two linguistic objects into a labeled linguistic object (to which MERGE can then apply). This point is particularly important for research seeking analogs of recursion in other species. A species may be able to comprehend or produce structures with a category embedded within another instance of the same category but lack the human language ability to generate phrases by means of a recursive procedure. Similarly, a very young child might be able to produce and comprehend strings such as "the story about cats" before having acquired the recursive procedure for generating an infinite set of language structures.[3]

Emerging Themes

The papers in this volume come from diverse perspectives and do not always agree, but there are several interesting themes that emerge from the papers, which outline an exciting set of research questions for the near future. Let's assume that human language phrases are built by a recursive procedure. What else do we have to say in order to have an explanatorily adequate theory of natural language?

Direct vs. Indirect Recursion and Labeling

In the current Minimalist Framework of Linguistics, a working hypothesis is that the only procedure needed to generate linguistic structures is the very simple operation Merge, which combines two words or phrases into a larger word or phrase. Merge is recursive because it can apply to its own output. Several papers in this volume point out the importance of distinguishing (at least) two types of recursive structures: direct and indirect recursion. Direct recursion takes as the input to MERGE two items of the same category. For example, a noun might combine with an NP to form a larger NP.

(ii) Direct recursion: $XP \rightarrow X\ (XP)$

Since MERGE is not sensitive to the category of its input, we should find ample instances of this kind of recursion. Arsenović and Hinzen (2012) as well as Hinzen and Juarrez-Daussà in this volume point out that there are no languages with only this kind of direct recursion, and in fact there are no languages in which this kind of direct recursion regularly occurs. More often, syntactic structures require

[3]Pullum and Scholz (2010) point out that it has never been empirically established that the set of possible expressions in any given language is infinite, and Langendoen (2010) proposes a way that this could be determined.

an intervening category. It is not clear why this is the case. Hinzen explores the consequences of this observation for the semantics-syntax interface, and Juarrez-Daussà explores its consequences for argument structure.

Related to the issue of direct vs. indirect recursion is the fact that sentence embedding seems to have a special status among recursive structures. As Chomsky and Roeper and Hollebrandse note, much of the controversy surrounding Everett's (2005, 2009) claims that Piraha lacks recursion is actually about the possible lack of sentential embedding. Since some research, such as de Villiers' paper in this volume, suggests a strong link between sentence embedding and the development and representation of the beliefs of others, the lack of sentential embedding in Piraha could be construed as an obstacle to the development of theory of mind. However, Everett argues against such conclusions. His view is that "... where the recursion is located in our language – in our stories, in our sentences, or both – has very little to do with our ability to reason recursively." (2012: 281) Although the goal of the UMass conference was to move beyond arguing over Everett's empirical claims, he was a valuable participant at the conference. He presented material which can be found in Chapter 11 of Everett (2012).

As he and others have made clear, as long as sentences are constructed at all there is Merge and the possibility of at least direct recursion. If Merge is the only procedure for natural language structures, we should not be surprised to find languages whose sentences are built via recursive MERGE but which happen to lack verbs that take sentential complements.

Constraints on Recursion: Functional Categories and the Lexicon

If nothing need be said about human language beyond the presence of recursion, we should find languages with lexical heads but no functional heads or no grammar-related features. Current research points strongly toward the view that all languages have functional heads. Although this point continues to be controversial, particularly for languages with extremely rich inflectional morphology and relative free word order, something beyond recursion needs to be said to account for the widespread presence of functional heads in at least some types of languages. To our knowledge, no analog of functional morphemes has been found in the communication system of any other species. Koopman's paper shows the centrality of such features in explaining restrictions on certain types of recursion. Juarros-Daussà suggests that restrictions on recursion are a characteristic property of the projections of lexical heads, and points out that it is surprising that the presence of functional heads does not override the restrictions on argument structure imposed at the lexical level. One possibility that emerges from both of these papers is that lexical restrictions are responsible for all cases where recursion is not completely free.

Conditions Imposed by the Interfaces

As several authors point out, many naturally occurring systems show discrete infinity that can be modeled using recursive Merge (numbers, the spirals in a shell, etc.). Perhaps what is unique about human language is that the objects that result from Merge interface with both the Sound/Meaning system and the Conceptual/Intensional system. Joshi and Stabler both draw attention to the equivalence of various ways of formalizing the core computational properties of language; what is necessary for an explanatory theory of language is an understanding of how relatively simple yet restricted structures map to these other cognitive systems.

As Krifka, Ludlow and Hinzen make clear, a prerequisite for this task is a concrete model of the other cognitive systems. Although we are far from having such models, these three papers show how studying discrepancies between compositional semantics and syntactic structures can lead to new insights about general cognition as well as about language.

The Papers in This Volume

Computational Questions

Joshi

Joshi explores a property termed "Indirectly Constrained Recursion," suggesting that beyond two levels of embedding, there are certain forms that cannot be generated. Indirectly Constrained Recursion is exemplified by three types of constructions:

- Constructions that have undergone "root transformations." These resist further embedding. For example, CPs that have undergone topicalization cannot be embedded within another CP: **Mary thinks that recursion, there are no restrictions on.*
- Relative clauses that have undergone extraposition. These resist further embedding. For example: **The woman said that language is recursive who my brother likes.*
- Complements in verb-final languages that have undergone scrambling. Examples come from SOV languages such as German.

Joshi's aim is to constrain recursion without limiting the potential infinity of language. He does so within a model of Tree Adjoining Grammar.

Stabler

Stabler's goal is to resolve a longstanding tension between the complexity of structures posited within linguistic theory and the much simpler structures that computer scientists claim are necessary for language processing. Contrary to what is assumed by some computer scientists, the simpler structures are not adequate to capture natural language grammar. Stabler observes that despite the apparent differences between various approaches to language computation, there is a significant consensus that linguistic structures are recursive and mildly context-sensitive, which may seem to conflict with the need of the processor to keep structures maximally flat.

Stabler shows that the presence of recursion need not conflict with the mechanisms of language processing if we are careful to distinguish the mechanisms for building structure from their implementation. The key idea is that the language processor recognizes similarities among linguistic constituents and operations. By making use of categorial frequency, lexical features, left edge/discourse effects and modular ranking of potential parses, processing can be efficient without a loss of syntactic complexity.

Syntactic Questions

Chomsky

As discussed above, Chomsky hypothesizes that the narrow language faculty (i.e. the generative procedures that allow us to construct sentences) arise from a very simple recursive procedure that merges two linguistic objects into a larger one. In his paper in this volume, he reviews the basic concepts behind the current "biolinguistic" framework, emphasizing the importance of studying language use along with language structure, while reiterating the crucial distinction between the two. He describes the basic task of the linguist as a reformulation of Otto Jespersen's program: "to investigate the true nature of the interfaces and the generative procedures that relate them, and to determine how they arise in the mind and are used."

The biolinguistic framework enjoins linguists to engage biologists and others in interdisciplinary research aimed at discovering what distinguishes the cognitive and communication systems of each species. In early Generative/Transformational grammar, the rules of grammar comprising the human language faculty were so intricate and language-specific that it was hard to see how they could possibly have analogs in other cognitive systems. Research over the past 50 years has gradually discovered ways in which the rules can be simplified, with the intricacies of grammar following from the interactions between a very simple computational component and the sensory-motor and conceptual/intensional systems. This "Minimalist Program" for linguistic research opens exciting new areas of interdisciplinary inquiry. In this paper, Chomsky discusses how minimalist assumptions might play out in specifying the mapping between minimal structures and the interfaces of language with the Sensory-Motor faculties and the Conceptual-Intensional faculties. He explores the following hypothesis, which he calls *Hypothesis T*:

(T) Language is optimized relative to the Conceptual/Intensional interface alone, with "externalization" a secondary phenomenon.

Chomsky's aim is not to argue conclusively for hypothesis T; rather, he is exploring precisely what would have to be explained in order to maintain this strong hypothesis. If hypothesis T holds, all constructions with properties that go beyond simple merger of words into phrases must be explained in terms of "externalization." Chomsky considers cases such as embedded control structures, ECP violations, reconstruction effects and structure-dependence, suggesting avenues that might be pursued to maintain Hypothesis T.

One may ask, then, what the "secondary phenomenon" of externalization looks like. If preserving Hypothesis T requires ad hoc "externalization" rules that simply recapitulate rules of control, binding, A' movement, etc., then obviously this does not constitute progress. This means that linguists working within a minimalist framework need to take seriously the interdisciplinary nature of the biolinguistic research program.

Chomsky proceeds to discuss the fact that human language seems to have evolved within a very small time window, which suggests that language evolution involved some slight rewiring of the brain that was adaptive for reasons perhaps unrelated to communication. He points out that this mutation may have taken place many times and discusses the conditions that would have to hold in order for human language to actually arise.

He then focuses on the particular properties of the narrow language faculty. In previous work he had claimed that External Merge, i.e. the merger of an item from the lexicon with the linguistic structure, is somehow more natural or 'perfect' than Internal Merge (i.e., merger of a "moved" constituent to a higher position). He notes interesting questions about the (imperfect) correlation of External Merge with argument structure and Internal Merge with discourse-related phenomena. He concludes by raising questions about the validity of the assumption that language structures are labeled and/or endocentric.

Koopman

Koopman reaches into the refined technical properties of grammar to show constraints on recursion, which in turn demonstrate the kind of tight fit between the formalism of generative grammar and empirically motivated constraints precisely in the domain of recursion. Completely free applications of recursion result in considerable overgeneration. Exploring and explaining constraints on recursion is thus central to the Minimalist program.

Koopman looks at particular cases where recursion is blocked in Dutch and German. Dutch verbal complexes are particularly interesting because their derivation yields a recursive syntactic output, but this recursion is restricted by +PH properties of particular lexical items. Dutch is generally head-final, and when an infinitival verb occurs embedded under just one verb, the higher verb follows the lower: ([[zwemmen] wil] 'wants to swim.' However, when there are more than two verbs, the head-final order is no longer possible. Instead, the order is head-initial: [zal [willen [zwemmen]]]. Significantly, it is only infinitival verbs that disallow recursion to their left.

Koopman argues that these are cases in which a lexical item imposes restrictions on the phonological spellout of its specifier. Interestingly, this conclusion is the same one reached by Juarros-Daussá based on properties of argument structure. Koopman's proposal is founded on assumptions about how merger of affixes works in a "single engine" theory (i.e., a theory that has no operations other than "Merge"). An affix merges as an independent item. The morpheme it affixes to then merges as the affix's specifier. This means that restrictions on the domain of affixation are restrictions on the affix's spec, which Koopman includes under a generalized extended projection principle. A bound morpheme is one whose specifier must be phonologically realized. Thus, the extended projection principle is not an idiosyncratic property of Subjects and/or Tense/Agreement; rather, it is a general property of all bound morphemes that they may impose restrictions on the phonological realization of their specifier. Koopman's proposal allows us to retain the Minimalist assumptions about Merge, and attributes constraints on recursion to systematic properties of individual lexical items.

Juarrez-Daussà

Juarrez-Daussà explores the problem of why recursion within lexical items is so limited. Although principles of affixation are recursive (nationalizationalization…), argument structures are strictly limited: As Hale and Keyser pointed out, there are intransitive, transitive and ditransitive verbs, but no language has verbs with three obligatory internal arguments plus an external argument. Verbs can have meanings that involve multiple thematic roles, so this cannot be a restriction on concepts. For example, the verb *send* can be used in expressing an action with a goal, theme and beneficiary, but only two of these can be internal arguments. (*We sent Mary a letter for John; We sent a letter to John for Mary; We sent Mary a letter for John; *We sent Mary a letter John.)* Explaining this restriction requires that Juarrez-Daussà establish clear distinctions between obligatory and non-obligatory arguments in a lexical representation. Using data from English, Spanish and Catalan, she shows that argument structure representations uniformly adhere to what she calls the "Two Argument Restriction (TAR)." She argues that this restriction follows directly from a specific prohibition against recursion within argument structure. Interestingly, the restriction only applies to argument structure, not to other possible instances of recursion within sub-lexical representations.

Interfaces, Acquisition and Evolution

Krifka

Krifka suggests that since semantics is compositional and syntax is recursive, semantics must also be recursive. Like many other authors, he observes that every category is recursive unless we impose restrictions, and takes up the question of how restrictions on recursion are to be explained. Krifka examines the status of one of the problematic constructions discussed by Joshi: root clauses and the restrictions on their embedding. He argues that the restrictions on embedding root clauses is due to semantics: Root clauses are independent assertions. As Hooper and Thompson (1973) have pointed out, root clauses that do embed are those that serve as the main assertion of the sentence. Krifka fleshes out this idea, with a model-theoretic reconstruction of speech acts that captures why speech acts generally cannot be embedded and explains why they sometimes can. In his theory, speech acts are events that trigger a change in the commitments of the speech act participants. Those cases where speech acts can be embedded share the property that the embedding verb does not itself introduce a new speech act. Thus, in Krifka's view speech acts are in principle different from propositions. While both compositional semantics and syntax are recursive, they compose different categories. Just as the syntactic features of certain categories may impose constraints on syntactic recursion, semantic and pragmatic properties may impose constraints on semantic recursion.

Ludlow

Ludlow addresses questions about the nature of the interface with the Conceptual/Intensional level. The minimalist system highlights the importance of the interfaces, but does not give much guidance as to their nature. Moreover, as Ludlow notes, a theory of the interface with the C/I system requires a concrete theory of the C/I system itself. He offers a family of proposals for a concrete theory of the C/I interface, in which semantic theory is in fact a theory of the interface. He argues that this task requires bridging the divide between so-called "use" theories of meaning and truth-conditional models. In particular, he outlines a strategy for constructing a formal use theory, using the tools of truth-conditional semantics along with insights from recent work on expressives. In the resulting framework, semantics is a "calculus of expressive attitudes." This theory predicts that second order attitudes are a precondition for the legibility of recursive structures, which explains the correlation between the emergence of recursion and false belief tests as well constraints on certain kinds of recursion within the complements of attitude predicates.

deVilliers, Hobbs and Hollebrandse

Like Ludlow and Krifka, deVillers et al. point out the close connection between complement recursion and propositional attitudes. They ask whether recursion might actually help the child acquire conceptual understanding of propositional attitudes and false beliefs. Noting that neither images nor external objects are adequate representations for attitude concepts, they adopt the hypothesis that sentences are in fact the appropriate medium. This assumption is supported by the extensive evidence that children necessarily acquire sentential complementation prior to acquiring the understanding of false beliefs. deVilliers et al. report on an ongoing set of language acquisition experiments that suggest that it is false complements rather than recursion per se that act as the ultimate trigger for the understanding of false beliefs. Truth contrasts indicate that certain clauses are complements rather than adjuncts, so truth contrasts enable children to recognize recursion more easily. This gives the child a way to represent false belief statements, which in turn leads to the understanding of false beliefs.

Hinzen

Hinzen points out that any system that displays discrete infinity can be modeled using recursive Merge, so something else is needed to capture the unique properties of human language. Hinzen's view contrasts in interesting ways with Krifka's, in that Hinzen proposes building the semantics in parallel to syntax, using Phase theory as a fulcrum. He begins with the observation that linguistic structures rarely allow direct embedding of a category within the same category ([X [X]]), and when such structures are found, such as in serial verb constructions or the argument structures investigated by Juarrez-Daussà, the embedded XP behaves differently from the embedding one. Hinzen claims that in order to understand the role of recursion in natural language, we need to understand the role of the cycle, or phase. He develops a model in which compositional semantics is not strictly a matter of composing the meaning of a sentence from the meaning of lexical terminals. Rather, at each point in the derivation where a phase is completed and sent to the CI level for interpretation, an intensional representation is created of the phasal head. Thus, the unique characteristic of natural language is not iterated Merge or category labels. Rather, it is the creation at each phase edge of a representation in which the individual parts put together by Marge are no longer visible, but map to an intensional representation of the meaning of the phasal head. This intensional representation, having been simplified, can then be embedded within a larger structure.

Hollebrandse and Roeper

Hollebrandse and Roeper use child language as a probe into the place of recursion in the grammar. Languages show variation in allowing recursion in (at least): adjectives, possessives, compounds, serial verbs, and sentential complements. Children acquire different types of recursion at different stages, and allow recursion in some cases where adults do not. They point out that Merge alone in the input isn't sufficient to trigger recursion, and ask whether children assume recursion as the default.

Hollebrandse and Roeper draw attention to the crucial distinction between direct recursion, where a category X merges directly with another instance of the same category (X ➔ X and X) (as discussed by Hinzen and Juarros-Daussa) and indirect recursion, where X merges with a phrase that *contains* another category X within it. (XP ➔ X ZP, ZP ➔ Z XP). Interestingly, despite the limited distribution of direct recursion in natural language, Hollebrandse and Roeper find that children resist indirect recursion, and that direct recursion, with a conjunctive interpretation, operates as either a first representation or a Default representation. They add that interaction with other kinds of syntactic constraints, such as cross-over principles (Rizzi), may be the primary factor in determining the acquisition path.

Fujita

Fujita takes up Hauser, Chomsky and Fitch's call to compare the human language capacity with language-independent human and nonhuman capacities. Focusing on recent claims that labeling is in fact the central innovation of human language, Fujita argues that labeling is in fact a form of Merge, and that Merge (including labeling and Move) developed from the cognitive capacity for concrete object combination. This evolutionary perspective interacts with recent discussion about exactly the status of labeling as a primitive of grammar or as a reflection of other cognitive capacities.

Acknowledgement We are grateful to the National Science Foundation, which funded the 2009 conference at UMass (grant #BCS 0834032).

References

Arsenović, B., & Hinzen, W. (2012). On the absence of X-within-X recursion in human grammar. *Linguistic Inquiry, 23*(3), 423–440.

Boeckx, C. (2012). Phases beyond explanatory adequacy. In A. Gallego (Ed.), *Phases: Developing the framework* (pp. 45–66). Berlin: Mouton de Gruyter.

Gärtner, H.-M., & Sauerland, U. (Eds.). (2007). *Interfaces + recursion = language? Chomsky's minimalism and the view from syntax-semantics*. Berlin: Mouton de Gruyter.

Gentner, T. Q., Fenn, M., Margoliash, D., & Nusbaum, H. C. (2006). Recursive syntactic pattern learning by songbirds. *Nature, 440*, 1204–1207.
Everett, D. (2005). Cultural constraints on grammar and cognition in Pirahã: Another look at the design features of human language. *Current Anthropology, 76*(4), 621–646.
Everett, D. (2009). Pirahã culture and grammar: A reply to some criticisms. *Language, 85*(2), 405–422.
Everett, D. (2012). *Language: The cultural tool*. New York: Pantheon Books.
Hauser, M., Chomsky, N., & Fitch, W. T. (2002). The language faculty: What is it, who has it, and how did it evolve? *Science, 298*, 1569–1579.
Hooper, J., & Thompson, S. (1973). On the applicability of root transformations. *Linguistic Inquiry, 4*, 465–497.
Langendoen, D. T. (2010). Just how big are natural languages? In H. van der Hulst (Ed.), *Recursion and human language* (pp. 139–146). Berlin: Mouton de Gruyter.
Pinker, S., & Jackendoff, R. (2005). The faculty of language: What's special about it? *Cognition, 95*, 201–236.
Pullum, G. K., & Scholz, B. C. (2010). Recursion and the infinitude claim. In H. van der Hulst (Ed.), *Recursion in human language* (pp. 113–138). Berlin: Mouton de Gruyter.

Minimal Recursion: Exploring the Prospects

Noam Chomsky

Abstract The most basic property of human language is that in each of its variants, it provides an unbounded array of hierarchically structured expressions that receive interpretations at two interfaces, sensorymotor for externalization and conceptual-intentional for mental processes. A central problem, then, is to determine the nature of the recursive procedures that have this property. General scientific norms direct inquiry to the simplest possible solution – minimal recursion – to deepen explanatory power and expedite the study of language acquisition, and to offer some hope for some eventual insight into evolution of the language capacity. Pursuit of this objective has led to elimination of many unwanted stipulations and also grounding some fundamental properties of universal grammar. Among other consequences, it provides evidence for an asymmetry of the interfaces, with externalization an ancillary procedure, a conclusion consistent with well-known properties of language and the very limited evidence about evolution.

Keywords Recursion • Interface • Merge • Order • Efficiency • UG • Minimalist program

For our purposes, we can think of recursion as enumeration of a set of discrete objects by a computable finitary procedure, one that can be programmed for an ordinary digital computer that has access to unlimited memory and time.[1] Taking

[1]Adopting what Soare (1996) calls the Recursion Convention, as is common practice, though in his careful conceptual and historical review Soare recommends replacing "recursive" by "computable" when the general framework is Turing computability, the usual case.

N. Chomsky (✉)
Department of Linguistics, MIT, Cambridge, MA, USA
e-mail: chomsky@mit.edu

T. Roeper and M. Speas (eds.), *Recursion: Complexity in Cognition*, Studies in Theoretical Psycholinguistics 43, DOI 10.1007/978-3-319-05086-7_1,

the recursive procedure P to be a function on the integers, its range $R = \{P(n)\}$, the set of objects enumerated by P. In the interesting cases, R is infinite, but it could be finite (or null). We are concerned with a special case of recursive procedures, generative grammars G_i, each of which enumerates a set of hierarchically structured expressions, assigning to each a symbolic representation at two interfaces, the sensorimotor interface SM for external realization ER and the conceptual-intentional interface CI for what is loosely termed *thought*: interpreting experience, reflection, inference, planning, imagining, etc. In this respect each G_i can be regarded as an instantiation of the traditional Aristotelian conception of language as sound with meaning (though sound is now known to be only a special case of ER).

The fact that the ER-meaning association is in principle unbounded has occasionally (but not often) been recognized explicitly in the long and rich history of study of language. Darwin wrote that "man differs [from animals] solely in his almost infinitely larger power of associating the most diversified sound and ideas," that is, in having a generative grammar G; "almost infinite" is a traditional phrase to be interpreted ambiguously as "very many" or "infinite." The unbounded character of the association is the basis for the "marvelous invention" that so entranced Galileo, the alphabet, which, in the words of the Port-Royal philosophers, provides the means to construct "from 25 or 30 sounds that infinity of expressions [that] enable us to reveal everything that we think, and all the movements of our soul." Recognition of the unbounded character of the language itself, and the deeper concern for the creative character of its normal use, was soon to become a core element of Cartesian science/philosophy. A century ago Otto Jespersen raised the question of how the elements of language "come into existence in the mind of a speaker" on the basis of finite experience, yielding a "notion of structure" that is "definite enough to guide him in framing sentences of his own," crucially "free expressions" that are typically new to speaker and hearer, over an unbounded range. The task of the linguist, then, is to discover these mechanisms and how they arise in the mind, and to go beyond to unearth "the great principles underlying the grammars of all languages" and by so doing to gain "a deeper insight into the innermost nature of human language and of human thought" – ideas that sound much less strange today than they did during the structuralist/behavioral science era that came to dominate much of the field, marginalizing Jespersen's insights (Falk 1992).

Reformulating Jespersen's program, the basic task is to investigate the true nature of the interfaces and the generative procedures that relate them, and to determine how they arise in the mind and are used, the focus of concern naturally being on "free expressions"; along with further questions about neural representation, evolution, and much else.

Returning to the Aristotelian formula "sound with meaning," from classical times there has been extensive study of sound in a broad sense of the term, including the entire process of *externalization* that converts the generated structured expressions to ER: phonology, morphology, prosody, etc.; and also study of the much more obscure topic of meaning. But the concept *with* has largely been ignored, partly because the tools to investigate it carefully were not readily available, partly because the matter was regarded as straightforward, just "the use of words." A common view during the cognitive revolution of the seventeenth to eighteenth century was that the order of words followed the order of thought. Hence there isn't much to say about the notion *with* beyond variations of order and grouping of words into larger units. We can now interpret *with* in terms of recursion, and as we now know all too well, it is by no means an easy complex to unravel.

The modern sciences have largely adopted Galileo's methodological guideline that nature is simple, keeping to the "easiest and simplest rules," as he put it: the task of the scientist is to try to establish the principle, despite the heterogeneity and diversity of phenomena. In our special case, the task is to determine how closely language approximates minimal recursion, within the boundary conditions set by interface satisfaction. That is why a recent collection of essays has the title *Interfaces + Recursion = Language* – followed by a big?, because there is so much that is not understood.[2]

Sometimes such efforts have been considered hopeless, or perhaps completely pointless, not just for language. Not many years ago a prominent molecular biologist could plausibly conclude that the variability of organisms is so free as to constitute "a near infinitude of particulars which have to be sorted out case by case" – "near infinitude" with the usual ambiguity. In much the same vein, Martin Joos (1957) reviewed the prevailing "Boasian tradition" that languages can "differ from each other without limit and in unpredictable ways" so that each language must be approached "without any preexistent scheme of what a language must be," without preconceptions. That of course could not be literally true, but it seemed at the time a reasonable general point of view. Adopting it, linguistic theory consists of procedures of analysis of data, basically segmentation and classification. In general biology, the picture now looks radically different; in linguistics too I think, though more controversially.

From its modern revival, generative grammar largely developed within what came to be called "the biolinguistic framework," which takes the faculty of language to be a component of the individual mind/brain, much like the modules of visual perception, organization of motor action, etc. The move from analytic procedures to the biolinguistic framework greatly enriches the variety of evidence that bears on the study of each individual language to include acquisition, neuroscience, dissociations, and much else, and also what is learned from the study of other languages, on the assumption that the capacity for language relies on shared biological properties. However, it took some time to make a clear transition from the prevailing procedural framework of structural/behavioral linguistics to the biolinguistic framework – in particular, to distinguish clearly between application of procedures of analysis to organize a corpus of data, on the one hand, and on the other, inquiry into language acquisition, the cognitive processes by which the language acquired arises from the data in which the child is immersed.[3]

---

[2]Sauerland and Gärtner (2007).

[3]One illustration, which has become influential in recent years (in rather misleading ways, I think), is "chunking" of running text into word-like elements. In my *Logical Structure of Linguistic Theory*

The extensive and very revealing studies of acquisition of language, particularly in the past 30 years, have shown that the process is no trivial matter. That is true even for the first step, not at all well understood: the extraction of language-relevant data from the "blooming buzzing confusion." There is recent evidence that the auditory system of chimpanzees interprets speech much as humans do.[4] Insofar as this is true, internal computational mechanisms must be substantially responsible even for the first step in language acquisition, as well as for later steps that lead to mature knowledge.

It seems to me well established (though much contested) that language "represents a domain-specific mental faculty, one that rests on structural organizing principles and constraints not shared in large part by other mental faculties, and in its processing and computation is automatic and mandatory," the formulation by Susan Curtiss (2013), who has done pioneering work in establishing these conclusions on the basis of dissociations and other evidence. Note that it is the processing and computation that are automatic and mandatory. Use of language extends far beyond these limits. There is no reason today to doubt the fundamental insight of Descartes that use of language has a creative character: it is typically innovative without bounds, appropriate to circumstances but not caused by them – or as far as is known by internal states – and can engender thoughts in others that they recognize they could have expressed themselves. We should bear in mind as well that von Humboldt's now oft-quoted aphorism that language involves infinite use of finite means refers to *use*. There has been great progress in understanding the finite means that make possible infinite use, but the latter remains largely a mystery.

Within the biolinguistic framework, the core concept of language is I-language – "I" signifying individual, internal, and intensional, in that we are concerned with the actual computational procedure, not just its range; other notions of language are derivative, involving considerations of varied kinds.[5] We can take an I-language to be nothing other than the computational procedure P that yields the set of generated expressions, including the paired interface representations.

LSLT (1955), I took for granted that this is done by statistical analysis of transitional probabilities, modifying a proposal of Zellig Harris's for detection of morphemes (not the right units, however, because morphemes are too abstract, lacking the required sequential properties; but a necessary assumption given the principled procedural ban on level-mixing). There is no reason to suppose that such analytic procedures relate at all closely to how a child analyzes discourse, making use of a variety of information, such as recognition of familiar words. It is also now known that the procedure does not work unless supplemented by prosodic information. See Yang (2004), Shukla et al. (2011).

[4]Fitch (2011). As is well-known, human vocal production involves different mechanisms, but the significance of this fact is limited in the light of the use of sign, also in languages created by children without input (Goldin-Meadow and Feldman 1977).

[5]In Chomsky (1986), I suggested that the term *I-language* be used instead of *grammar*, in one of the senses of this ambiguous term. I added that any other concept of language might be called a variety of *E-language* ("E" for external). The latter term has come to be used to refer to a (finite)

As in the case of all organic growth and development, several factors interact in acquisition of language: (1) external data, (2) genetic endowment that converts the data to experience and then guides the course of acquisition of I-language, (3) general principles of development that hold more broadly and may even be extra-organic. For a computational system like I-language, we would expect the third factor to include general conditions of computational complexity. While these are not entirely well-defined antecedently, some of their aspects are clear: e.g., less is better than more, minimal search is better than deeper search, etc. Even such simple notions carry us quite far.

The genetic factor (2) in turn has several components, among them (a) those specific to language (Universal Grammar, UG), (b) other cognitive processes, e.g., methods of statistical analysis of data, (c) determinants of structure and organization of the brain and other systems relevant to language. Exactly how these interact is the topic of the study of language acquisition.

Normal scientific rationality requires that at every stage of investigation we adopt the simplest assumptions, unless empirical evidence requires complication, thus weakening explanatory force and setting up still more barriers to an eventual account of evolution of language, if this proves to be a feasible endeavor.[6] This has always been a guiding principle in serious work on language, even if only tacit, including such traditional questions as mode of transcription and search for symmetries in structuralist phonology; and within generative grammar, among other examples elimination of stipulations and complex technology in the course of reduction of phrase structure grammar to X-bar theory and then to bare phrase structure, parallel simplification of transformational technology finally to Move-α, and then unification of the residues of phrase structure and transformational grammar to the simplest computational operation, as discussed below.

In the early days of modern generative grammar, these investigations were generally framed in epistemological terms, as a search for simplicity of theory – not to be confused with the construction of evaluation measures for I-languages, theory-internal empirical hypotheses about what counts as legitimate linguistic generalization. As the distinction between procedural analysis and the biolinguistic

---

corpus of material, or to some infinite object generated by the I-language, usually a set of well-formed formulas WFF. This usage is unfortunate, and should be abandoned, I think. A corpus is not a language in any sense. A set of WFFs is a *formal language*, determined by some finite generative procedure. Apart from its derivative character, it is not even clear that such a set can be coherently identified for human language; see *LSLT*, Chomsky (1957), and much since. In actual practice dimensions of deviance have been a rich and productive topic of inquiry since the early days of generative grammar.

[6]That much can ever be learned about evolution of language, or of cognition generally, is by no means obvious. For well-argued (and unfortunately, largely ignored) skepticism on the topic, see evolutionary biologist Richard Lewontin (1998). There is a huge and growing literature on evolution of language, but it seems to be almost entirely devoted to a different topic, speculations about evolution of communication; and in fact the alleged target, language, is rarely even characterized. Study of evolution of the eye can proceed only so far as we know what an eye is. The same holds for language or anything else.

framework became more clear, the epistemological version was reinterpreted in metaphysical terms: as a search for what is true of cognitive systems, language in particular. In a seamless continuation of such endeavors, this search was reformulated as the minimalist program MP about 20 years ago. The primary innovation of MP was to suggest a new research paradigm, sometimes called "approaching UG from below." Pursuing this approach, we ask what the optimal solution would be for I-language, then formulating the *Strong Minimalist Thesis* SMT that holds that each language actually is an optimal solution for the interface conditions. We can then inquire into the (naturally vast) gap between the SMT and what linguistic data appear to require, asking whether the gap can be overcome by a deeper reanalysis of the data, or by reformulating the assumptions about the nature of minimal computation (and other postulated third factor principles, empirical matters that can in principle be tested independently). The MP is entirely theory-neutral; one can choose to pursue these questions or not whatever one's ideas about the nature of language and cognition. MP is not a theory; rather, a research program that falls strictly within the bounds of normal science. Despite repeated efforts at clarification, the MP has often been misinterpreted as a new theory of language. We read, even in professional literature, that the MP has been refuted, or is not minimalist enough, etc. No such critique is intelligible. Research programs may be poorly implemented, or premature, or misguided, but not refuted.

One far-reaching thesis about language acquisition, with many important consequences, is the *Continuity Hypothesis*, which implies that "children are free to try out various linguistic options (compatible with Universal Grammar) before they set these parameters in the same way as the particular human language that is spoken to them."[7] It follows that child language may (and in interesting ways does) exhibit properties different from those of the adult language, though found in other languages compatible with UG. A related view is the *Maturation Hypothesis*, holding that certain principles of UG are available only at later stages of maturation (Borer and Wexler (1987). Both have substantial empirical support and significant consequences.

As noted, that UG interacts with other cognitive processes in language acquisition has been taken for granted since the origins of modern generative grammar. There is a substantial literature on how such interaction takes place (Yang 2002). An alternative view, widely held since the 1950s, is that other cognitive processes alone suffice to account for acquisition of language, so that language does not really exist as an object of independent study. I will put all of this aside here, simply assuming

[7]Crain (2012), focusing primarily on "logical nativism, which is the proposal that humans are biologically equipped with the tools for logical reasoning," a contingent property that allows for very limited variation among languages and is not derivable from assumptions of rationality. The thesis traces to Pinker (1984). A contrasting thesis is Kenneth Wexler's "*Very-Early Parameter-Setting (VEPS)*, holding that from the earliest observable ages (around 18 months), children have set their parameters correctly" (Wexler 2002); also well-supported empirically.

(as I think is well established) that the non-existence thesis is no more tenable today than it was in the past.[8]

A finitary computational procedure P will have buried in it in some form an operation – call it *Merge* – that takes objects already constructed and forms from them a new object, beginning with a set of atomic objects (which may have internal structure). To first approximation, we can take the atomic objects to be lexical items drawn from the lexicon, though this is not an innocent move. We can therefore think of P as having available a *work space* consisting of the lexicon (or some subpart extracted from it for the purpose of the computation) and objects that have already been formed by P. The optimal assumption is that Merge takes the simplest form: a binary operation that applies to X, Y, forming $Z = \{X, Y\}$, with X and Y unchanged by the operation (the No-tampering Condition NTC), and also unordered.

This optimal assumption too is far from innocent, and has substantial import. Take linear order. If Merge leaves elements unordered, then it is natural to expect that linear order and other forms of arrangement are reflexes of SM, where they are required, and therefore do not enter into the derivation of the CI interface, that is into core syntax and semantics. The general architecture of P, then, would consist of a *narrow syntax* NS that yields hierarchically-structured but unordered expressions that are transferred to CI, leaving atomic elements unchanged (by virtue of NTC) and with no new elements added (by virtue of sole reliance on Merge – the inclusiveness condition INC). An ancillary procedure of externalization EXT maps the expressions generated by NS to SM, not affecting the mapping to CI.[9]

There is a still broader thesis that is strongly suggested by these conclusions, call it the thesis T:

(T) Language is optimized relative to the CI interface alone, with externalization a secondary phenomenon.

There are other considerations tending in the same direction, conforming to a traditional view that language is primarily an instrument of thought, with other uses secondary. If so, we should revise the Aristotelian picture of language as sound with meaning; rather, it should be regarded as meaning with sound, a very different conception.

There is familiar evidence supporting a slight revision of INC: unvalued features (including structural case and ϕ-features of verbs and adjectives) are valued in the narrow syntax, a prerequisite for mapping to CI as non-deviant (see note 5). But other than that, there seems to be no strong reason to modify INC. As for ordering, the conceptual argument for restricting it to EXT seems strong, but empirical

[8]For detailed critical discussion, see among others Crain, Wexler, *op. cit.* On the failure of the most careful efforts to establish the non-existence thesis, see Berwick et al. (2011); also below.

[9]Note that this picture could in principle accommodate the more complex and widely explored thesis that CI considerations enter into externalization; as usually formulated, that Spell-Out accesses LF. Questions then arise about cyclicity and other matters.

arguments have been offered to the contrary, particularly in recent work of Richard Kayne's. I will keep to the simplest assumption here, with this caveat.

A Merge-based system is compositional, in the Fregean sense: interpretation of X is a function of the interpretation of its parts and the way they are combined. There is ample evidence in all aspects of language that the principle holds quite generally, with cyclic computation in all components of grammar. A more restrictive condition is *strict cyclicity*: an object already computed can no longer be modified. This condition plainly yields substantial computational saving, and has many consequences. A specific version of the condition is *phase theory*, which designates specific stages in the narrow-syntactic computation where material is transferred to the interfaces, and thus induces limited (but insufficient) compositionality to interface computation. I leave this interesting topic aside here.[10]

If Merge is binary, then Merge(X, Y) has only two possible cases: either X and Y are entirely distinct (*external Merge EM*), or one is a *term* of the other (*internal Merge IM*, where a term of W is a member of W or a term of a member of W). IM yields the ubiquitous displacement property of natural language, in the form that has been called "the copy theory of movement." If Y is a term of X, then by virtue of NTC, IM(X, Y) = {X, Y}, with two *copies* of Y, one the copy that was already part of X and the other one merged to X. For example, skipping details, if X is (1) and Y is *which boys*, then (2) has two copies of Y, one externally merged within X and the other internally merged to X:

(1) the girls expect which boys to see each other
(2) which boys [the girls expect which boys to see each other]

Further applications of EM yield (3):

(3) I wonder [which boys [the girls expect which boys to see each other]]

Externalization to SM imposes order and also deletes the hierarchically lowest copy, yielding (4):

(4) I wonder which boys [the girls expect to see each other]

If the bracketed clause of (4) appears alone, then *the girls* is the antecedent of *each other*, but not in (4); what is transferred to CI for interpretation is (3), not (4), so that the local antecedent is "which boys," a case of *reconstruction* (the term deriving from earlier more complex and stipulative approaches in which the fronted copy is reinserted by a new operation). The deletion of the internal copy follows from additional principles of minimal computation, presumably third factor properties: mapping to SM is radically simplified by copy-deletion, as is external realization by SM. The hierarchically most prominent copy remains, or there is no evidence that the operation took place.[11] Far more intricate cases proliferate, but

[10]See Gallego (2012), for various developments of the general idea.

[11]In embedded contexts, under some conditions, a residue of the raised copy remains, providing additional support for the copy theory; see Trinh (2011) for review and analysis. Reflexes

minimal computation – optimal Merge and copy deletion – seems to yield the basic properties.

Even such simple sentences as (4) have a number of interesting properties. One is that they pose problems for perception: it is necessary to locate the "gap," which may be a difficult task in moderately complex constructions, a familiar problem for parsing systems. If the copy were retained, the problem would be largely overcome. We therefore have a conflict between computational efficiency and efficiency of use (and its subpart, communicative efficiency). In this case, computational efficiency prevails, universally. There are other such cases, and to the extent that they are understood, the same holds: ambiguities, garden path sentences, islands. Throughout, these provide further reasons to suppose that the overarching thesis T is correct.

A further question about (4) is why we do not interpret it to mean something like (5):

(5) I wonder for which boys, the girls expect that those boys will see the other girls

That interpretation, which is a fine thought, would preserve Principle (A) of the binding theory in its simplest form, applying directly to the surface representation. Thus the antecedent of the reciprocal would be the closest appropriate choice in the expression that reaches the ear or eye, whereas the actual principle applies to the abstract form including the copy, visible only to the mind but not to SM. But language design does not permit that simpler option: what reaches SM, again, is of secondary significance for the workings of language.

There are many other examples illustrating the same conclusion. Consider ECP violations, such as (6), close to gibberish, as contrasted with (7), far more acceptable:

(6) (a) how many mechanics do you wonder if fixed the cars
(b) which mechanics do you wonder which cars fixed
(c) which mechanics do you wonder fixed which cars

(7) (a) how many cars do you wonder if the mechanics fixed
(b) which cars do you wonder which mechanics fixed

Here too the thoughts that cannot be expressed by (6) are perfectly straightforward ones, though language design requires some circumlocution. This again provides evidence that computational efficiency trumps efficiency of use (production, perception, communication, etc.), if – and it's a big "if" – the ECP island condition receives an explanation in terms of optimality of design (as may well be possible; there are some suggestive proposals).

---

of movement through intermediate positions have also been discovered, including verb-raising (Esther Torrego, Spanish), agreement (Sandra Chung, Chamorro; Chris Collins, Ewe), deletion of verbal affix (Doug Saddy, Indonesian). Some accounts of covert movement allow for the internal copy alone to remain after externalization.

An even simpler case is provided by the much-debated principle of structure-dependence – an informal term, referring to the fact that minimal distance is computed on structural not linear proximity, as in (8):

(8) can eagles that fly swim

The meaning is (9), not (10), though (10) again is a fine thought:

(9) is it the case that eagles that fly can swim
(10) is it the case that eagles that can fly swim

The computational procedure that yields the association ((8), (10)) is far simpler than the one that yields the correct association ((8), (9)), but language does not permit it, in this or any construction in any language in which the issues arise.

The most straightforward explanation for these universal properties of language is that linear order does not enter into core syntactic/semantic computations. That again supports the architectural assumptions sketched earlier, and the broader thesis T. In the case of structure-dependence, a cottage industry has developed within computational cognitive science attempting to show that the facts follow from other cognitive processes, but with no success.[12]

The simplest account of displacement, in terms of IM and NTC, requires a distinction between copies and repetitions. The intuitive basis is clear. In just about every utterance there likely to be repetitions: of articles, prepositions, verbs, etc. These are unrelated because they are drawn independently from the lexicon (workspace). In the accusative construction (11a) there are two repetitions of *John*; in the unaccusative (11b), two copies:

(11) (a) John saw John
(b) John arrived John

Here too the cases differ in how items are extracted from the workspace; in (11a), *John* is extracted twice, and in (b) only once. Repetitions are unrelated in interpretation; copies in contrast can be regarded as a *discontinuous element* at CI, interpreted as a single element (though the separate copies can make distinct contributions to the unitary interpretation).[13] If IM takes place only at the phase

[12]See Berwick, *op. cit.* The critique there is understated, failing to make clear a more fundamental point: the proposed methods, or very similar ones, would work for a pseudo-language that relied on linear order, but no such language exists, surely not an accident. There is evidence from neuroscience that while invented systems conforming to UG (with structure-dependence) activate normal language areas, those that use linear order do not, and are presumably treated as non-linguistic puzzles. See Musso et al. (2003). Note that the problem with (10) cannot be reduced to a principle that predicates do not tolerate gaps; they easily do, as in "(it is a shame to see) how angry John is at the world," etc. More generally, there are, to my knowledge, no accounts of any non-trivial linguistic phenomena in terms of non-linguistic cognitive processes alone, though it is common to claim that they alone suffice. See note 8.

[13]Separate question have to do with referring and binding theory, where questions arise relating to modes of conceptualization ("cognitive value"). See Heim 1998.

level, then the distinction between copies and repetitions is readily established at the point of transfer, as required. The functioning of discontinuous elements has interesting aspects, which I will have to put aside here.[14]

The architecture suggested above, and the overarching thesis T, correlate with familiar general properties of language. Mapping to the CI interface is generally assumed to be invariant, or virtually so. It is hard to see how it could be otherwise, given the paucity of evidence available for acquisition. There is by now substantial evidence that narrow syntax may also allow only limited variety, virtually none if parametric variation is restricted to the lexicon, or even to functional elements of the lexicon. And simple principles of computational complexity, such as NTC and INC, appear to hold rather generally of narrow syntax, maybe completely. In contrast, externalization radically violates these and other natural principles of efficient computation. It is in general complex, varied, and easily subject to diachronic change, again suggesting that it is an ancillary phenomenon.

The general picture also correlates well with the very little that is known about evolution of language. It appears that there has been no evolution of language (or virtually none; or of cognitive capacities generally) since our ancestors left Africa, perhaps about 50,000 years ago. There are individual differences, but no known group differences. An infant from a tribe in New Guinea that has been isolated for tens of thousands of years, if raised from birth in Boston would have the language of its Boston peers, and conversely.

If we go back roughly 50,000 years before that, there is little evidence that human language existed at all; archaeological evidence suggests that language, and with it complex cognition, emerged within this very narrow window, in what Jared Diamond called a "great leap forward."[15] Doubling the numbers or more changes little; the window remains very narrow in evolutionary time, and many millions of years after separation from other surviving species. These facts suggest that at some point within this narrow range some slight rewiring of the brain occurred yielding the core property of language: Merge with its output linked to the CI interface. Mutations occur in an individual, not a group. The individual endowed with this rewiring would therefore have had a "language of thought" LOT: a means to interpret, reflect, plan, etc., in principle unbounded.[16] An argument would be needed to support the thesis that there is a separate LOT, over and above what the CI interface yields as a mode of representation.

Since there would have been no external pressures, the system arising this way in some individual would have been optimal, determined just by third factor properties, rather like a snowflake, hence conforming to SMT. The mutation could be partially transmitted to offspring, and if it carries some selectional advantage, as seems plausible, it could proliferate through a small hunter-gatherer society. When the

---

[14]See Chomsky (2013).

[15]See Tattersall (1998) for general discussion.

[16]It is perhaps worth noting that the general picture seems plausible, maybe even necessary, to leading evolutionary biologists; for quotes from Nobel Laureates, see Chomsky (2005).

capacity is shared, there is some point to externalizing it for interaction, including communication as a special case. But externalization poses a hard cognitive task: it is necessary to link the internal system, perhaps approximating SMT, to a sensorymotor system that had been present for hundreds of thousands of years, so fossil evidence indicates. Accordingly, the problem can be solved in many different ways, each of them complex, and subject to modification. That seems to correspond fairly well with what we know about language. In fact, it is hard to imagine a course of evolution that does not include at least these steps; and it is not easy to find arguments supporting others.

Survival of a beneficial trait is a low-probability event, so this might have happened many times and been extinguished, but at least once it was retained, with our ancestors. One might also suspect that the mutation that yields Merge might have taken place many times, not just in the hominid line. But unless it is linked to CI, it would have been of little use, hence unlikely to survive. There have been claims that something of the sort might have happened with songbirds, but simpler interpretations of the data that have been advanced (see Beckers et al. 2012). There are other cases of recursive generation in human cognition, the most famous being the natural numbers, a concern that goes back to Wallace and Darwin. But this and other examples that have been suggested might well be piggy-backing on language.[17]

Gallistel (2011) reviews interesting evidence that other primates (and perhaps other species) have representations of the predicate-argument form of human language, and suggests that the "great leap forward" might have been externalization, not recursive generation (let us say just Merge). The latter conclusion seems to me questionable. To yield a LOT that provides the means for free expression and interpretation of thought in the required sense, Merge-generated expressions must be linked to elementary pre-existing mental structures of some kind, perhaps these; perhaps actor-action schemata and others.[18] Hence two steps are necessary to yield externalized human language: Merge and Externalization. The first step carries bounded elementary representations to unbounded and structured LOT. If the first step has been taken, it would be rather surprising if externalization does not follow (as for example happens spontaneously with children acquiring language without input; Goldin-Meadow and Feldman, *op. cit.*). Easy use of other modalities shows

[17]Reliance on internal computational capacity may yield infinite digital outputs, but that of course does not entail that the output system has its own recursive procedures. Thus externalization of generated I-language expressions does not entail that the vocal apparatus incorporates a recursive procedure. Same with use of the alphabet, or the number system, if it is an offshoot of I-language; or infinite digital visual displays, or other examples that can easily be constructed.

[18]It is worth noting that the most elementary concepts of human languages, such as those used for denoting, seem to be radically different from those found in animal communication systems, raising very serious problems for evolution of language and thought. Perhaps insuperable problems, Lewontin's arguments suggest. Note also that these comments do not touch on far broader and in many respects convoluted issues of what should be regarded as constituting thought.

that the problem does not reduce to limitations of vocal apparatus. The speculations above seem to me a more parsimonious and plausible account.

It is commonly held that IM is more complex than EM, requiring the operations Form-Copy, Remerge, Copy identification, and Copy deletion. But that is unwarranted. There are no operations Form-Copy or Remerge, just Merge in its simplest form, satisfying NTC and hence yielding copies. Copy identification is based on a straightforward property, easily detectable at the phase level. Copy deletion follows from elementary third factor considerations.

Prior to the recognition that EM and IM are unified under Merge, it was generally assumed (by me too) that EM and its precursors are somehow natural and that displacement, though ubiquitous, is an "imperfection" of language design, something that requires explanation. But that belief too does not seem tenable. One could on the contrary claim that IM is simpler, because EM requires search through the work space, including the lexicon, while for IM search is restricted to a single object, hence far more economical.[19] Blocking either of the two logically possible cases is a stipulation that requires empirical justification, along with additional justification for whatever new technology is introduced to deal with problems that have a free solution in terms of EM, IM.

In a well-designed system, we would expect that the two logically possible kinds of Merge would correlate with interface effects. At the SM level, that is obvious; IM correlates with displacement, EM with adjacency. At the CI level the distinction correlates closely with duality of semantics: EM yields generalized argument structure, IM yields everything else, specifically discourse-related structure like new/old information, focus, and so on. The correlation is close, but not fully established, and sometimes denied, as in the proposals to reduce control to A-movement.[20] An interesting task is to determine whether the correlation is perfect; if so why, and if not, why not.

As noted at the outset, a recursive procedure need not yield an infinite set of objects (or any at all). We can invent cases in which the range of the I-language would be finite. Imagine, for example, the pseudo-language "truncated English" TE, lacking all the devices of English that permit unbounded length. TE lacks connectives, iterable modifiers ("old old man," "very very old," etc.), possessor constructions ("father of father of Bill"), and so on. A child learning TE will, of course, be unaware of its radical expressive restrictions, and will proceed to acquire it just as if it were normal English based on unbounded Merge, discovering at some point that there is not much to say.[21] It's doubtful that TE could exist. Probably the child would simply invent new modes of expression using the resources of UG

[19]For arguments on early resort to IM in acquisition, see Roeper (2013).

[20]For extensive review and discussion, see Landau (2013).

[21]Merge is either available by virtue of UG, or unattainable. In general, there no way to learn a system of unbounded computation: addition, Merge or any other. That has been clear at least since David Hume. Separate questions have to do with the stage of development at which these properties are exhibited by the child in performance, either perception or (often much later) production.

as in cases already mentioned: isolated acquisition without any input, or the many examples of early acquisition of languages differing from the adult language but conforming to UG. But it is a theoretical possibility. From TE we would learn nothing about UG, or about recursion in language, or about the language faculty generally. It would just be an oddity. It has been proposed that the Amazonian language Pirahã is similar to TE, though the conclusion is contested.[22] The proposal has received enormous publicity, and has been taken to establish far-reaching results about language and cognition, but right or wrong, it is not clear that anything general follows, any more than from TE.

The earliest versions of generative grammar assumed two components – phrase structure grammar PSG and transformational grammar TG – and identified a number of general properties to be accommodated by the generative system: composition, order, projection, and displacement. The first three were assigned to PSG, the last to TG (along with long-distance morphological relations). PSG was abandoned for substantial reasons in the 1960s, replaced by X-bar theory, which greatly simplified the generative system but imposed new conditions: specifically, universal endocentricity. That aside, the distribution of labor was retained.

We have seen reasons to separate order from compositionality, and to assimilate displacement with composition under Merge. That leaves projection – or in more recent terminology, labeling – which differs from the others in that it is a theory-internal notion, while the others are virtually detectable in the data. It also raises interesting questions, some related to the endocentricity assumption (questionable in my view), but I will put these aside here.[23]

This is the barest sketch of what seem to me to be reasonable conclusions today on the status of minimal recursion, clearly a desideratum for a theory of language, with broad ramifications that should be worth exploration and careful examination.

References

Beckers, G., Bolhuis, J., Okanoya, K., & Berwick, R. (2012). Birdsong neurolinguistics: songbird context-free grammar claim is premature. *Neuroreport, 23*, 3.

Berwick, R., Pietroski, P., Yankama, B., & Chomsky, N. (2011). Poverty of the stimulus revisited. *Cognitive Science, 35*, 7.

Borer, H., & Wexler, K. (1987). The maturation of syntax. In T. Roeper & E. Williams (Eds.), *Parameter setting*. Dordrecht: Reidel.

Chomsky, N. (1955). *Logical structure of linguistic theory*. 1956 revision (partial) published by Plenum and Chicago, 1975.

Chomsky, N. (1957). *Syntactic structures*. The Hague : Mouton.

[22]A further proposal is that this alleged property of Pirahã derives from the cultural features of the community, but these, if they exist, would not be known to a child acquiring the language, hence could have no bearing on UG, just as in the case of TE. On these matters see Everett (2005), Nevins et al. (2009), and subsequent interchange.

[23]For discussion, see Chomsky (2013).

Chomsky, N. (1986). *Knowledge of language*. New York: Praeger.
Chomsky, N. (2005). Three factors in language design. *Linguistic Inquiry, 36.1*, 1–22.
Chomsky, N. (2013). Problems of projection. *Lingua. 130*, 33–49.
Crain, S. (2013). *The Emergence of Meaning*. Cambridge: Cambridge University Press.
Curtiss, S. (2013). Revisiting modularity: Using language as a window to the mind. In M. Piatelli-Palmerini & R. Berwick, (Eds.), *Rich languages from poor inputs*. Oxford: Oxford.
Everett, D. (2005). Cultural constraints on grammar and cognition in Pirahã: Another look at design features of human language. *Current Anthropology, 46*, 621–646.
Falk, J. (1992). Otto Jespersen, Leonard Bloomfield, and American structural linguistics. *Language, 68*, 3.
Fitch, T. (2011). Speech perception: A language-trained chimpanzee weighs. *Current Biology, 21*, 14.
Gallego, A. (Ed.). (2012). *Phases: Developing the framework*. Berlin/Boston: Mouton de Gruyter.
Gallistel, C. R. (2011). Prelinguistic thought. *Language Learning and Development, 7*, 4.
Goldin-Meadow, S., & Feldman, H. (1977). The development of language-like communication without a language model. *Science, 197*, 401–403.
Heim, I. (1998). Anaphora and semantic interpretation: A reinterpretation of Reinhart's approach. In U. Sauerland & O. Percus (Eds.), *The interpretive tract* (MIT working papers in linguistics 25). Cambridge, MA: MIT Press.
Joos, M. (Ed.). (1957). *Readings in linguistics*. Washington, DC: American Council of Learned Societies.
Landau, I. (2013). *Control in generative grammar: A research companion*. Cambridge: Cambridge University Press.
Lewontin, R. (1998). The evolution of cognition: Questions we will never answer. In D. Scarborough & S. Sternberg (Eds.), *Methods, models and conceptual issues: An invitation to cognitive science* (Vol. 4). Cambridge: MIT Press.
Musso, M., Moro, A., Gluache, V., Rijntjes, M., Bücheli, J. C., & Weiller, C. (2003). Broca's area and the language instinct. *Nature Neuroscience, 6*, 774–781.
Nevins, A., Pesetsky, D., & Rodriguez, C. (2009). Pirahã exceptionality: A reassessment. *Language, 85*, 2.
Pinker, S. (1984). *Language learnability and language development*. Cambridge, MA: Harvard University Press.
Roeper, T. (2013). There-insertion: How internal merge guides the acquisition path. In M. Becker, J. Grinstead & J. Rothman (Eds.), *Generative linguistics and language acquisition: Essays in honor of Nina Hyams* (pp. 251–270). Amsterdaman: John Benjamins.
Sauerland, U., & Gärtner, H.-M. (2007). *Interfaces + Recursion = Language?* Berlin: Mouton de Gruyter.
Shukla, M., White, K. S., & Aslin, R. N. (2011). Prosody guides the rapid mapping of auditory word forms onto visual objects in 6-mo-old infants. *Proceedings of the National Academy of Sciences of the United States of America, 108*(15), 6038–6043.
Soare, R. (1996). Computability and recursion. *Bulletin of Symbolic Logic, 2*, 3.
Tattersall, I. (1998). *The origin of the human capacity* (68th James Arthur Lecture on the Evolution of the Human Brain), American Museum of Natural History. New York.
Trinh, T. (2011). *Edges and linearization*. MIT Phd dissertation. Cambridge, MA.
Wexler, K. (2002). Lenneberg's dream: Learning normal language development and specific language impairment. In J. Schaffer & Y. Levy (Eds.), *Language competence across populations: Towards a definition of specific language impairment*. Mahwah: Erlbaum.
Yang, C. (2002). *Knowledge and learning in natural language*. New York: Oxford.
Yang, C. (2004). Universal grammar, statistics, or both? *Trends in Cognitive Sciences, 8*, 10.

[illegible]

Recursion Restrictions: Where Grammars Count

Hilda Koopman

Abstract Since recursion is a fundamental property of human languages, it is puzzling that we regularly find cases where recursion is impossible or restricted. In this paper we argue that these restrictions follow from an independently necessary property associated with individual lexical items, which encodes sensitivity to phonological properties. These restrictions must be stated on the output of the syntactic derivation, when syntactic structures are transferred to phonology as expected in current late spell-out models. The main idea is that phonological properties can be "grafts" on the structure-building requirement of a lexical item, referred to as an epp property, which can then be viewed as a repository of the finely grained knowledge speakers have of the phonological properties associated with local syntactic environments. In this view, restrictions on recursion, though accidental, can be straightforwardly and simply accounted for as arising from the way that independently necessary properties interact in specific local syntactic environments. This accounts for a number of well-known effects, including left branch restrictions, restrictions on center embedding, and complexity effects.

Keywords Recursion • Interface • Spell-out • EPP • Syntax

This paper builds on my presentation at the Recursion conference in Amherst, May 2009, and is a substantial revision of my 2010 paper *On restricting recursion and the form of syntactic representation* lingBuzz/001139. I would like to thank Natasha Abner, Melanie Bervoets, Chris Collins, Uli Sauerland, Ed Stabler, Dominique Sportiche and Anna Szabolcsi, for feedback and comments.

H. Koopman (✉)
Department of Linguistics, UCLA, Los Angeles, CA, USA
e-mail: koopman@ucla.edu

T. Roeper and M. Speas (eds.), *Recursion: Complexity in Cognition*, Studies in Theoretical Psycholinguistics 43, DOI 10.1007/978-3-319-05086-7_2,

1 Introduction

Since recursion is a fundamental property of human languages, it is puzzling that we regularly find cases where recursion is impossible or restricted. This paper explores how to investigate and understand such restrictions. It argues that they follow from an independently necessary property associated with individual lexical items, a property that encodes sensitivity to phonological properties. These restrictions must be stated on the output of the syntactic derivation, as expected in current late spell-out models, and they effectively rein in recursion in very specific local syntactic environments. This accounts for a number of well-known effects, including left branch restrictions, restrictions on center embedding, and complexity effects.

This paper makes a proposal about the nature of the phonological knowledge that is integrated into lexical properties, and argues, on the basis of a case study of verbal complexes in Dutch and Hungarian, that the grammar requires a limited amount of counting to capture the recursion restrictions.

More particularly, building on Koopman and Szabolcsi (2000) and Koopman (2002), I will show that recursion restrictions are to be accounted for at the syntax-phonology interface, when syntactic structures are transferred to phonology under current late spell-out models. The main idea is that phonological properties can be "grafts" on the structure-building requirement of a lexical item, referred to as an epp property, which can then be viewed as a repository of the extremely finely grained knowledge speakers have of the phonological properties associated with local syntactic environments. In this view, restricted recursion, though by itself accidental, can be straightforwardly and simply accounted for as arising from the interaction of independently necessary properties. The present account integrates finely grained phonological knowledge with local structural syntactic properties of syntactic atoms, yielding a more complete view of how to account for an individual speaker's knowledge, and providing some insight into individual variation.

The paper will proceed as follows. Against a general theoretical background where there is no separate morphology, it first addresses what needs to be said about individual atoms of syntactic structure that were previously dealt with in morphology. It argues that these LIs, or heads, can impose various types of phonological properties on their second merged objects (formerly called 'Specs'). Such properties, I argue, should be localized as part of the structure building epp of an LI, and depend on and vary according to the individual atom. In particular, these properties encode the maximum size of phonological material allowed in a specific location, as measured on the output of the syntactic derivation, by (limited) counting of the number of nodes dominating pronounced material, and separating it from the sisternode with which it is merged. This essentially phonological property reins in recursion, as we will show for several different cases. This discussion sets the stage for a particular case study in the domain of verbal complexes in Dutch and Hungarian, building on Koopman and Szabolcsi (2000) and Koopman (2002). Here we find two instances of recursion restrictions. One case is particularly interesting and revealing, because the restriction on recursion, which must be stated on the output of the syntactic derivation, is found where phonologically light material is syntactically embedded, allowing to rule out a possible prosodic account. It is here that the particular syntactic derivations lead to a uniform understanding of the restriction, and where it can be shown that the grammar needs to allow for a limited amount of counting, so as to allow for the observed variability between speakers, dialects, and across languages. If this analysis is on the right track, it provides insight into the type of syntactic derivations we need to adopt.

2 Some Theoretical Background

The discussion in this paper is cast within a particular version of the Minimalist Program, and takes the following set of axioms as a point of departure:

- There is a single computational system for building structure, and that is syntax. It departs from Distributed Morphology, and earlier accounts of morphology, and assumes that there is no structure-building in either post- or pre-syntactic components (cf. Koopman 2005a, b). The difference between words and phrases is one of output size, not one of modules, or atoms, or ways of composition (head or phrasal movement).
- Syntactic atoms (LIs) are tiny and can be semantically or phonologically meaningful: they are *not* the fully inflected lexical items familiar from much work. These small atoms enter into the structure building component (Merge, Bare Phrase Structure, Chomsky 1995). Syntax is fundamentally decompositional, and antisymmetric (Kayne 1994, 2000).
- Lexical properties of LIs, including phonological properties, must be satisfied strictly locally, under sisterhood (first or second Merge Sportiche 2005). Agree plays no role in the properties of lexical items in question in here (and perhaps not for agreement either—see Koopman (2006)).

Since lexical properties drive derivations and determine which representations converge, it becomes crucial to establish what these properties are. Starting from "bound morphemes" which often are out of syntacticians' sight, Sect. 2.1 argues that syntactic atoms can impose very specific phonological requirements on their second merged objects (which I sometimes refer to as 'Specs', for convenience), and proposes that these are "grafts" on the structure building epp. These finely grained and varying phonological properties can yield restrictions on recursion, in effect filtering out well-formed syntactic derivations that do not pass the phonological conditions imposed by specific LIs. The ensuing ungrammaticality is thus a result of unsatisfied, specifically phonological, properties of lexical items.

It is against this general background that I will turn to the domain of verbal complexes in Dutch and Hungarian, building heavily on the syntactic derivations motivated in Koopman and Szabolcsi (2000) and Koopman (2002), where we recast the complexity and splitting filters as phonological properties of the epp. The domain of verbal complexes lends itself well to this investigation: we find restrictions in recursion, as well as micro- and macrovariaton, which restrict both the possible analyses of the variability, and the factors that must enter into the account. The Koopman and Szabolcsi (2000) syntactic derivations that underlie the particular phenomena are highly determined, general and uniform, and yield non-trivial results for Dutch, German and Hungarian. The syntactic derivations provide syntactic vocabulary that allows us to directly relate the observed restrictions on recursion to the derivational output. The restrictions will be shown to be (1) sensitive to the presence (but not to the absence) of phonological material in a designated Spec at spell-out, (2) sensitive to the depth of embedding of pronounced phonological material, even if that material is phonologically light on the surface (say a light foot, or a CV syllable), (3) sensitive to syntactic category, and (4) are variable within a particular language or language area, and across languages. Restricted recursion, though largely arbitrary, can be accounted for straightforwardly and simply as a result of independently necessary properties, provided of course we have the right kind of syntax.

2.1 On the Notion **Bound** *and the epp*

Under the single engine hypothesis, all word formation is syntactic structure building: there is no difference between derivation, inflection, compounding, or syntactic phrase formation. This implies that properties traditionally taken to be morpheme-specific are potentially properties of any LI. It is important in this respect to consider the properties of "bound morphemes", and what these reveal about general properties that could, in principle, characterize any individual LI.

As an atom, an LI merges in the spine with a constructed syntactic representation (its complement). English bound morphemes additionally require a particular category to their left (Williams 1981 RHHR), i.e. they require second merge of a specific category. At transfer, LIs can be spelled out (i.e. past tense *–ed*), or not. The English little v_{cause} that merges with change of state complements famously lacks any phonology ([clean $[_v]$]), as does the root question LI Q, which triggers T to C in English. This second merge requirement of bound morphemes is familiar: it looks just like what syntacticians have called the epp (Koopman 2002, 2005a, b), a requirement a head can have to build structure of a particular category (epp_{cat}). Thus the entry for English past *-ed* can be written as $epp_{[V]}$; v has epp[V], and Q has $epp_{[Tfinite]}$. Some LIs may require "phonological", i.e. [+ph] material in their Specs, a property which I will code as follows:

(1) $epp_{cat[+ph]}$

These Specs will end up containing [+ph] material; this is simply what being bound means.

It is in this sense that a phonological property can be "grafted" on the epp_{cat}. The epp forces a particular piece of structure, but in addition, it can further require the presence of phonological material in that category at spell-out.

As another example, finite T (or some head in the T region) and *'s* D both require a subject of the D category (epp_d): yet Spec TP can be silent as a result of wh-movement (*who did they believe—would sign this petition?*), but Spec DP may not (**who did you like —'s brother*).

For some LIs then, the specifier ends up containing pronounced phonological material at spell-out; others allow subsequent extraction. For the purposes of this paper, I simply code this distinction as follows, awaiting further understanding.

(2) a. T_{finite} epp_d
b. D_s $epp_{d[+ph]}$

Can [+ph] be further specified?

What Kinds of Phonological Properties Can Be Imposed on [+ph]?

As is well known, LIs (individual affixes) seem to be able to impose a variety of phonological properties. These are typical of suffixes (and perhaps never of prefixes). Since suffixes have a second merged element on their left, these heads have an epp structure-building feature.

Affixes can be sensitive to syllable structure, impose particular metrical structure, as well as minimal and maximal size requirements. For example, the spell-out of Korean affixes routinely depends on whether an affix follows a consonant or a vowel. Nominative is spelled out as *-i* after a consonant, but *-ka* after a vowel, hence spell-out needs to see the syllable structure of the item in the Spec.

Affixes sometimes impose a *minimum* size requirement, say at least two syllables. To meet such requirements, languages insert expletive materials. Sometimes however affixes seem to impose a *maximum size* requirement ("no larger than x"). For example, the English comparative *–er* is often said to merge with an A no larger than a 'light foot' (*happy-er vs *yellow-er, *intelligent-er*).[1]

How exactly should we understand "no larger than x"? As we will see, this notion cannot always be measured in terms of the phonological vocabulary, number of syllables or feet, or phonologically coherent constituents ('words'). As will be argued in this paper, there is a notion of maximum "phonological size" that is measured on a local syntactic output. I will argue that it is this notion that plays a crucial role in restricting the possible outputs of syntactic derivations. Restrictions on recursion thus are to be understood as representations that fail to converge because they violate properties of lexical items, in particular the maximum [+ph] size requirement.

[1] Throughout this paper, the question of interspeaker variability arises. Some speakers are perfectly happy with yellower, but others are not. This variability is expected (see Sect. 3 for further discussion), and should make fine predictions about correlating patterns of judgments and systematic interspeaker differences.

2.2 *More on Maximum Phonological Size and Restrictions on Recursion*

This section forms the background for the discussion of verbal complexes in Sect. 3, and establishes the following:

(3) (i) $epp_{cat[+ph]}$ can vary as to the maximum allowed number of syntactic nodes dominating the phonological form (notated as $epp_{cat[ph\ \mathbf{max}:(number)]}$ for an individual LI).
(ii) $epp_{cat[+ph\ \mathbf{max}:(number)]}$ leads to restrictions in recursion.
(iii) different LIs can set different $epp_{cat[ph\ \mathbf{max}:(number)]}$, which leads to messy surface patterns.

So here we are not directly interested in what is regular and beautiful, but rather in finding out where exactly the beautiful design can break down on the surface. Insofar as there is a unified theme here, it will instead inform the boundary conditions on the syntactic derivations that allow us to state these in a uniform way.

Let us start with two English bound morphemes: past tense *-ed* and adjectival –*ing*. Both (second) merge with a V_{cat} constituent, which must contain [+ph] material at spell-out, $epp_{V,[+ph]}$, yielding local structures like (4). Note that bare phrase structure labeling will be adopted everywhere, except for cases where confusion might arise.

(4) a. [$_T$ [$_V$ [$_N$father] [$_V$e]] [$_T$ed]]
b. [$_{\text{'ing'}}$[$_V$ [$_V$melt] [$_V$e]] [$_{\text{'ing'}}$ing]]

While this example suggests that *-ed* and *-ing* each require epp_V, this is not sufficient to account for their differences. As the following examples show, the constituent that [$_T$ed] merges with must be comparatively[2] small in size; *–ing,* on the other hand, seems to be able to merge with a bigger structure (taking compositionality into account):

(5) a. *the person who [[$_A$sad] [$_V$seem] ed]
b. [[[$_A$sad] [$_V$seem]] ing], (cf *good-looking, foul-smelling*)

If we look at the syntactic representations, (5a) is excluded with T *-ed,*[3] but fine with *-ing*. T *-ed* is only compatible with small constituents, but –ing is not restricted in this way.

[2]Comparatively, because we know from decomposition of the VP domain that such transitive structures are in fact syntactically quite large.

[3]Richard Kayne (personal communication) finds *they homeschooled all their children, they wrongfooted us twice last night*, perfectly acceptable. This may imply that the structures in (5a) are different w.r.t. size and perhaps category. The general point made here is that there are differences in the relative sizes of structures and that restrictions on relative sizes can be an arbitrary property of individual heads.

(6)

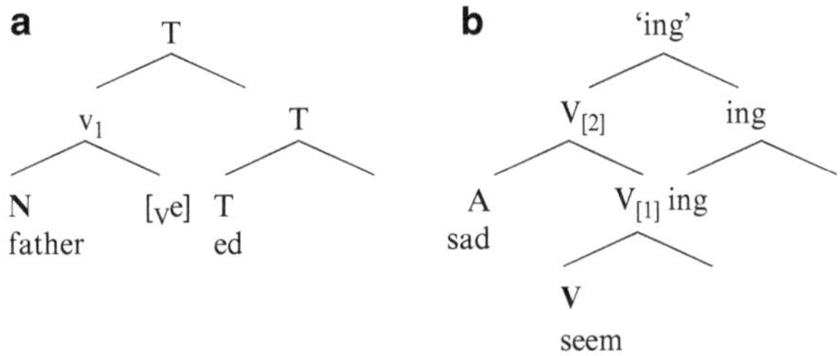

The size can be measured on the output of the syntactic structure by the following procedure:

(7) locate the most deeply embedded category with [+ph] material (boldfaced above); count the nodes that separate it from the sister of T or *-ing*, (annotated in the representations for convenience).

This gives a rough measure of depth of embedding, which will be sufficient for our purposes. What matters here is the difference in depth of embedding, not the exact number of words, though in many (but not all) cases, further embeddings will lead to more phonological material. Here is how this will be annotated on [+ph] (max[n])

(8) a. T_{ed}, $epp_{catV,[+ph], max[1]}$
b. ing, $epp_{catV, [+ph], max[2]}$

(8a) will filter out (5a), since it will not satisfy the lexical requirement. Other imaginable outputs are excluded as well, like piedpiping a big verbal projection for example. The net effect is that the output condition forces a derivation which obeys (8a), i.e. the verbal structure must be split up.

Next, we broaden the discussion, bringing in the head spelled out as the phrasal affix' 's with epp $_{d\,[+ph]}$. This affix does not care so much[4] about the size and depth of embedding of phonological material in its Spec. It allows coordination of the second merged DP, which neither *-ed*,*-ing*, or plural allow, as well as possessor recursion:

(9) a. *he [$_V$[close [$_v$e]] and [$_V$open [$_v$e]]]-ed the doors every day
b. * [$_V$[$_V$sad seem] and [$_V$crazy look]]-ing
c. [$_{DP}$[$_{DP}$The king of England] [and [$_{DP}$the queen of France]]]'s offspring
d. [$_{DP}$[$_{DP}$John], [$_{DP}$Mary] [or [$_{DP}$Bill]]]'s house
e. [$_{DP}$[$_{DP}$[$_{DP}$my] father]'s friend]'s car

Coordinated structures are basic recursive structures, embedding a node of the same type under the mother node [$_x$ [$_x$] [and [$_x$]]]. (9a) and (9b) illustrate contexts

[4]Though speakers vary in their tolerance to right-branching relative clauses (e.g. %the queen that you adore's minister).

in which recursion fails, (9c–e) contexts where it does not. Regardless of what we say about coordination, the structures in (9a) and (9b) are immediately excluded by their [+ph max: n] settings: the number of nodes that dominate ph material in these examples necessarily exceeds the number in (7). The max size setting thus rules out coordination, and any type of recursion. LIs which lack such a setting, or which allow a much bigger size setting, like English's, allow recursion under coordination or other types of recursion.

[+ph max] is specific to individual LIs. It varies with the individual LI, as *-ing* and *-ed* show, and some LIs may lack such restrictions altogether. The settings must be established on the basis of primary data, and thus are expected to vary slightly between speakers and across regions.

They can also varies cross linguistically. The size restriction of English *-ed* is not shared by Japanese or Korean past tense Ts, which allow coordination and disjunction of the verbal constituents on their left, and behave in this respect like "phrasal affixes".

Another of our examples, English *'s*, which allows recursion, differs from the German head that licenses prenominal genitives, which does not. German prenominal genitives are largely restricted to bare proper names or pronouns (Longobardi 2001, p. 567), and coordination of these elements is not allowed. No such restriction holds for the postnominal genitive position, suggesting this is a particular property of a D head on the left edge of the DP, not a property of genitive case itself.

(10) *Marias sorgfältige Beschreibung Ottos*
Maria's careful description of Otto

(11) *??Des Zeugens/*Dieser Frau/*Meiner Schwester/sorgfältige Beschreibung Ottos*
The witness's/this woman's/my sister's /careful description of Otto

(12) * *[Marias und Susies] sorgfältige Beschreibung Ottos*
Maria's and Susie's careful description of Otto

(13) Marias sorgfältige Beschreibung des Zeugen/ dieser Frau/ meiner Schwester
Maria's careful description of the witness/this woman/my sister

How can we capture this distributed restriction in German? Here only names and pronouns survive the derivation. These are the elements independently known to be associated with the D region. This makes the approximate structure in (14) a very reasonable candidate for the syntactic configuration at transfer to the phonology. If we consider the boldfaced D and count the number of nodes dominating the pronounced material, we find roughly the $epp_{d\ [+ph\ max\ [2]]}$:

(14)

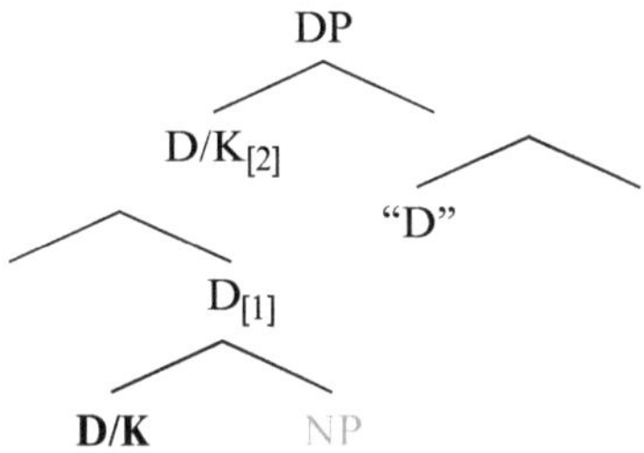

The starred example in (11) violates the [+ph max size] of the epp of "D", regardless of the finer details. Inserting [+ph] material as the complement of D (i.e. pronouncing N) will obligatorily lead to additional hierarchical structure dominating N, and violate the [+max] setting. This may yield the impression that number of words or syllables could be relevant. I address this issue in Sect. 3, where I will show that the determining factor here must be the syntactic configuration. Similarly, coordination of names (12) or other recursions are excluded because they violate the max [+ph] properties of this particular German LI.

Under this view, it is expected that maximum phonological size can be set differently for different speakers and regions, as it must be deduced on the basis of primary data, mapping sound structures onto syntactic structures. Confronted with the examples above, one hears sometimes that there are varieties of German, or speakers of German who accept such sequences, and hence the data are somehow invalidated. Note that this cannot be taken to mean that speakers who have restrictions have bad judgments, or that judgments are unreliable. This is simply a consequence of the fact that speakers end up with different settings, reflecting their different linguistic exposures. Variability in max size is expected to correlate with distributional properties elsewhere in very precise ways for individual speakers, and exploring such correlations thoroughly will be important in the future to understand the finer details of the syntactic representations.

Finally, max [+ph] size requirements can be "grafted" on the epp of any LI, regardless of the height at which the latter seems to enter the derivation. Indeed, for some LI merged early in the derivation, it could be argued it is compatible with just a small phonological size because that is all that is present in the derivation at that point (assuming constituents are built up from small parts—cf. Sportiche's split D hypothesis (1998, 2005), Kayne's adpositions (2000, 2005)). It is important to point out in this respect that size restrictions routinely show up for LIs that are generally taken to be merged *high* in the derivation, like particular Ds or Cs. This seems to be the case for the German prenominal genitives discussed above, or, for example, the root C that hosts the finite verb in Dutch verb second.

2.3 When Max Size Is Set Differently for Individual LIs: English C, T and V

If individual LIs can have different max [+ph] associated with them, we expect to find distributional effects. Only those syntactic derivations in which the max size is respected will converge. Other in principle well-formed syntactic outputs, including recursive structures, will be 'filtered' out, because they will violate lexical requirements. Such factors seem to be at work in the distributional properties of English modals, finite verb forms of *have* and *be*, and the auxiliary *do*, versus all other verb forms, as we sketch below.

English root questions require T to C movement, driven by the need to satisfy Q, [$epp_{Tfinite\ [+ph]}$]. Within Bare Phrase structure, this epp property could in principle be satisfied by second merging any T constituent with C, subject to locality. However, merging TP will not converge in English root yes/no questions (though it will in superficial head-initial languages with final question particles). Instead, the English C_Q shows effects of a size restriction (related of course to Germanic T to C) with a small max size 2: [[$_{T[1]}$**V T**] $Agr_{[2]}$] [C]. Only those Vs which are merged highest in the Cinque hierarchy (1999), i.e. the tensed modals, the last resort auxiliary *do,* all finite instances of *be*, and finite forms of perfect *have*, survive in this environment. If all other English finite verbs are necessarily "bigger", i.e. have further nodes embedding them because of the syntactic hierarchy, they will never survive in this C environment.

But English verbs are compatible with finite T ($T_{[hab]}$ or $T_{[past]}$). This is non-problematic, as in this environment, only the [+ph] size requirement of these Ts (wherever these might be merged) needs to be satisfied: C does not enter into the picture. It is only when T is forced to merge with other heads, like Neg and Focus which, perhaps not accidentally, happen to share the properties of C_{+Q}, that we find the size restriction, because now the [+ph] max of the epp of Neg and Focus come into play. This view attributes much of the mystery of the distribution of Vs in English to an independently motivated and necessary property of the epp. Specific phonological knowledge, stored as part of the epp, can filter out otherwise perfectly well-formed syntactic structures. Such representations do not converge because they violate particular lexical properties of the syntactic atom.

Under the single computational engine, any type of LI with an epp property could have [+ph max] restrictions associated with it. These are expected to vary between speakers (as primary data vary), regions, and languages, and to be instrumental in the slow, incremental, and LI-specific induced syntactic drifts and changes.

Against this background, we now turn to the question whether [+ph max] could yield a further understanding in specific cases of recursion, and what variability we could expect to find and not find. The remainder of this paper explores these issues for verbal complexes in Dutch and Hungarian, which provide a particular window into recursion restrictions, how these arise and what they may tell us about the grammar. Here in turn we can find support for specific theories of syntactic structure building, as the outputs of the derivations yield the structures on which the relevant phonological restrictions can be simply stated.

3 Verbal Complexes: Restrictions on Recursion and Variability

Here we turn to recursion in the domain of verbal complexes in Dutch, German, and Hungarian, drawing heavily from Koopman and Szabolcsi (2000), and Koopman (2002).

In (standard) Dutch, recursion of infinitives is restricted on the left of the finite verb (one recursion is fine, but not two). To the right of the finite verb, sequences of three or four embedded infinitives are routine. German infinitives, on the other hand, happily recurse to the left of the finite verb. This suggests the effects of a particular LI in Dutch with a [[+ph],max] restriction, which the equivalent German LI may simply lack. If so, we expect to also find intermediate cases of [[+ph],max] restrictions, where a slightly bigger size is allowed, say, two recursions are fine, but three are excluded. As we will see below, this state of affairs seems to hold of a particular variety of Hungarian.[5]

In order to get to the relevant syntactic output structures, the discussion will need to become more technical and precise. Indeed, as the simpler examples discussed so far show, it is the syntactic output structures that provide insight into how to understand the recursion restrictions found, and these will enable us to investigate further predictions about the actual patterns of variation. If successful, the study of these seemingly exceptional and messy patterns holds valuable information about the form of the output of the syntactic derivations, and hence about the way these outputs must have been built by the derivations.

3.1 Dutch Verbal Clusters and Restricted Recursion

In (standard) Dutch, an infinitival verb can appear to the left or right of a finite verb. However, Dutch verbal clusters show restrictions on recursion of infinitives to the left of the (finite) verb, but not to the right. Note that these variable orders do not correspond to any interpretative differences. They seem, as Jean Roger Vergnaud once told me, "to be a pure reflection of the workings of the computational system". The surface patterns are given below (with the terminology used by Koopman and Szabolcsi (2000) in italics). Numbers refer to the hierarchical order, with 1 (*wil* 'want') selecting 2 (the infinitive *zwemmen* swim) as its complement:

[5]Other varieties of Hungarian appear to lack such restrictions.

(15)	a.	...	zwemmen wil	*Inverted orders/Roll-up*: 21
		...	swim.inf want+T	
	b.	...	wil zwemmen	*English order*: 12
		...	want swim.inf	

However if 2 itself selects for an infinitival complement (3), only the 123 order is acceptable; 321 is excluded, as are the other four possible orders. Why would this be the case?

(16)	a.	*...	zwemmen willen zal	*Inverted orders/Roll up: *3-2-1*
			swim.inf want.inf will	
	b.	*...	zal zwemmen willen	**132*
			will swim.inf want.inf	
	c.	*...	zwemmen zal willen	*Climbing orders: *312*
			swim.inf will want.inf	
	d.	*...	zal willen zwemmen	*English orders: 123*
			will want-inf swim.inf	

As the patterns in (15) show,[6] infinitivals can form embedded 21 orders, but an infinitive cannot be to the left of another infinitive (that selects for it), even when separated by a finite verb. These patterns are exceptional from a Dutch-internal perspective, as well as a micro- and macro-comparative perspective.

Within the broader Dutch paradigm of complex verbs, the starred 312 patterns are exceptional: only infinitivals disallow this pattern, particles, adjectives, bare nouns, directional PPs, and participles all yield the 312 (or for that matter 4123, 51234) patterns[7]. An example with a participle is shown below:

(17)	...	geschreven	*wil hebben*	$3_{Part}\ 1_{fin}\ 2_{inf}$
		ge.write.part	want have.inf	

Second, while these strings are excluded in standard Dutch, they do occur in certain Dutch dialects (cf. Barbiers 2002). Third, both 312 and 321 patterns surface in similar clustering languages like German and Hungarian (abstracting from the position of the tensed verb in Hungarian). In particular, take the 312 pattern in (16c): this order shows up in all German varieties with *zu*-infinitivals and participles, as well as in the so-called IPP environment in Southern varieties of German (Den Besten and Edmondson 1983; Wurmbrand 2004).

[6] The *213 pattern is left out of consideration here. Barbiers (2002) argues this order is universally impossible. Koopman and Szabolcsi (2000, p. 173) list all the conditions that must hold at the same time for a derivation to even yield this pattern. This leads to the expectation that this order will surface very rarely.

[7] I have nothing to say about this mysterious fact. It could be a historical accident or perhaps it is not. It seems to also be found in English in the form of the double-ing constraint (Ross 1972). See Appendix 1 for further discussion.

(18) a. *ohne singen zu wollen* $3_{inf}\ 1_{zu}\ 2_{inf}$
without sing.inf to want.inf
'without wanting to sing'

b. *singen hat wollen* $3_{inf(ipp)}\ 1_{fin}\ 2_{inf}$
sing.inf has want.inf
'... has wanted to sing'

This pattern is also found in neutral clauses in Hungarian (clauses without any focus or negation) (Koopman and Szabolcsi 2000: p. 74 (108)):

(19) *Dolgozni fogok akarni* $3_{inf}\ 1_{fin}\ 2_{inf}$
work.inf will-1sg want.inf
'I will want to work'

But, within Hungarian dialects, there seems to be one region in which such structures show further restrictions on recursion: while 312 is fine, adding another infinitive clustering predicate in this dialect yields *4123. (cf. Sect. 3.3 for further discussion.)

How do these restrictions fall out? The proposal here is that these can be accounted for in a unified way: they reflect different settings of $epp_{[[+ph]max]}$ calculated on the output of the syntactic derivation in a specific local configuration.

The next sections introduce the basic background for the syntactic derivations, just enough to show how the restrictions on recursion follow from the [[+ph], max] setting of a specific linguistic atom. The interested reader is referred to Koopman and Szabolcsi (2000) for fully specified derivations.

3.2 *Verbal Clusters: What Must Be Captured and How to Capture It*

Starting from the general properties that any analysis of verbal clusters should capture, we outline the particular implementation given by Koopman and Szabolcsi (2000). This analysis is, as far as I know, the only theory which provides insight into how to capture the variable orders and deal with the restrictions on recursion. This will allow just enough detail to focus on the output of the syntactic representations that underly the different orders at the point of transfer to phonology. These can be ordered in terms of [max size], which in turn allows the relevant restrictions to be stated quite simply. The 312 orders will turn out to be particularly important and revealing, as these allow us to test if syntax indeed provides the vocabulary that is responsible for stating the restrictions. Then we discuss suggestive data that show that the predictions of variation made by these representations are indeed borne out.

What an Analysis Should Achieve

1. Analyses for verbal clusters should extend beyond infinitives and participles. Verbal complexes with infinitives and participles are part of a more general pattern of complex verb formation (for example, Dutch verb particle constructions, e.g. *opbellen* 'up call', bare adjectivals, e.g. *schoon maken* 'clean make.INF', bare nominals, e.g. *piano spelen* 'piano play.INF', directional small clauses, e.g. *naar huis gaan* 'to home go.INF', and other small clauses). Such complexes are ubiquitous in clustering languages.
2. Analyses should account for the fact that the two parts of a cluster can be manipulated separately, yielding discontinuous surface constituents (for example, Dutch *op te bellen* 'to call.INF up', verb second *bel hem nu op* lit. 'call him now up').
3. The different surface orders should fall out from the derivations, without any additional stipulations.
4. (At the very least) the following two properties of the relevant verbs should be captured: first, the verbs participate in cluster formation (not all verbs do), and second, their individual needs must be satisfied (some verbs select for infinitives, some for participles, some for *te* complements).
5. Analyses should capture the distribution of the (bare) infinitival, *te*-infinitival and participial morphology.
6. Analyses should yield insight into where and how languages differ.
7. And finally, they should yield insight into the restrictions on recursion, and do so by using independently motivated mechanisms.

Koopman and Szabolcsi (2000)

Koopman and Szabolcsi (2000) propose a unified analysis attempting to achieve these goals, with movement (second merge) the only means to satisfy the lexical properties locally. The analysis is summarized below. Note that within bare phrase structure, there is no difference between head and phrasal movement, the only distinction will be one of size.

1. We stipulate that there is a general complex verb formation configuration that governs all complex verb formation. This is mediated by a head (arbitrarily named V+), which merges with V as its complement, and requires a small clause to second merge with it. This yields a configuration in which the V complement can in principle be separated from the small clause in V+, depending on what attracts it. But nothing in principle prevents them from internal merging (i.e. moving) together as a unit, and both possibilities are attested within the same internal grammars. Depending on which option is taken, the surface patterns will be different.
2. Clustering verbs (restructuring verbs) form complex verbs (again a stipulation). This means they consist of V+ and V, with V+ attracting a V+ from the complement (i.e. forming a complex verb). The derivation must succeed in

satisfying the configuration outlined above. Given locality of selection, a V+ constituent must undergo second merge with the V+ of the restructuring verb at some stage in the derivation. This can be achieved either by V+ extraction from the complement, or by pied-piping, subject to the extension condition (Chomsky 1995).

3. Clustering verbs select for particular types of complements (infinitives, participles, *te*, etc.). Note that the single computational engine forces the projection of infinitival morphology (Inf), participial morphology, and *te* to be in the (narrow) syntax: there are no other places to build structure.
   (a) Assuming Inf first merges with V+, there are two options to satisfy the needs of Inf morphology (it selects for V as a specifier): either V second merges with Inf (stranding V+), or V + second merges along with V. Both options are available in the grammars of the three languages, and lead to different surface constituencies and orders.
   (b) Infinitival complements are always "CPs": C needs to be typed as infinitival: this forces Inf to merge with C.
4. Surface constituency is the result of external and internal merge (this is all that is available), and derivations obey the extension condition and locality. In the course of the derivation, there are very few options[8]: they concern whether the Spec extracts (splits the structure into smaller parts), or whether pied-piping occurs (keeps local parts together). Spec extraction keeps the derived structures small, but creates discontinuous structure, pied-piping keeps local parts together, but yields heavy phonological constituents.
5. Individual heads can have what we called complexity filters associated with them; here these are renamed as epp[[+ph] max]. These are an important locus of variation, both between speakers and between languages, and these are responsible for reining in recursion.

The Output of the Syntactic Derivation Yielding 12, 21, 123, 312 and 321 Orders

Given these derivations, we can examine the linear orders, and match these to their output syntactic structures (elements will be pronounced in positions where all properties have been satisfied). Thus, we can zoom in on the particular configurations that yield 21, 312 or 321 structures (which were of interest because of the restrictions on recursion). These structures can be arranged in terms of the [[+ph] max] size. This ranges from 0 in Dutch, for the 12 and 123 patterns (there is no local phonological material present at spell-out—the infinitive is pronounced at the edge of the infinitival CP, and distributes like a CP), to the smallest number n of possible

[8]Other choices leading to variability depend on the height of merger of functional material and order.

nodes, which gives the 21 orders, to $n + 1$, the next smallest possible amount of structure embedding, which is associated with the 312 pattern, to the maximum size in this location, $(n + 1) + x$, which gives the 321 pattern. The following table gives the possible patterns for Dutch.

(20) Dutch: V+[epp, Inf[[+ph] max]]

	12	123	21	312	321
	+ph max:[0]	+ph max:[0]	+ph max:[n]	+ph max: [n+1]	+ph max [[n+1]+x]
epp_{inf} [+ph max: [0-n]]	√[9]	√	√	*	*

What is relevant here is that the output of the syntactic derivations provide the vocabulary to state the restrictions, reflecting an independently necessary algorithm which captures the sensitivity of the distribution of phonological material to depth of syntactic embedding. Indeed, this is expected if syntactic structures are directly transferred to phonology.

More on the 312 Order

The 312 orders are particularly important. Under the current analysis, these are excluded because of a [+ph] max size associated with the category Inf, as shown below.

(21) *... *zwemmen zal willen*
swim.inf will want.inf
'... *will want to swim*'

(22)

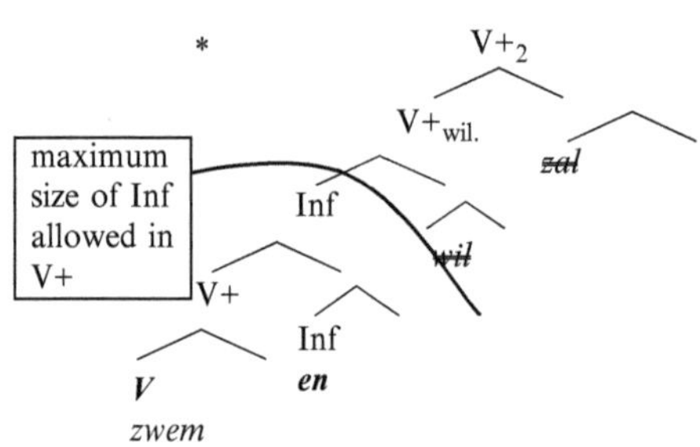

[9]Suppose there is a language that is just like Dutch, but where the epp of V+ requires [+ph], i.e. the presence of phonological material in V+. This will rule out *V1 V2 orders, but allow for 312 orders, (found in some German varieties (Wurmbrand 2004) (with 312 occurring in the IPP context.

As expected 312-forms are possible in certain contexts in German and in neutral contexts in Hungarian: this follows from the idiosyncratic absence of a [[+ph] max] restriction.

This particular derivation yields an output in which an infinitive is syntactically embedded in a specifier. On the surface however, it just looks like there is a simple infinitive. So access to the syntactic structure must be the determining factor for ruling out this pattern.

Similarly, for certain Dutch verbs, the infinitival and participial forms are indistinguishable segmentally and prosodically. Yet, these forms are strongly excluded when the syntax calls for an infinitive, but fine in contexts where the syntax calls for a participle.

(23) a. Dat hem toch zoiets (*overkomen) zou moeten/ zou moeten overkomen!
that him PART so-something (overcome.inf) would must.inf/would must.inf overcome.inf
'That something like this should have to happen to him!'
b. Dat hem toch zoiets overkomen zou zijn/zou zijn overkomen!
that him PART so-something overcome.part would be /would be overcome.part
'That something like this should have happened to him!

Thus, to exclude the 312 context, reference to the syntactic category infinitive is crucially necessary, and that is exactly what the present proposal expresses.[10]

These examples are important in other ways as well; they show that purely surface prosodic constraints play no rule in excluding these examples, though they might look relevant in other cases. Under the account pursued in this paper, the contrast between (23a) and (23b) is entirely expected as it is dependent on the depth of syntactic embedding dominating Inf, which plays a role in accounting for other excluded cases as well.

3.3 *Do Grammars Really Count? Where Embedding Twice Is Fine, but Thrice Is Too Many*

If the proposal here is on the right track, we should also find cases of languages or dialects which have a slightly larger cut off for recursion, and allow recursion twice, but not a third time. That is, we expect to find grammars that allow a limited amount of counting. What would such a case look like? As Koopman and Szabolcsi (2000, p. 124) discuss, in the context of Cinque's (initial) restructuring proposal,

[10]See Koopman (2002) for further discussion of problems with the prosodic account in Den Besten and Broekhuis (1989), and the processing account in Broekhuis (1992).

some Hungarian speakers simply "do not accept climbing across two infinitives, regardless of their relative order" (Koopman and Szabolcsi: p. 124 (31) and (32)):

								Hungarian "B"	Hungarian "A"
(24)	a.	Haza	fogok	akarni	menni		4-1-2-3	OK	OK
		home	will-	want.inf	go.in				
		'I'll want to go home'							
	b.	Haza	fogok	akarni	kedzeni	Menn i	5-1-2-3-4	OK	*
		home	will.I	want.inf	begin.inf	go.inf			
		'I'll want to begin to go home'							

Within the perspective of the current paper, these facts immediately suggest an answer: in the 4123 configuration in (24a), 4 (a particle) formed a complex verb with 3, which subsequently extracted and combined with Inf, 3 is embedded under 2 which has subsequently extracted, etc., ultimately yielding the representation: [[[**4**] ~~3~~] ~~2~~] ~~1~~] [**1** [**2** [**3**]]]. But suppose that for some dialect, (which turns out to be spoken in the Széged area, see Szendrői and Tóth (2004), this output also sets [+ph max] of the LI that has attracted the remnant constituent containing 4. This rules in (24a) but rules out (24b), as the additional cycle will yield a violation of [+ph max].

This in turn leads to predictions about the correlations we expect to find. Restrictions on recursion come about as a special phonological property of the epp. These should show up only in cases where V+ is filled, and be sensitive to the size of the constituent at the end of the derivation. Within the context of Hungarian, where the three possible orders (roll up, climbing, and English orders) are allowed by all speakers (albeit in different contexts), we expect the judgment patterns in (25), based on the first column, where V1 has extracted to T. Of particular relevance are the (c) patterns of Hungarian Széged.

(25) a. If V1 [V3 V2 ~~V1~~]] is √ *then* √: V3 V1 V2 *and* √ V1 V2 V3
b. If V1 [V4 V3 V2] is √ *then* √: V4 V1 V2 V3 *and* √ V1 V2 V3 V4
c. **If V1 [V4 V3 V2] is * *then* *: V4 V1 V2 V3 but √ V1 V2 V3 V4**

At this point, the actual correlating patterns found within speakers have not been investigated, but the following data from Szendrői and Tóth (2004) are suggestive enough to warrant further inclusion here. In their questionnaire-based study of the regional distribution of verbal complexes in Hungarian, Szendrői and Tóth (2004) find that speakers in this particular region neither allow climbing nor four verb inverted clusters, as expected. Judgments on English orders improve comparatively as expected, Szendrői and Tóth however suggest that these should not be considered acceptable in this area either, and state that "speakers of this region seem to have a general aversion to verbal clusters".

This conclusion appears to be too hasty, as we need to understand what to measure improvement against. In fact, their data show the predictions go in the right directions. The raw questionnaire data (on a three point scale) show a 100 % rejection of climbing and inverted orders, but a 20 % improvement or acceptance of the English order (their Fig. 10). This may look like a minor effect, but if we look at the acceptance rates of such cases in the overall study we see the same pattern. Throughout the Hungarian region, there was a very high number of negative answers for three or four verb clusters in the English orders (40 % no, 30 % maybe, 30 % yes) with inverted orders degrading 20 % (60 % no, 20 % yes, 20 % maybe). Given the fact that speakers in the Szeged region in question start out with 100 % no for inverted and climbing orders, an 80 % no, which is a 20 % improvement. This might well be all that is expected for English orders anyway. Interestingly, throughout the Hungarian region, the acceptability rates of the three orders with 3/4 clusters follow the predictions: full roll-up inversion is consistently judged worse than climbing orders, and English orders are judged most acceptable. This is expected from the thesis in this paper, where maximal phonological size is measured on the output of syntactic structures, with fully inverted structures of the biggest phonological and syntactic size, climbing structures next on the scale, and English structures lightest. These data then seem to make a reasonable case for the suggestion that grammars indeed must allow a limited amount of counting, where this is restricted to matching phonological material with syntactic outputs as a function of the epp[+ph max] of specific LIs. While further investigation should confirm if the predictions about distributions indeed hold up, it is remarkable that the proposed syntactic derivations together with independently motivated interface condition that is sensitive to structure dominating phonological material not only create order in understanding where and why recursion fails.

4 Conclusion

This paper focused on how to understand recursion restrictions given current theoretical understanding. It argues that such restrictions are to be accounted for by independently necessary phonological properties, which can be "grafted" on the structure building epp. These have the effect of reining in recursion. We discussed, in particular, cases in which recursion is restricted in the domain of verbal complexes. Here, specific syntactic derivations provide the vocabulary to understand particular idiosyncratic cases of recursion, which result from the sensitivity of phonological material to syntactic embedding in highly localized contexts. The epp of any atom thus serves as a repository of the finely grained knowledge speakers have of the phonological properties associated with local syntactic environments. If this hypothesis is on right track, recursion restrictions are expected to be limited to particular syntactic configurations (not in first merge but second merge cases). They are expected to not be absolute, but to vary according to region or perhaps generation, as the [+ph max] setting is set on the basis of primary input data.

In effect, then, the study of the messy and variable patterns leads to factoring out the general workings of syntax from the specific interactions with phonological insertion. But this is only possible in light of fully explicit and detailed syntactic derivations since only these will lead to insights into the treatment of variability and specific predictions that can be verified or not.

While the paper focused on the [+ph max] restrictions and their relation to recursion, this study can start to make some sense of the data reported in Bresnan and Ford (2010). In their study of double object construction in American and Australian English varieties, Bresnan and Ford find that speakers of each variant have different probabilistic knowledge of the constructions that corresponds to corpus data (where available). Importantly, in Australian English but not American English, the length and phonological weight of the first object in the double object construction plays a significant role. Within the present paper, this suggests a different settings of the [+ph max] settings for the head hosting the first object, with Australian English allowing a small max, and American English a much bigger one. This looks comparable to the difference between German pronominal genitives and English pronominal genitives, or the difference between the different sizes Japanese T and English T can host (Sect. 2.2). This may reflect an absolute size restriction, or statistical knowledge about with the [+ph] properties associated with the epp, i.e. about particular phonological shapes. Thus, the [+ph] properties grafted on the structure-building epp could be the very locus where all such knowledge is represented and stored.

Appendix A: A Possible Linearization Account for the Recursion of Infinitives?

We add here a short discussion of Norvin Richards (2006) linearization account. Richards Proposes a general constraint on linearization, which imposes distinctness of linearization of categories within the same phase, a kind of modern incarnation of the OCP, and applies it to doubl-ing. Adapting parts of Kayne's (1994) LCA, his proposal in essence prohibits two categories within the same phase from being spelled out, because they would yield a conflicting linearization statement. He applies this to a number of different phenomena (distinct markings on DPs, the doubl-ing filter (Ross 1972, but see Koopman 2002 for an explicit proposal along the lines in this paper). Applied to Dutch, this proposal would prohibit the linearization of two infinitives within the same phase, but not across a phase boundary. It is easy to adapt Richards (2006) to Dutch, given the syntactic analysis developed in Koopman and Szabolcsi (2000) and outlined here. Indeed, inverted infinitives end up occurring in the same V+ constituent (hence within the same phase), but infinitives in the VT1 Vinf2 Vinf3 order are each in CPs, hence separated by a phase boundary [$_C$ V_{inf} [$_C$ $V_{inf.}$. This would allow some understanding as to why Inf is subject to the restriction on recursion but other categories are not, which under the account in this paper is purely a historical accident. However, the proposal does not extend (easily) to any of the other observed and excluded patterns. It will not extend to exclude the 3 1 2 pattern with infinitives in Dutch (**zwemmen wil kunnen*), as here the infinitives do not seem to be in the same phase as the selecting infinitive. Nor will it allow a simple understanding of variation, i.e. what happens with Dutch dialects or speakers that allow such patterns, or with German and Hungarian speakers? How do we let the right cases in? If Richards needs Koopman and Szabolcsi (2000) to understand Dutch, the problem then is how to allow German and Hungarian which allow surface constituents of this type. Richards proposal could force a double structural analysis where preverbal infinitives in German and Hungarian are or can be in their own CPs or within smaller constituents, but this would lose all the basic insights of Koopman and Szabolcsi's analysis (see Koopman and Szabolcsi for an argument against this option), and any hope to model the observed variability between individual grammars. This analysis will have nothing to say about the Hungarian cases where it seems you can invert twice but not three times, and where the restriction on recursion does not seem sensitive to a particular category, just to dept of embedding in a specific location. And finally, it will have nothing to say about other cases of restrictions on recursion, and the general notion of [+ph] size that is independently needed, and that seems widely applicable.

References

Barbiers, S. (2002). Book review: Koopman, H. and Szabolcsi A. 2000. *Verbal Complexes: Journal of Comparative Germanic Linguistics, 6*, 53–79.
Bresnan, J., & Ford, M. (2010). Predicting syntax: Processing dative constructions in varieties of American and Australian english. *Language, 86*(1), 168–213.
Broekhuis, H. (1992). *Chain government: Issues in Dutch syntax*. Leiden: HIL.
Chomsky, N. (1995). *The minimalist program*. Cambridge, MA: MIT Press.
Cinque, G. (1999). *Adverbs and functional projections: A cross-linguistic perspective*. New York: Oxford University Press.
den Besten, H., & Broekhuis, H. (1989). Woordvolgorde in de werkwoodelijke eindreeks. *GLOT, 12*, 79–137.
den Besten, H., & Edmonston, J. (1983). The verbal complex in Continental West Germanic. In W. Abraham (Ed.), *On the formal syntax of Westgermania* (pp. 155–216). Amsterdam/Philadelphia: John Benjamins Publishing Company.
Kayne, R. (1994). *The antisymmetry of syntax*. Cambridge, MA: MIT Press.
Kayne, R. (2000). *Parameters and universals*. New York: Oxford University Press.
Kayne, R. (2005). *Movement and silence*. New York: Oxford University Press.
Koopman, H. (2002). Derivations and filters. In A. Alexiadou, E. Anagnostopoulou, S. Barbiers, & H. Gaertner (Eds.), *Dimensions of movement: From features to remnants* (pp. 151–189). Amsterdam: Benjamins Publications.
Koopman, H. (2005a). On the parallelism of DPs and clauses in Kisongo Maasai. In A. Carnie (Ed.), *Verb first* (pp. 281–303). Amsterdam/Philadelphia: John Benjamins Publishing Company.
Koopman, H. (2005b). Korean (and Japanese) morphology from a syntactic perspective. *Linguistic Inquiry, 36*, 601–633.
Koopman, H. (2006). In defense of the Spec head configuration. In C. Boeckx (Ed.), *Agreement systems* (pp. 159–199). Amsterdam/Philadelphia: John Benjamins Publishing Company.
Koopman, H., & Szabolcsi, A. (2000). *Verbal complexes* (Current studies in linguistics 34). Cambridge, MA: MIT Press.

Longobardi, G. (2001). The structure of DPs. In M. Baltin & C. Collins (Eds.), *The handbook of contemporary syntactic theory* (pp. 562–604). Malden/Oxford: Blackwell Publishers.

Richards, N. (2006). *A distinctness condition on linearization*. ms, MIT. http://web.mit.edu/norvin/www/papers/Distinctness2.pdf

Ross, J. (1972). Doubl-ing. *Linguistic Inquiry, 3*, 61–86.

Sportiche, D. (2005). Division of labor between reconstruction and merge: Strict locality of selection and reconstruction paradoxes. http://ling.auf.net/lingBuzz/000163

Szendrői, K., & Tóth, I. (2004). Hungarian verbal clusters: Results of a questionnaire survey. In K. Kíss & H. van Riemsdijk (Eds.), *Verb clusters. A study of Hungarian, German and Dutch* (pp. 87–121). Amsterdam/Philadelphia: John Benjamins Publishing Company.

Williams, E. (1981). On the notion 'Lexically related' and 'Head of a Word'. *Linguistic Inquiry, 8*, 101–139.

Wurmbrand, S. (2004). West Germanic verb clusters: The empirical domain. In K. Kíss & H. van Riemsdijk (Eds.), *Verb clusters. A study of Hungarian, German and Dutch* (pp. 43–87). Amsterdam/Philadelphia: John Benjamins Publishing Company.

Deriving the Two-Argument Restriction Without Recursion

Eva Juarros-Daussà

Abstract It is a universal property of argument structure that verbs cannot take more than three arguments –one external and two internal (Hale K, Keyser SJ, On argument structure and the lexical expression of syntactic relations. In: Hale K, Keyser J (eds) The view from Building 20: essays in linguistics in honor of Sylvain Bromberger. MIT, Cambridge, 2002, henceforth H&K); any additional (pseudo)arguments are introduced via (possibly recursive) syntactic or discursive procedures, not lexical. This is the Two (internal) Argument Restriction (TAR), which is part of the basic architecture of natural language. It is argued that a modified version of the theory presented in H&K can account for this restriction. In this proposal, the TAR crucially depends on the lack of recursion in the lexicon (in building argument structures), thereby restricting the possibility of using this formal mechanism to the domain of sentential syntax (building of sentences). Argument structures are thus built in a cyclic and directional way, and without recursion.

Keywords Argument structure • Lexicon • Recursion • Selection • Cyclicity • Argument • Adjunct • Ditransitivity • Lexical syntax

---

Thanks to Chisato Kitagawa, Jean-Pierre Koenig, Tilman Lanz, Tom Roeper, Peggy Speas, Ellen Woolford, and two anonymous reviewers for their generous help.

E. Juarros-Daussà (✉)
Department of Romance Languages and Literatures, University at Buffalo, Buffalo, NY, USA
e-mail: ejuarros@buffalo.edu

T. Roeper and M. Speas (eds.), *Recursion: Complexity in Cognition*, Studies in Theoretical Psycholinguistics 43, DOI 10.1007/978-3-319-05086-7_3,

1 Properties of Argument Structure

Most linguistic frameworks accept that each verb in natural language is associated with a set of arguments, which are not systematically predictable from the verb's meaning, and which form its argument structure. While argument structures can also include adverbial or other semantic information associated with the predicate, arguments are commonly taken to be those elements that are realized syntactically as the projected sentence's subject, direct and indirect objects. Other participants or anchor elements appear in the sentence as adjuncts, typically completing and further specifying the meaning of the predication. The distinction between arguments and adjuncts has thus been long recognized in linguistic theory (e.g., Dowty 1979, 1991, 2001; Chomsky 1986), even when no set of necessary and sufficient criteria has yet been proposed as the basis for the ontological distinction between these two kinds of elements (cf. Koenig et al. 2003; Sybesma 2004), and the precise differences in the way they are introduced in the sentence's structure remains a subject of debate (e.g., Borer 2005; Babby 2009; Randall 2009).

At the processing level, most frameworks also accept as a working hypothesis the distinction, supported by psycholinguistic evidence, between some participants that are implicitly introduced by the verb as soon as it is recognized (lexically encoded arguments), and those participants that, even when they might be part of general world knowledge associated with the verb and frequently appear in the same context as the verb, are not necessarily activated each time the verb is encountered, and are thus adjuncts (Boland 2005).

Throughout this paper, thus, argument structure is taken to be the level of linguistic representation which includes a predicate and all of its lexically encoded arguments. In this context, this paper explains the basic fact (*apud* Hale and Keyser 2002) of argument structure that verbs cannot take more than three lexical arguments –more precisely, one that will become the sentential subject, or *external* argument, and a maximum of two additional arguments that will be realized as the sentential objects, or *internal* arguments. This maximal argument structure corresponds to that of the English verb *give* (1).

(1) [The LinguistList] gave [nice prizes] [to the winners of the challenge]

Proof of the restriction is that introducing a conceivable additional participant without the help of a lexical preposition (such as *for* in (3) below) which contributes its own argument-taking abilities, results in ungrammaticality (2):

(2) * [The LinguistList] gave [nice prizes] [to the winners] [the students]
(3) [The LinguistList] gave [nice prizes] [to the winners] [for the students]

In previous work (Juarros-Daussà 2010b), I showed that this limitation in fact concerns the number of *internal* arguments associated to a predicate, and thus formulated the above empirical observation as the *Two-Argument Restriction* (TAR):

"A single predicate can have at most two internal arguments."[1] I claimed that the TAR is an *unrestricted universal* (in the sense of Croft 1990), and I argued that the commonest valence-increasing operations, which result in applicative and causative constructions, while they present a challenge, however do not to violate the TAR.[2] I further showed that, since there is no known processing or discursive reason not to lexically associate more than three (strictly, two) participants to a predicate, the TAR is syntactic in nature, and it is one of a family of architectural constraints that determine and limit possible attainable languages (in this case possible argument structures).

Following that work, in this article I propose an explanation of how (an extension of) the framework of lexical syntax put forth by Hale and Keyser (2002) is especially suited to derive the TAR at the theoretical level. Crucially, in my proposal, deriving the TAR involves negating the existence of a recursive function in the domain of argument structure. The paper is structured as follows: first I show how H&K's theory as it is encounters the problem of generating a recursive structure in order to account for the maximal argument structure (the one in *give* above). I claim that, since recursion threatens the restrictive power of the theory, it should be avoided. I argue that an account within H&K's framework that avoids recursion can be achieved once we accept a restriction in the combination of lexical categories, the *Uniqueness of Selection Hypothesis (USH).* I then show that such view of argument structure derivation, together with a cyclical, directional view of lexical Merge very much alike the one advocated in Chomsky (2001) and Collins (2002), and captured in the proposed *Cyclicity of Selection Hypothesis (CSH)*, can both capture the maximal argument structure (that of a predicate with two internal arguments), and explain its maximality.

---

[1]The TAR has been implicitly accepted in many frameworks. Rarely, one finds explicit clam of the restriction. Un such rare examples is Levin and Rappaport Hovav (2005: 233); discussing template augmentation in ditransitives, these authors note "Such verbs cannot undergo event composition, which adds another subevent, since they already have a maximally articulated event structure." However, I have not found the restriction the matized anywhere.

[2]Still unexplained is the counterexample presented by what Juarros-Daussà (2010b) called bet-verbs, which seem to systematically accept a forth argument, as in

(i) [I] bet [you] [fifty dollars]

While verbs like *bet*, taking two phrasal internal arguments and one sentential internal argument, are rare, they potentially present a legitimate counterexample. Even more worrisome for the universalist claim of the generalization is the fact that certain verbs like promise or guarantee can also enter into what seems a four-argument construction, as in (ii) (Lyn Frazier, p.c.):

2 Deriving the TAR: Rationale of Eliminating Recursion in Argument Structure

Ideally, the TAR, as an empirical generalization, would follow from independent principles of grammar. To that purpose, it is useful to recall that, in essence, what we are deriving here is the fact that argument structures are finite in a way that sentential structures are not: they have a limit on the number of linguistically encoded participants, where at the level of sentences there is not such restriction. Now, it is well known that in many grammars recursion is the formal mechanism used to attain this kind of infinitude at the sentential domain.[3] Therefore, one obvious way to explain the lack of infinitude in argument structures is to simply claim that recursive mechanisms are not available in this grammatical domain. This apparently simple move only makes sense within a theory that otherwise assumes the same principles to apply for the levels of argument (lexical) structure and sentential structure – otherwise, the finitude of argument structures would be a fact incommensurable with the infinitude of sentences (they would just form two different grammatical systems, and their properties following from the same mechanisms would be no more than a coincidence). Hale and Keyser (2002) present such a theory, and I next show that, when assuming it, the claim of there be no recursion in the building of argument structures precisely derives the facts captured by the TAR.

2.1 *Hale and Keyser's (2002) Theory of Lexical Syntax*

Lexical syntax, as it is compiled from a series of formerly distributed articles in Hale and Keyser 2002 (henceforth H&K), is a theory of the grammatical level representing the relationships between the verb and its arguments, without taking into account their extended or functional projections. It assumes that a verb's argument structure is not a mere enumeration of its arguments, but rather a structural representation that follows the same principles as sentential syntax.

Crucially, H&K claim that verbs are not syntactically simplex items, but rather they all contain heads (roots) with diverse projection and incorporation properties. For example, even an otherwise seemingly simplex (from the point of view of its internal argument structure) unergative verb like *laugh* is created by incorporating

[3](ii) [We] promise [our clients] [our first born] [that we will deliver your order in time]

In strings captured by rules in which an element contains another element of its same type. For example, NP → N PP and PP → P NP, which results in potentially infinite NPs.

the complement of a transitive-like structure into a light-verb-like head –a structure equivalent to the analytical counterpart "do laugh"[4]:

(4)

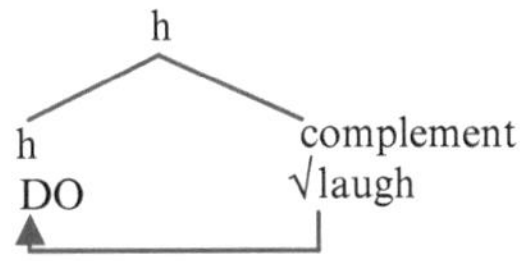

In defining the lexical categories that form the building blocks of argument structures, two structural relations are taken as basic: whether the lexical head (*h* below, the root) has a specifier (*spc*) or not, and whether it has a complement (*cmp*) or not. Based on these parameters, H&K (2003: 13, adapted) distinguish the following four lexical structural categories (α is an unprojected head):

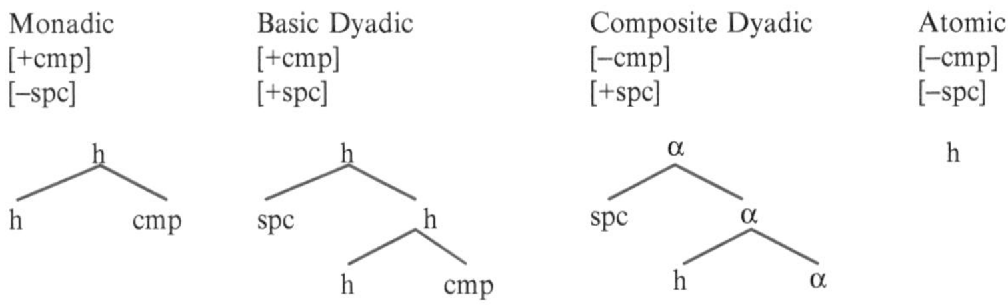

These structures contain information about the projection properties of the word. Obviously, structural information of this kind might coexist in the lexical entry with information of semantic character that also plays a role in the syntactic behavior of the word, such as aspectual type, certain features on the arguments such as definiteness or number, or features on the root itself, such as boundedness.[5] It is the interaction between structural and semantic information what ultimately determines the syntactic behavior of the word.

Under this view, all words are morphologically derived from one lexical type or other: there are no verbal, nominal or otherwise, basic stems, but the category of the word depends on the structure the stems appear in (much in the spirit of Distributed Morphology). Since all possible argument structures are formed by combinatorial merging of these primitive structural categories, the goal of the theory is to determine the zero-relatedness of all predicates, i.e., to identify their argument structures in terms of the types allowed by the basic lexical categories above.

---

[4]A vision supported by those languages that present surface transitive structures (oftentimes based on a light verb) for many of their verbs which would otherwise be crosslinguistically unergative, such as Basque and Persian (see, for example, Folli et al. 2005).

[5]See the relevant work of Heidi Harley (1995, 1996), Folli and Harley (2005), and Folli et al. (2005).

In the fashion illustrated for unergative verbs above in (4), the above configurations correspond prototypically to the following taxonomy in the verbal domain:

(a) monadic: unergative (*John laughed*) or (pseudo)transitive (*John made a fuss)*
(b) basic dyadic: transitive (*I shelved books*) or ditransitive (*I put books on the shelf*)
(c) composite dyadic: unaccusative (*Many guests arrived; The screen cleared*)
(d) atomic: presumably weather verbs (*It rains*), although they are not discussed in H&K.

Other argument structures are obtained by combination of the basic structural categories. For example, a transitive verb like *shelve*, in *Ayse shelved the books*, is formed by combination of a basic dyadic structure with a monadic head:

(5)

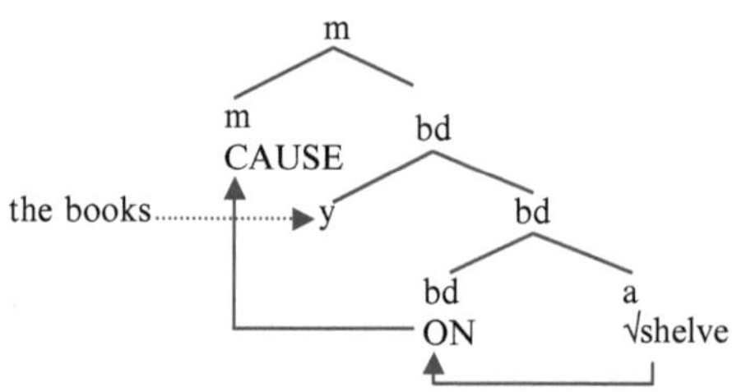

In this structure, a basic dyadic head *bd* (roughly glossed as the homonymous preposition *ON*) projects a structure with the properties of [+complement, +specifier]. The [+complement] requirement is satisfied by the morphological constant (the root), itself an atomic structure. Because of *bd* being phonologically empty, a (*shelve*) is automatically incorporated (conflated) onto it. The required specifier is projected, and is occupied by a lexical variable, here represented by *y*, acting as a placeholder for the object to be inserted in sentential syntax. The whole basic dyadic structure is itself the complement of a monadic element, which is responsible for the introduction of the external argument in sentential syntax (hence its gloss as CAUSE), thus making the resulting verb necessarily transitive (cf. **The books shelved*.) Again, the emptiness of *m* requires the incorporation of the complex [shelve + p], resulting in the transitive verb *shelve*.

In this fashion, H&K claim that the constrained nature of argument structure[6] follows from the structural properties of the existing lexical categories. In a nutshell, we have a limited typology of argument structures because of the structure limitations of the primitive building blocks that form those argument structures. Despite of this claim, a problem arises when, in building argument structures for different kinds of verbs, H&K allow for unrestricted, recursive combination of the

[6]In Hale and Keyser (1993), what they mean by the constrained nature of argument structure is the limited absolute number of theta roles.

lexical categories. For example, in their argument structure for *give*, as in I *gave the bottle to the baby*, a basic dyadic category contains another basic dyadic category, a classical example of recursion (H&K 2003: 163; some re-labeling applied):

(6)

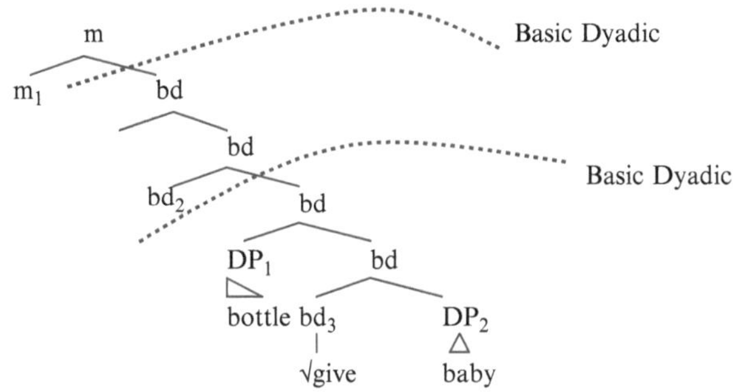

In H&K, nothing prevents recursion of the lexical categories in this fashion; in fact, it seems that H&K's model actually predicts that, in order to build the argument structures of verbs with two internal arguments (such as *give* above), recursive merging of the basic lexical categories is actually required. The result is that their system produces a potentially infinite number of argument structure types. Moreover, driven to its ultimate consequences, this situation has the potential to create argument structures with an unlimited number of arguments, each one introduced by recursive combination of the basic lexical categories. Such structures would obviously be violating the TAR –something that does not occur in natural language.

2.2 *Maximal Verbs*

Many verbs have[7] the maximal argument structure of a verb with two internal arguments: obligatory ditransitives like *give, send*, etc., optional ditransitives like *write, kick, pass*, etc. (*write a book* and *write a sentence on the blackboard*; *kick a ball* and *kick me a ball*), prepositional verbs like *put, splash, pour*, etc.; and within the latter, many denominal verbs that have received special attention in the literature: *butter, nail, spit*, etc. (Jackendoff 1990; Randall 2010; Juarros-Daussà 2003). In this article I will take the verb *spit* as an example, following my previous work (2003, 2010), as a representative of this kind of verbs[8] that are formed by the

[7]Or, should we say, "frequently or even regularly have" (Jespersen 1927: 278, *apud* Mukherjee 2005).

[8]Levin (1993: 218) calls them verbs of "bodily process" of the "breathe" group, and includes the following: *bleed, breathe, cough, cry, dribble, drool, puke, spit, sweat, vomit* and *weep*. Although semantically unified (they are all verbs of excretion, related to emitting a substance from the body), the group in Levin (1993) is not unified as for syntactic behavior. Hence, for example, the verbs in (i) form good sentences, but the ones in (ii) do not:

maximal argument structure found in natural language. Crucially, the analysis of the argument structure of these verbs presented here does not need recursion, but only one single level of selection among lexical categories. Due to the fact that the argument structure of this type of verbs is taken as the most complex one that we find in natural language (a predicate with two internal arguments), and given that we can account for it avoiding recursion, a general principle banning recursion in the lexicon altogether is deduced, as it makes sense in the theoretical framework of H&K (the *Uniqueness of Selection Hypothesis,* or *USH*). The rationale is that if the USH allows for the structurally most complex case to be accounted for, simpler cases might be straightforwardly explainable within the limits allowed by the USH. The implications of this move for syntactic theory are also considered; mainly, it is claimed that a derivational conception of syntactic processes is not only favored, but necessary for the idea to work.

The verb *spit* presents the following argumental possibilities:

(7) a. When llamas get upset, they spit [$_{VP}$ $V_{unergative}$]
 b. The volcano started spitting ash and lava [$_{VP}$ $V_{transitive}$ NP]
 c. The baby spat porridge on the table [$_{VP}$ $V_{resultative}$ NP PP]

The latter example is one of the verb with two internal arguments. The challenge is to present an analysis of this possibility that makes use of no recursion in the sense discussed above.

The first task is to determine the zero-relatedness of *spit*-verbs in terms of the lexical categories above. I claim that the properties of these verbs are compatible with them being related to a composite dyadic (*cd*) category. We will see that, for this, we have to revise the structure of *cd* lexical categories, which are the hardest to interpret in H&K's system. In Juarros-Daussà (2003). I presented a revision of this structure, in order to make it more compatible with the tenets of their own theory. I here present a summary.

According to H&K (2003: 16), a synthetic unaccusative verb like *clear* in the inchoative sentence *The sky suddenly cleared*, has the complex lexical projection structure of a composite dyadic element (*cd*):

(8)

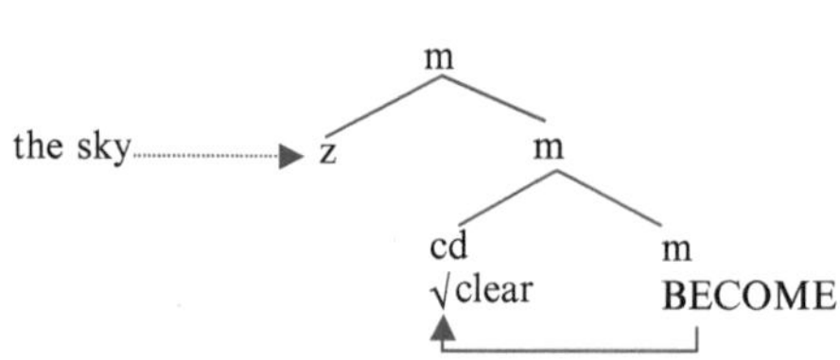

In this structure, a head *cd* (the root √clear) projects a composite dyadic structure,[9] by definition a [−complement, +specifier] category. In orderto do so,

[9](i) The president spat/vomited/puked on the floor
(ii) * The president breathed/coughed/drooled/sweated on the floor

it appears as the complement of a monadic element, *m* (roughly glossed as the light verb *BECOME*), hence satisfying *m*'s lexical requirements. H&K then say that *m* acts as a host that projects the specifier required by *cd*, hence providing structural space for a variable (*z*) that acts as a place holder for syntax to contribute the derived subject argument (*the sky*). Conflation of phonologically empty heads onto phonologically full ones produces the target inchoative unaccusative verb once the default morphosyntactic realizations of the lexical categories for English are instantiated (mainly, m = Verb, cd = Adjective).[10]

The whole basic dyadic structure can optionally be constructed as the complement of a monadic element m_2, which would be responsible for the introduction of an external argument in syntax, thus forcing the internal argument to become the verb's object. This is the structure corresponding to the transitive sentence *The wind cleared the sky*. Again, the phonetically null character of *m* requires the incorporation of the complex [[BECOME + √clear] onto [CAUSE].

(9)

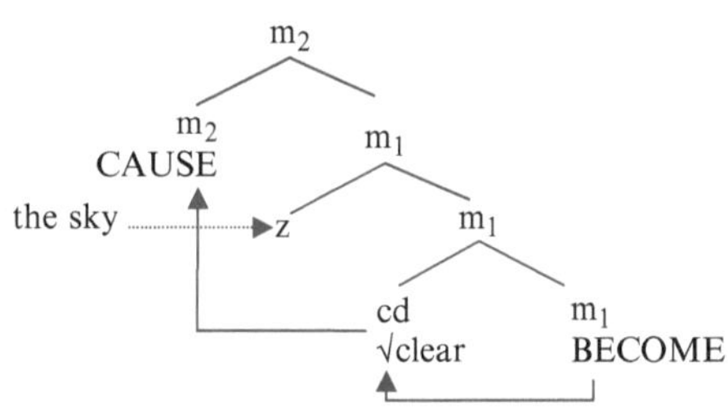

Since composite dyadic verbs are predicted to participate from this kind of diathesis alternation, were we to defend now such an analysis for *spit*-verbs based on a composite dyadic structure, we will have to offer an explanation for the lack of alternation in these verbs:

(10) a. The volcano was still spitting ash and lava
b. *Ash and lava were still spitting

I will do so in the next section, relating the absence of inchoative form to the Source role of the subject, following a suggestion first presented in Hale and Keyser (1993), and refined in their (1998) paper. But first let us just see in detail how a derivation of *spit*-verbs as composite dyadic element works out under the assumptions presented above.

For the reasons explaining why clear projects a composite dyadic structure instead of a basic dyadic one, please read Hale and Keyser (1993, 2002).

[10]This structure will be revised below.

Following the model of (8), the root $\sqrt{spit}$, itself a composite dyadic element, merges with a monadic element in order to project a specifier for its argument:

(11)

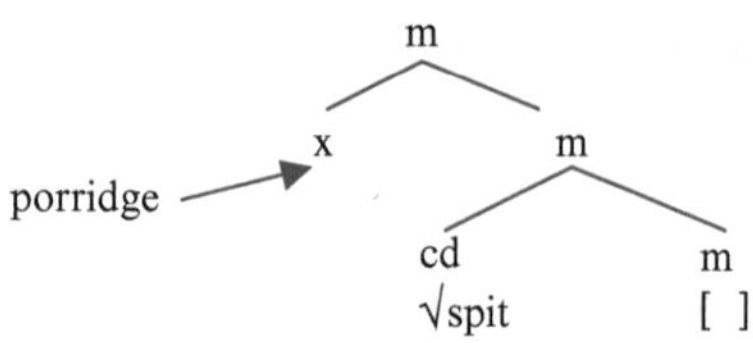

Let us next direct our attention to the PP argument in *The baby spat porridge on the table* (7c above). Arguably (Randall 2010, Juarros-Daussà 2003), when it appears, the PP is part of a secondary resultative predicate, with a head (roughly glossed as CAUSE, apud Stowell 1983, Bowers 1991, cf. Kratzer 2005[11]) that takes the Theme argument (*porridge*) as its subject and the PP (*on the table*) as its complement. I here propose that a basic dyadic structure containing both the Theme argument and the resultative PP be inserted in specifier position of a composite dyadic structure, where the Theme (overt or not) would otherwise be usually inserted:

(12) The baby spat porridge on the table

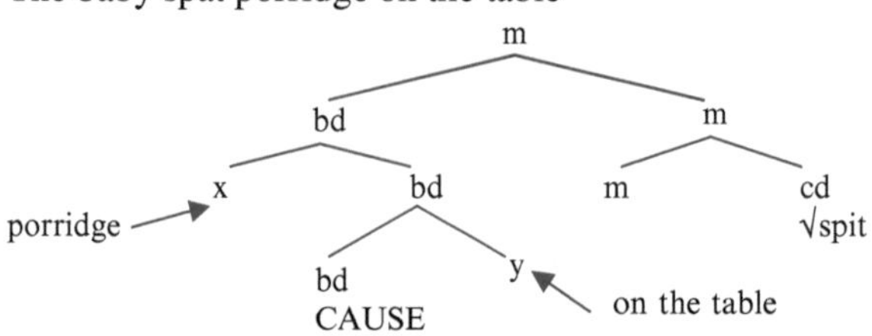

While in (12) all relevant predication relations are met, there is still one problem: the relation between the surface verb and noun *spit* (resulting from the default morphosyntactic realization of lexical categories in English) is not captured. Although H&K present de-adjectival verbs (*clear, narrow, thin, redden,* and the like) as the clearest example of unaccusative (composite dyadic) verbs, this does not mean that unaccusatives are limited to this class (cf. Levin and Rappaport Hovav 2005). *Spit*-verbs are de-nominal verbs, and still they are composite dyadic. To capture this claim, I propose that in the structure in (12), it is actually the *cd* element in (12) which is phonetically empty, the morphological constant (the root *spit*) being an atomic head, whose default morphosyntactic realization is a N, instead of a monadic head, whose default morphosyntactic realization is a V. The complex

[11]Note that Kratzer (2005) outlines an alternative analysis of resultatives as uniformly raising constructions, in the context of arguments being introduced by functional categories (see also Borer 2005).

thereby becomes (by default assignation of morphological categories) a lexical noun (with the rest of arguments unmodified)[12]:

(13) The spit of porridge on the table

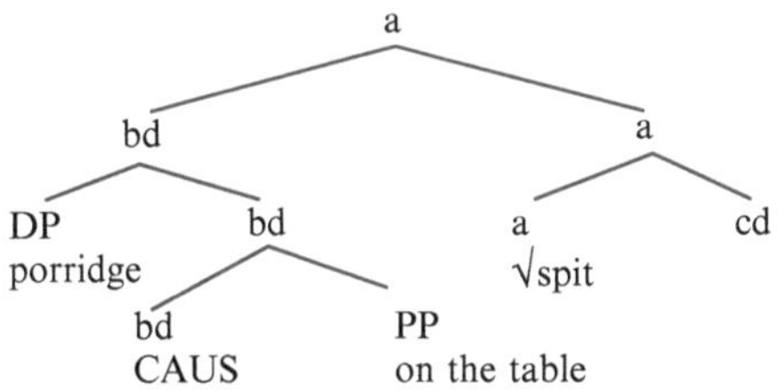

The whole complex can itself be embedded under a *monadic* head, to form either the activity nominal or the verb *spit*:

(14) (The baby) spat porridge on the table/ (The) spitting of porridge on the table

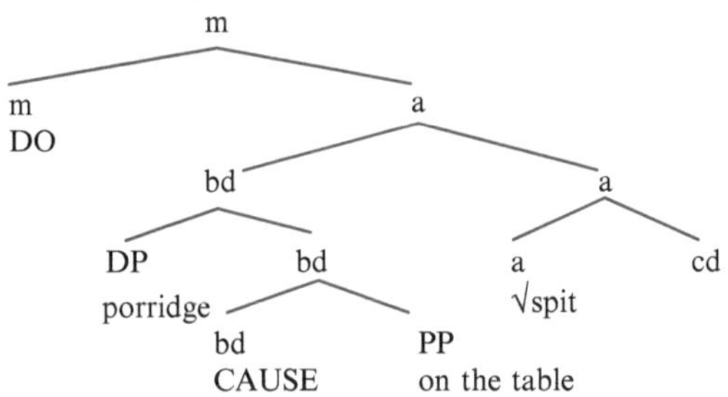

The following is the structure showing the default assignations of morphosyntactic categories in English[13]:

(15) (The baby) spat porridge on the table

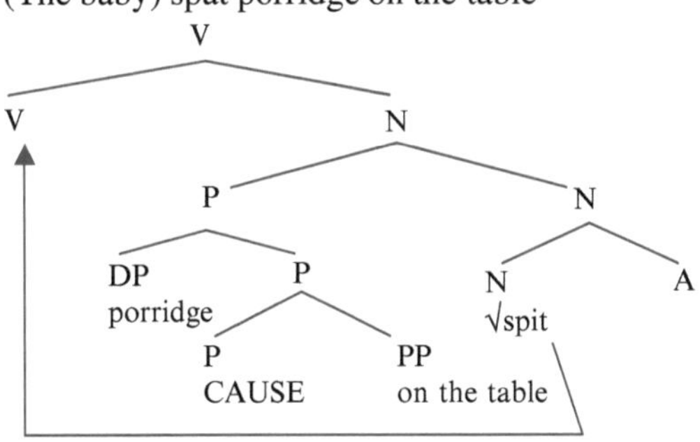

---

[12]The now phonetically null element *cd* can be identified as the abstract clitic position argued by Keyser and Roeper (1992). This is the position in which a particle can be inserted, to form the verb *spit out.*

[13]I am following H&K's system strictly here, with regards to default morphosyntactic labeling of structures. Obviously, there is another possibility, that of lexical structures to be completely category-free, and receiving their morphosyntactic label by merging with the corresponding functional category, D for Nouns, T for Verbs, and so on (Marantz 2000; Alexiadou 2001; McGinnis 2001; Pylkkänen 2002; Juarros-Daussà 2003; Borer 2005).

The structures in (13)–(15) present a problem, however. As they stand, the atomic head appears to have the composite dyadic category as its complement. Recall that this fact was not problematic when a monadic head was merged in this position, since the monadic head and the composite dyadic head were in a relationship of mutual satisfaction, in which *cd* satisfied the [+ complement] requirement of *m*, and *m* provided *cd* with a specifier. If an atomic category appears in this position, the problem arises because the atomic category is [−complement], [−specifier], and (13) and (14) thus seem incompatible with the lexical requirements of *a*, since *cd* cannot be in complement position of an atomic category.

I propose that this problem however disappears when (13) and (14) are considered derivationally, rather than representationally, in the following fashion. Let us assume that a lexical element can only select and be merged with another lexical element (or, in some cases, lexical variable) whose requirements [± complement, specifier] have been satisfied (*checked off*, in minimalist terms, a process attained via *Merge*). In other words, a lexical category has to first be *saturated* (in Speas 1990's terms), or *complete*, before it becomes the target of merge. We can express this idea with the hypothesis in (16):

(16) *Cyclicity of Selection Hypothesis (CSH)*
A lexical category *x* must be complete before another lexical category *y* can be merged with *x*

The view of Merge represented in (16) has two crucial characteristics: on the one hand, it is cyclical: a category y can only be the target of Merge if it is completed, since once targeted and merged, it cannot further target any other category. On the other hand, Merge is directional, since a category x merges with a category y to check x's features. Both properties are consistent with recent syntactic work (Chomsky 2001; Collins 2002). Moreover, the idea of a derivational approach in which (syntactic) requirements are established in the structural description of a transformation (rule) is in line with recent work by Epstein and his colleagues (see Epstein et al. (1998), and Epstein and Daniel (2001), for example). This view however contradicts H&K's own proposal. Hale and Keyser (1998), for example, use representational terminology in their definition of the structural relations *complement* and *specifier*. However, very often they adopt a derivational language, in the step-by-step construction of their lexical structures. For now, I will just remark that the assumptions made so far about the status of the arguments that appear in *spit*-verbs, together with the adoption of (16), have allowed us to construct a lexical structure of these verbs without any degree of recursion, a result that didn't seem possible when we followed H&K's theory as it was stated. In this sense, and given that recursion is a natural enemy of a restrictive theory of argument structure, we can say that the adoption of (16) serves progress.

Below I closely analyze the structure of composite dyadic verbs (such as *spit*), in an attempt to show the changes that the elimination of mutual satisfaction in favor of (16) introduce, and the advantage to choose one over the other. The structure is repeated below for convenience, and their derivation follows:

(17) (The baby) spat porridge on the table

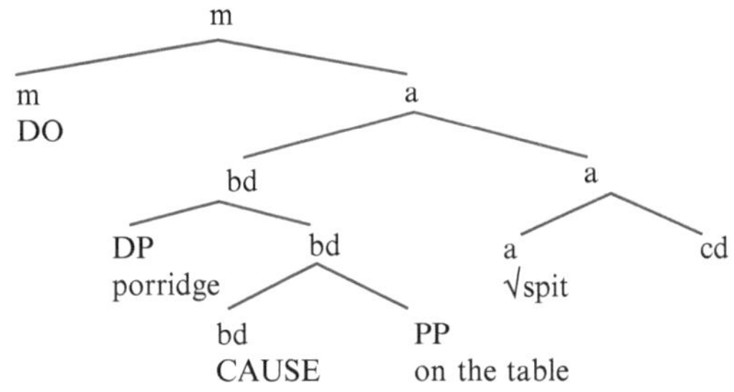

The composite dyadic head, *cd*, as the selector, dictates the conditions of the building operation; in this case, *cd* demands that it be merged with a category that will project a specifier, but will not "claim" such specifier as its own; moreover, such category cannot be constructed as the complement of *cd*. When *cd* merges with the complete atomic category *a*, in order not to construct *a* as the complement of *cd*, *cd* **adjoins** to the maximal *a*, creating segments, not projections, of *a*. Hence, the [−complement] requirement of *cd* is met (since *cd* itself does not project in merging, *a* is not its complement). Moreover, since *a* is complete, its lexical requirements are not modified as a consequence of being selected and merged. When the complex [*a bd*], in which *a* is the head and *bd* the selector, creates another segment, the resulting specifier can only belong to *cd*, the only non-complete lexical category. A basic dyadic category is then merged in the specifier position.

Therefore, following (16), and despite of what might look representationally, when understood derivationally, neither *cd* nor *bd* are complements of *a*, and *a*'s requirement of [−complement] is met. Likewise for *cd*.

Now, let's briefly come back to the general composite dyadic structure that H&K propose for deadjectival unaccusatives, repeated below for convenience:

(18)

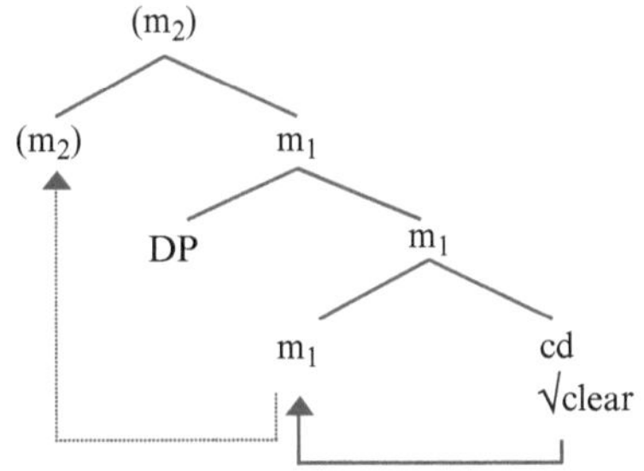

Recall that H&K's account of this structure, consists in mutual satisfaction occurring when *cd* is merged with *m*: on the one hand, the [+complement] requirement of *m* is met by the merging of *cd* in the complement position of *m*; on the other hand, *m* is able to project a specifier by its merging with *cd*, hence satisfying the [+specifier] requirement of *cd*. Yet recall also that, if we are to adopt (73), *cd* cannot be the complement of *m*, since *m* must be completebefore being selected by *cd*. For *m*, that means that its lexical requirements of [+complement], [−specifier] be met prior to its merging with *cd*. The structure in (18) should be modified to the one in (19), in which *m* has satisfied its [+complement] requirement (say, by merging with an atomic category) prior to its being the target of merging with *cd* by adjunction[14]:

(19)

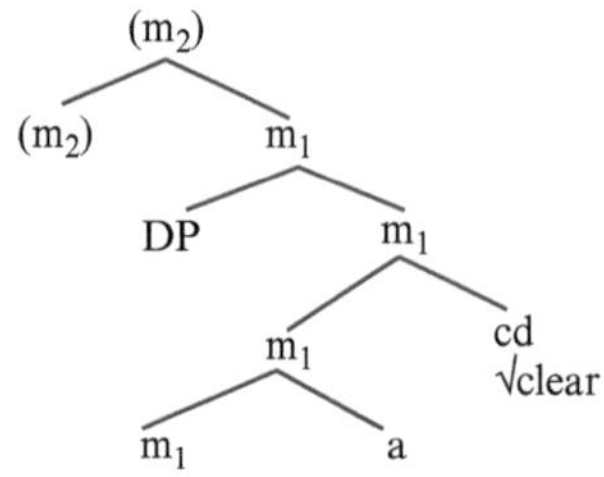

Addendum: The Source Role of the Subject, and the Lack of Inchoative

We are now to address the last of our concerns about the lexical structure of *spit*-verbs, mainly the lack of transitivity alternation that these verbs present, unlike most of composite dyadic verbs studied by H&K. In my answer to this problem, I will use a version of the "manner indices" fist introduced in Hale and Keyser (1993) and reformulated in binding theoretic terms in Hale and Keyser (1998, 2002). The authors notice the contrast between two kinds of verbs, both taking a PP complement: the ones in (20) participate in the transitivity alternation, while the ones in (21) don't:

(20) a. I splashed/got mud on the wall
b. Mud splashed/got on the wall

(21) a. I smeared/put mud on the wall
b. *Mud smeared/put on the wall

Alleging that it has been frequently noticed in the literature that certain aspects of the meaning of lexical items have are relevant to their syntactic behavior, H&K propose an account of the contrast in (20)–(21) based on what they call "manner component indices", which they indicate by {i}. While the status of these indices is not totally clear, syntactically speaking they seem to act as binding entities. For example, in *splash* and *get* the manner component index {i} is "patient-manner", meaning that it must be bound internally.[15] Both in the transitive (20a) and in

[14]The structure in (19) does not look completely satisfying to me. Although a right justification would take more space and time than what I have available for the purposes of this paper, a raising structure in which the lower *m* selects a lexical variable (instead of an empty *a*), which raises to the specifier position of *cd* and becomes the sentential complement makes more sense to me. For a related idea about resultatives as uniformly raising constructions, see Kratzer (2005).

[15]Hale and Keyser (1998: 12) define patient-manner as follows: "an adverbial semantic "feature" which identifies the physical motion, distribution, dispersal, or attitude, f the entity denoted by the

the intransitive (20b), where the object has raised to subject position), the object binds the index from its specifier position. Since the object is an internal argument, the index {i} is always internally bound, and its requirement is met. Hence the grammaticality of the transitivity alternation.

On the other hand, *smear* and *put* have a manner component index {i} that is "agent-manner", meaning that it must be bound externally.[16] In (21a), {i} is bound by the subject, which, as external argument, satisfies {i}'s requirement. But if we raise the object *mud* to subject position (to form the intransitive (21b), the object binds the index from its new position. Since *mud* is an internal argument, the index {i} will be internally bound, against its requirements, and the intransitive will crash.

In order to explain the lack of the inchoative form of *spit*-verbs (recall 10), I propose that these verbs are like *smear* and *put*: they have an "agent-oriented" index that must be externally bound and therefore prevents the inchoative from surviving. Furthermore, it could be that the index of these verbs could well bear the specification *Source* for the subject, since the agent of the action must also be the source of the substance expelled from the body.[17] This way we explain the Source role of the subject, where a simple Agent would be expected.

Now, as pointed out above in (7), *spit* does present a transitive/intransitive alternation of a different sort, namely, what looks like an unergative/transitive alternation:

(22) a. When llamas become upset, they spit
b. The volcano spat even more ash and lava

We can refer to this alternation that *spit* presents between a transitive and an intransitive form as the transitive/unergative alternation. Levin (1993: 33, and references given therein) refers to a very similar alternation that she calls "Unspecified Object Alternation". According to her, in the intransitive variant a typical object of the verb can be omitted, or left implicit. She further claims that this alternation appears with activity verbs. Some clear examples are given below:

(23) a. Mike *ate* the cake
b. Mike *ate* (→ Mike ate a meal or something one typically eats) (Levin 1993: 33)
c. She *is cooking* again this weekend, and of course you are invited

argument (the "patient") occupying the specifier position in the P-projection which functions as the complement of the verb".

[16]Hale and Keyser (1998: 13), for agent-manner: "an adverbial feature which describes the actions of entities denoted by their *external* arguments –to "smear X on Y" requires an "agent" which executes the gestures which, in accordance with the lexical encyclopedic entry, are necessary in performing the action so named".

[17]As evidenced by the following scenario: if Elliott wipes out from the floor the spit produced by one of my dogs, and in his way to the kitchen his mop drips some of the substance on the floor, we cannot however say that Elliott (and even less the mop) spat on the floor.

This is consistent with spit being related to an atomic element; the unergative counterpart occurs when the root appears as a complement of a monadic category:

(24)

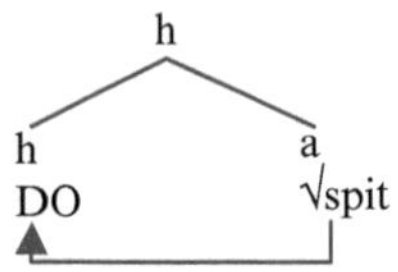

Notice that other unquestionably composite dyadic verbs do not present this possibility (in the relevant interpretation, where the subject is the Agent and not the Theme):

(25) a. *I cleared
b. * I broke

This is further encouragement to abandon the analysis of *spit* as the head of a composite dyadic category, in favor of it being the head of an atomic one. In addition, the unergative/transitive alternation exhibited by *spit* is consistent with what we know of the syntax of verbs in English, where the addition of a goal phrase to an arguably unergative predicate like *run* results in the predicate showing transitive characteristics (26 a, b, from Levin and Rappaport Hovav 2005)[18]:

(26) a. Fred ran
b. Fred ran Alex to the emergency room

Summary: Lexical Projection Structure of *spit*-Verbs

In sum, the lexical projection structure of *spit*-verbs argued for in this chapter is the following:

(27) (The baby) spat porridge on the table/(The) spitting of porridge on the table

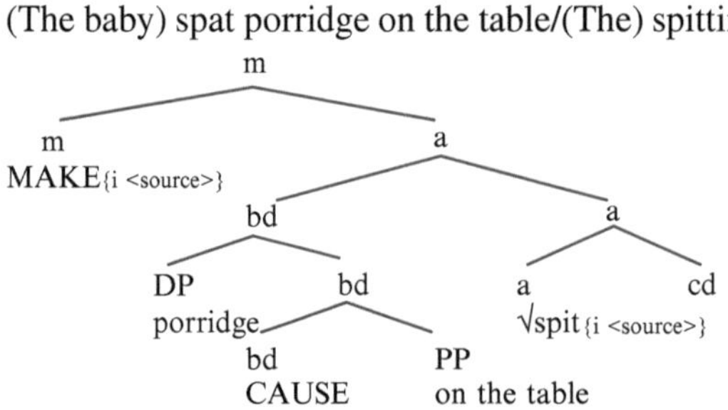

[18]See Juarros-Daussà (2010a) for an analysis of expanded unergatives within Hale and Keyser's framework.

The crucial feature of (27) is that, assuming the USH, it exhausts the combinatorial possibilities of the lexical categories: there is one instance of each. This makes (27) the largest argument structure allowed in natural language. Since it is a structure in which two arguments are associated to one single root, the TAR is thus derived.

2.3 *Cyclic Merge and the Uniqueness of Selection Hypothesis*

The conception of Merge implied by the CSH as a cyclical, directional operation, poses a restriction on the construction of lexical structures, by limiting the number of possible argument structures to those that are obtained by combining completed lexical categories. In this final part of the paper, I explore one more consequence of restricting lexical structures in this fashion: that of calculating the actual absolute maximal *number* of possible argument structure types allowed in natural language. The reasoning goes as follows: if argument structures are obtained by mere combination of the basic lexical categories according to Merge, and if we restrict the number of categories that can be present in a single lexical entry, we can actually calculate, in an abstract manner, the number and shape of all possible argument structure types allowed by natural language.[19] Of course, it is expected that some of the resulting structures be eliminated by other factors (such as patterns of conflation, for example, and crosslinguistic factors), and it is possible that some others be added (for example, by adding additional factors such as the obviative/proximate indexes, as we saw before), but the important thing is that the basic number of abstract structures predicted by the theory can actually be calculated. Once this is done, the task is to match these structures with the empirical facts of each language regarding verb types, while trying to account for any divergences.

Notice that in the above reasoning one further condition has been added, mainly, that of restricting the number of basic lexical categories (in H&K's sense) that can be present in a single lexical entry. This is necessary to restrict the combinations, but it is not an obvious condition. After all, if language allows for a single argument structure to be composed of more than one basic element, any random restriction based on the number of segments would be mere stipulation. Imagine though that the restriction is not on the number of basic elements that form an argument structure per se, but on there not being two lexical categories of the same type contained in a single argument structure. We can formulate this hypothetical restriction, which I call Uniqueness of Selection Hypothesis (USH), as follows:

(28) *Uniqueness of Selection Hypothesis (USH)*
An argument structure can contain two lexical categories x, y, only if $x \neq y$

[19]Since there are four lexical category types, and only one is allowed in each structure, the calculation in gross terms is: 1P: n = **4**; 2P: n(n − 1) = **12**; 3P: n(n − 1) (n − 2) = **24**; 4P: n(n − 1) (n − 2) (n − 3) = 4! = **24**; Total: **64**. I thank Gert Hoffmann for mathematical advice.

Presumably, the intuition behind the USH in (28) has to do with the identification of lexical categories with roots. If each lexical category of H&K's type represents a root, and some form of Late Insertion of roots into structures is assumed, if the case presents itself of two roots competing for the same insertion slot, the derivation crashes. Of the most relevance to the topic of this volume is that this principle amounts to negating the existence of a recursive function in the domain of argument structure, relegating such functions to other grammatical domains.[20] Moreover, it predicts the set of possible argument structures restricted to a given number. While the actual number of structures is not important, the fact that we can predict a restricted typology of argument structures is a very interesting consequence of the particular extension of H&K's theory that I am proposing.

3 Conclusion

This article assumes that, at the representational level of argument structure, there is a Two (internal) Argument Restriction (TAR) that is part of the basic architecture of natural language; any additional (pseudo)arguments are introduced via (possibly recursive) syntactic or discursive procedures, not lexical. As shown, a modified version of H&K can account for this restriction. In this proposal, the TAR crucially depends on the lack of recursion in the lexicon (in building argument structures), restricting the possibility of using this formal mechanism to the domain of sentential syntax (building of sentences). Argument structures are thus built in a cyclic and directional way, and without recursion.

References

Alexiadou, A. (2001). *Functional structure in nominals*. Amsterdam: John Benjamins.

Babby, L. H. (2009). *The syntax of argument structure*. New York: Cambridge University Press.

Boland, J. E. (2005). Visual arguments. *Cognition, 95*, 237–274.

Borer, H. (2005). *In name only. Structuring sense*. Oxford: Oxford University Press.

Bowers, J. (1991). The syntax and semantics of nominals. In *Proceedings from SALT I* (Cornell working papers in linguistics 10, pp. 1–30).

Chomsky, N. (2001). Derivation by phase. In M. Kenstowicz (Ed.), *Ken Hale: A life in language* (pp. 1–52). Cambridge, MA: MIT Press.

Collins, C. (2002). Eliminating labels. In S. Epstein & S. Daniel (Eds.), *Derivation and explanation in the minimalist program* (pp. 106–132). Malden: Blackwell.

Croft, W. (1990). *Typology and universals*. New York: Cambridge University Press.

Dowty, D. (1979). *Word meaning and Montague grammar*. Dordrecht: Reidel.

[20]While if the TAR is right, no language would make use of recursive mechanisms of this type in the domain of argument structure, languages could differ on the extent to which they use recursion in other domains.

Dowty, D. (1991). Thematic proto-roles and argument selection. *Language, 67*(3), 547–619.
Dowty, D. R. (2001). The semantic asymmetry of 'argument alternations'. *Groninger Arbeiten zur germanistischen Linguistik 44*, Groningen: CLC.
Epstein, S., & Daniel, S. (Eds.). (2001). *Derivation and explanation in the minimalist program.* Malden: Blackwell.
Epstein, S., Groat, E., Kawashima, R., & Kitahara, H. (1998). *A derivational approach to syntactic relations.* New York: Oxford University Press.
Folli, R., & Harley, H. (2005). Flavors of v. In P. Kempchinsky & R. Slabakova (Eds.), *Aspectual inquiries* (pp. 95–120). Dordrecht: Springer.
Folli, R., Harley, H., & Karimi, S. (2005). Determinants of event type in Persian complex predicates. In L. Astruc & M. Richards (Eds.), *Cambridge occasional papers in linguistics 1* (pp. 100–120). Cambridge: University of Cambridge.
Hale, K., & Keyser, S. J. (1993). On argument structure and the lexical expression of syntactic relations. In K. Hale & J. Keyser (Eds.), *The view from Building 20: Essays in linguistics in honor of Sylvain Bromberger.* Cambridge: MIT Press.
Hale, K., & Keyser, S. J. (1998). *Bound features, merge, and transitivity alternations,* ms. MIT Lexicon Workshop.
Hale, K., & Keyser, S. J. (2002). *Prolegomenon to a theory of argument structure.* Cambridge, MA: MIT Press.
Harley, H. (1995). *Subjects, events, and licensing.* PhD dissertation, MIT, Cambridge, MA.
Harley, H. (1996). If you have, you can give. In *Proceedings of WCCFL XV.* Stanford: CSLI.
Jackendoff, R. (1990). *Semantic structures.* Cambridge, MA: MIT Press.
Jespersen, O. (1927). *A modern english grammar on historical principles (Part III: Syntax, second volume).* Heidelberg: Carl Winters.
Juarros-Daussà, E. (2003). *Argument structure and the lexicon-syntax interface.* PhD dissertation, University of Massachusetts, Amherst.
Juarros-Daussà, E. (2010a). Obviation in argument structure: Spanish prepositional arguments and expanded unergatives in Italian and English. In C. Borgonovo, M. Español-Echevarría, & P. Prévost (Eds.), *Selected proceeding of the 2008 HLS* (pp. 38–49). Cascadilla Press.
Juarros-Daussà, E. (2010b). Lack of recursion in the Lexicon: The two argument restriction. In H. van der Hulst (Ed.), *Recursion and human language.* Berlin/New York: Mouton de Gruyter.
Keyser, S. J., & Roeper, T. (1992). Re: The abstract clitic hypothesis. *Linguistic Inquiry, 23*(1), 89–125.
Koenig, J.-P., Mauner, G., & Bienvenue, B. (2003). Arguments for adjuncts. *Cognition, 89*, 67–103.
Kratzer, A. (2005). Building resultatives. In C. Maienborn & A. Wöllstein-Leisten (Eds.), *Events in syntax, semantics, and discourse* (pp. 177–212). Tübingen (Niemeyer).
Levin, B. (1993). *English verb classes and alternations. A preliminary investigation.* Chicago: University of Chicago Press.
Levin, B., & Rappaport Hovav, M. (2005). *Argument realization.* Cambridge: Cambridge University Press.
Marantz, A. (2000). Words. Ms., MIT, Cambridge, MA.
McGinnis, M. (2001). Variation in the phase structure of applicatives. In J. Rooryck & P. Pica (Eds.), *Linguistic variations yearbook* (pp. 105–146). Amsterdam: John Benjamins.
Mukherjee, J. (2005). *English ditransitive verbs. Aspects of theory, description and a usage-based model.* Amsterdam/New York: Rodpi.
Pylkkänen, L. (2002). *Introducing arguments.* PhD thesis, Massachusetts Institute of Technology, Cambridge, MA.
Randall, J. (2009). *Linking: The geometry of argument structure.* Dordrecht: Springer.
Randall, J. H. (2010). *Linking. The geometry of argument structure.* Boston: Springer.
Speas, P. (1990). *Phrase structure and the lexicon.* Dordrecht: Kluwer Academic Publishers.
Stowell, T. (1983). Subjects across categories. *The Linguistic Review, 2*, 285–312.
Sybesma, R. P. E. (Ed.). (2004). *Arguments and structure: Studies on the architecture of the sentence.* Berlin: Walter de Gruyter.

Embedding Illocutionary Acts

Manfred Krifka

Abstract Speech acts have sometimes been considered as not embeddable, for principled reasons. In this paper, I argue that illocutionary acts can be embedded under certain circumstances. I provide for a semantic interpretation of illocutionary acts as functions from world/time indices to world/time indices, which provides them with a semantic type, and allows for operators that take them as arguments. I will illustrate this with three cases: First, with illocutionary acts as arguments of verbs like *tell*, second, as semantic objects modified by speech act adverbials like *frankly* and third, with Austinian conditionals. By these exemplary cases, I show that illocutionary acts (or rather, speech-act potentials) become part of the recursive structure of language.

Keywords Speech act verb • Semantics • Illocutionary acts • Adverbials • Conditionals

---

This paper had a long gestation period. It is part of ongoing work on the nature of speech acts and the interaction of It profited tremendously from comments of audiences on presentations of related material at Stanford University, University of California at Santa Cruz, the World Congress of Linguists in Seoul, the recursion conference in Amherst in 2009, the conference on sentence types and illocutionary acts at ZAS Berlin in 2010, and a talk at New York University, April 2013. The current paper focuses in particular on the issue of recursion of speech acts within the more general topic. There are too many colleagues that, in one way or other, had influence on the points to be presented here, but I should at least mention Chris Barker, Arik Cohen, Hans-Martin Gärtner, Andreas Haida, Sophie Repp, Anna Szabolcsi, Hubert Truckenbrodt, and Tom Roeper. Work on this topic is supported by a grant of the Bundesministerium für Bildung und Forschung (BMBF, Förderkennzeichen 01UG0711) to the Zentrum für Allgemeine Sprachwissenschaft; the responsibilities remain with the author.

M. Krifka (✉)
Center for General Linguistics (ZAS), Humbolt University, Berlin, Germany
e-mail: krifka@rz.hu-berlin.de

T. Roeper and M. Speas (eds.), *Recursion: Complexity in Cognition*, Studies in Theoretical Psycholinguistics 43, DOI 10.1007/978-3-319-05086-7_4,

1 Introduction

The literature on recursion typically focuses on syntactic aspects of recursion. Syntactic rules are recursive if they generate structures in which an expression β occurs within an expression α of the same category c, either directly or indirectly. Adjectival modification as in [$_N$ *old* [$_N$ *man*]] is a case of direct recursion; clausal complements as in [$_S$ *John* [$_{VP}$ *thinks* [*that* [$_S$ *Mary left*]]]] are cases of indirect recursion. As the elements of a syntactic category have the same label and the syntactic rules that combine syntactic categories cannot look into their internal syntactic composition, unlimited recursion is allowed by default, as in [$_N$ *dirty* [$_N$ *old* [$_N$ *man*]]], or [$_S$ *Bill* [$_{VP}$ *suspects* [*that* [$_S$ *John* [$_{VP}$ *thinks* [*that* [$_S$ *Mary left*]]]]]]. If such structures are to be excluded from the generative capacity of the language, recursion would have to be blocked explicitly, leading to a more complex grammar.

Now, syntax is there for a purpose: to guide the construction of semantic representations. This is done compositionally, following the Frege principle: The meaning of a complex expression is derived from the meaning of their immediate syntactic parts and the syntactic rule that combines them.[1] Interpretation must be compositional, given that the number of possible expressions is very large or (in case of recursive rules) infinite, and speakers must be able to learn, in a finite and actually quite short time, how to interpret these expressions. If the syntactic rules are recursive, the corresponding compositional semantic rules must be recursive as well. For direct recursion we must allow for functions of a type[2] **(σ)σ** that take arguments of type **σ** and deliver values of type **σ**. For example, for attributive adjectives we assume functions of type **(et)et** like $\lambda P \lambda x[P(x) \wedge \text{OLD}(x)]$, that can be applied to arguments of type **et** like $\lambda x[\text{MAN}(x)]$, and deliver values of type **et** like $\lambda x[\text{MAN}(x) \wedge \text{OLD}(x)]$. This kind of recursivity is standardly assumed in semantics.

Recursive syntax does not mean that every syntactic category is recursive, in the sense that every syntactic category can contain expressions of the same category, directly or indirectly. One kind of systematic exceptions are lexical categories, the categories employed in terminal rules in phrase-structure grammars. Semantically,

---

[1] There are additional factors that influence the meaning of a complex expressions. First, the meaning of the constituents may be shifted by semantic operators, e.g. the type shift of names to quantifiers in cases like *John and every girl* (cf. Partee 1987), the aspectual type shift of semelfactives to iterative activities in *the light blinked for hours* (cf. Moens and Steedman 1988), and metonymic type shifts as in *begin the book* discussed in Pustejovsky (1995) or *the ham sandwich* discussed in Nunberg (1977). Second, the linguistic and extra-linguistic context may influence the meaning, as in suppressing implausible interpretations of ambiguous expressions and in specifying the values of variables and indexical expressions. Third, the availability of other expressions in the language may lead to pragmatic optimization, also ruling out certain interpretations that otherwise would be available; e.g. an indefinite noun phrase suggests non-uniqueness as otherwise a definite noun phrase would be used.

[2] In naming semantic types, I follow the convention that **(σ)τ** denotes the type of functions from meanings of type **σ** to meanings of type **τ**; if **σ** is a simple type, parentheses are omitted. Simple types are **e** for entities and **t** for truth values; see below for additional simple types.

these are expressions whose meanings have to be learned (which does not preclude that the meanings of some syntactically complex expressions, like idioms, have to be learned as well). But it also has been claimed that certain complex categories are systematically excluded from recursive syntax. We will consider here so-called root clauses.

While clauses can be embedded as complement clauses, adverbial clauses, or relative clauses, it has been claimed by Emonds (1969) and Ross (1970) that there are clauses that allow for certain syntactic configurations, so-called root transformations, that are not available for other clauses, and do not occur embedded in other clauses, except perhaps in coordinations. As an example for such root transformations that lead to clauses that cannot be embedded, consider a clause that underwent Left Dislocation:

(1) a. *This room, it really depresses me.*
b. *They put so much furniture in here that*
(i) **this room, it really depresses me.*
(ii) *this room depresses me.*

Why should there be such a systematic restriction of recursion? Certainly, this would not be due to a conceptual necessity of syntax, as in the case of the terminal categories. In traditional phrase-structure grammars, there is no need that the starting symbol S cannot occur as the output of a rule. But we may be able to find a reason for this systematic exception of embeddability in semantics, as suggested in Hooper and Thompson (1973): Root clauses are independent assertions – more generally, speech acts – and speech acts generally cannot be embedded. This is corroborated by those cases in which root clauses actually do embed, as in the following example:

(2) *Carl told me that* [*this book, it has recipes in it*].

In such cases, Hooper & Thompson claim, it is the embedded clause that constitutes the main assertion, whereas the embedding clause has the role of a parenthetical expression, in spite of the complementizer *that*. In a sense, there is a mismatch between overt syntax and function; the embedded clause is not embedded after all.

The phenomenon of restrictions of embeddability of clauses and their exceptions grew into an important topic over the years; cf. the survey in Heycock (2006). More syntactic phenomena that characterize root clauses were found (e.g., modal particles that are specific for root clauses, or verb-second syntax in Germanic languages), and additional types of exceptions where identified.

It is quite likely that the various root transformations of early generative grammar actually are a mixed bag, that is, that the various restrictions and their exceptions do not fall under a uniform description. But at least for a substantial subset of cases, a version of Hooper & Thompson's explanation seems to be right. According to this, root clauses have a feature that allows them to express assertions, or perhaps also other kind of speech acts, and due to this feature they cannot be embedded, if it were not for those exceptional cases that do allow for the syntactic embedding of speech acts. This view is supported by a widespread assumption in natural language philosophy, that speech acts are of a nature that makes it impossible that they become part of semantic recursion (e.g. Stenius 1967; Green 2000, to be discussed below).

In this paper, I will try to argue that this line of argument is semantically sound. In particular, I will try to elucidate what speech acts are, from which it follows that they typically cannot be embedded. But the main focus will be on those cases that have been treated as exceptions, for which I will argue that they involve the embedding of speech acts. I will propose a model-theoretic reconstruction of speech acts, leading to a framework in which both truth-conditional semantics and speech-act theory can be expressed. This theoretical reconstruction will allow for the embedding of speech acts in certain cases. I will be able to discuss this with a few examples only, and within a fairly simple, stripped-down semantic theory. Nevertheless, I hope that in this I will go beyond a mere proof-of-concept, and that some of the proposed analyses will be insightful in their own right.

In Sect. 2 I will develop a representation for truth-conditional semantics and for speech acts. The crucial idea, which goes back to Szabolcsi (1982), is that illocutionary acts are not propositions that are evaluated at word-time indices. Rather, they trigger a change in the world in which they are performed; hence they map world-time indices to other world-time indices. In this they generate events, a point also stressed by Recanati (1987). In particular, they change the commitments of the participants of conversation. Such changes are modeled by restricting the possible continuations in a branching-time model. I will show how assertions can be treated within this framework, and I will show how it allows for the representation of speech act reports and of explicit performatives. In Sect. 3, I will turn to two exemplary cases of speech act embedding: First, a case of indirect recursion, the embedding of an assertive act under *tell*, and secondly, a case of direct recursion, the embedding of an assertive act under *frankly*. Section 4 concludes this paper with an outlook on other kinds of embeddings.

2 The Formal Representation of Speech Acts

2.1 The Nature of Speech Acts

The literature on speech acts is huge, and I cannot begin to do it justice in this paper. Here, I will concentrate on the nature of speech acts and on the ways how it should be modeled formally.

It is safe to say that there are two broad perspectives on speech acts, as exposed in Lewis (1970). One view, most clearly expressed in Stenius (1967) but already present in Frege's distinction between thoughts and judgements (Frege 1879), considers speech acts as communicative actions. Speech acts make use of semantic objects like propositions, but transform them to something of a different nature. This view distinguishes between sentence radicals, which denote propositions, and illocutionary acts that are formed when illocutionary operators are applied to sentence radicals. Speech acts are moves in a language game in the sense of Wittgenstein (1958). For example, if a speaker asserts *Mary has left* to an addressee, the speaker uses the proposition 'Mary has left' for a particular game. Stenius called this the Report Game which follows the rule "Produce a sentence in its indicative mood only if its sentence-radical is true." Under this view, speech acts are actions, not propositions.

Lewis himself favors another view, which he calls "paraphrased performatives."[3] This view considers both sentence radicals and speech acts to be propositions. In our example, if the speaker asserts *Mary has left* to the addressee, then this can be captured by 'the speaker tells the addressee that Mary has left', which is itself a proposition, hence a semantic object. This leads to the problem that the assertion *Mary has left* would necessarily be true whenever uttered by a speaker to an addressee, and so for all performative sentences. For this reason, Lewis assumes the method of paraphrased performatives only for speech acts other than assertions ("declaratives"), and disregards that non-assertive speech acts have spurious truth values – e.g., the questions *Has Mary left?,* analyzed as 'the speaker asks the addressee whether Mary has left', is true as soon as the speaker asks the question. If we want to treat speech acts in a homogenous way, we might assume the method of paraphrased performatives for assertions as well, and simply disregard the truth value of the whole paraphrase, as it is always true. The method of paraphrased performatives is related to the performative hypothesis of Katz and Postal (1964) and Ross (1970), which assigns the sentence *Mary has left* a deep structure of the form *I tell you* [*Mary has left*], which would be interpreted as a paraphrased performative in Lewis' sense. Hence, such deep structures can be seen as syntactified versions of para-phrased performatives.

The two perspectives on speech acts differ in their consequences concerning the role of semantic representations in syntactic recursion. According to Stenius, speech acts are distinct from regular semantic objects. As regular semantic recursion is defined over entities, truth values, worlds, times, contexts and the functions one can build from that, we should not expect that speech acts can themselves be arguments of semantic operators. Once an illocutionary operator has been applied and has transformed a semantic object into a speech act, there is no chance for it to be embedded again. According to Lewis' method of paraphrased performatives, on the

---

[3]The only argument that Lewis gives is that the sentence-radical view would not allow for a treatment of constituent questions like *Who came?*, and encouragements like *Hurrah for Mary!* But this is clearly not the case. Constituent questions can be treated like polarity questions if we assume that their sentence radical denotes a set of propositions or a structured proposition. Encouragements can be seen as speech acts that require a person-denoting referential expression as radical; in our example, *hurrah* can be treated as an illocutionary operator applied to *Mary*.

other hand, speech acts are propositions, regular semantic objects, and there is no intrinsic reason to assume that speech acts cannot be embedded.

What we find is that speech act embedding occurs, but in a restricted way. We can take this as indicating that the Stenius view is right: speech acts are not just propositions, otherwise they would participate more fully in recursion. But then we must explain how the Stenius view can be reconciled with the embedding of speech acts that we do find. For this, we first have to develop a theory of speech acts in which they differ from regular semantic objects, but still can be folded back into semantic meanings.

2.2 A Dynamic Interpretation of Illocutionary Acts

In this and the next two subsections, a semantic framework will be developed that is able to account for standard semantic phenomena, and which is designed to accommodate speech acts.

For the denotational part of the semantics of natural language, the model-theoretic approach of Richard Montague has proved to be extremely fruitful. It is the natural choice of a semantic theory on which to build a more general theory of communication that encompasses speech acts. Montague (1973) provided a framework which allows for evaluating the truth value of a sentence with respect to an index (a world and a time). Kaplan (1978) extended this framework by introducing contexts, thus allowing for a principled treatment of deictic expressions referring to the speaker, the addressee, and the world and time of the utterance. The basic explanandum remained the same: the derivation of the truth values of sentences. This static picture changed with Stalnaker (1974) and Karttunen (1974), who modeled the communicative impact of expressions as a change of the common ground of speaker and addressee. Smaby (1979), Kamp (1981), Heim (1982), Staudacher (1987) and Groenendijk and Stokhof (1987) extended this dynamic view, allowing for a treatment of pronouns referring to entities mentioned in the prior discourse. The resulting picture is one of a dynamic conversation and a static world: a sentence changes the common ground and the set of available discourse referents, but the world and time of the utterance stay the same. This contrasts with the notion of speech acts as seen by Stenius: Speech acts are not true or false at a world and a time, but rather create new facts, after which the world is different. Communication does not just change the common ground of interlocutors, it changes the world itself. It turns out that this view is not quite novel. Szabolcsi (1982), in a paper ahead of its time that was not taken up by semanticists or speech act theorists, sketches exactly this view of speech acts as an index changing device.[4] What follows can be seen as an execution of her idea.

---

[4]Thanks to Hans-Martin Gärtner who directed me towards that paper.

In the following subsection, I will introduce a model frame that provides for an interpretation of regular semantic expressions, but also accommodates speech acts as world changers. This model frame will be minimal in the sense that it should illustrate how a dynamic representation of speech acts works. I will not attempt to integrate discourse referents, or events, or even all aspects of speech acts, like the utterance act (part of the locutionary act) and the perlocutionary act (the achievement of what the speaker intended by the speech act). Rather, I will concentrate here on the illocutionary act.

What do illocutionary acts change? Speech-act theory has been characterized from two distinct perspectives (cf. Harnish 2005): First, the idea that the speaker expresses some attitude, like a belief or desire; for example, in an assertion, the speaker expresses a desire that the addressee believe the content. This Gricean view is prominent in Bach and Harnish (1979). Secondly, the idea that in an illocutionary act the speaker takes on certain commitments; for example, in an assertion, the speaker takes on the liability that the asserted proposition is true, which involves, for example, the obligation to provide evidence for the truth of the proposition (cf. Alston 2000).

I will follow here the second approach, for which I think there is convincing evidence. In particular, certain speaker intentions can be derived from the commitment view of speech acts, but not vice versa. First, if a speaker commits himself to a proposition, then it is likely that he has reasons to believe it to be true, otherwise it would be difficult for him to come up with evidence to support it. Secondly, public commitment to a proposition can also be construed as indicating an intention to make the addressee believe that proposition. This is due to the intersubjectivity of reasoning; if one person (the speaker) considers himself to have sufficient evidence to express public liability for a proposition, then this often can be construed as evidence for another person (the addressee) to believe that proposition as well. It is important to see that these attitudes concerning beliefs and intentions of the speaker are not necessary for a successful assertion. First, there are lies (the speaker knows that the asserted proposition is false), and there is bullshit (cf. Frankfurt 1986; the speaker does not have evidence for the asserted proposition). This is perhaps not a serious problem for the intentional view of assertions, as one could explain this as the speaker giving a false impression that he believes that a proposition is true. Secondly, there are assertions that explicitly express that the speaker declines interest in whether the addressee forms a belief about the proposition, cf. (3):

(3) *Believe it or not, I never cheated on you.*

(4) *#I don't believe it, but I did not cheat on you.*

One might ask, then, why assertions like (4) are self-defeating (Moore's paradox), if the expression of speaker's belief is secondary. The reason is that in this case, the speaker indicates that he does not have any grounds for the public commitment to the proposition, thus defeating this commitment itself.

So we take it that with an assertion, the speaker takes on commitment This change of a commitments is momentaneous – it doesn't take time. In terms of aspectual classes (cf. Vendler 1957) it is an achievement, just as the events expressed by the verb *arrive*. A linguistic reflex of this is that explicit performatives do not occur in the progressive form, which would express ongoing events. But this refers to the illocutionary act only. The utterance act, which is part of the locutionary act, does take time. We can see the utterance act as an event by which the speaker brings about the state change, using the rules of language. Here, I will disregard utterance acts, and concentrate on illocutionary acts, which are seen as changes of commitment states.

With this exclusive representation of the illocutionary act understood as a change of commitment states, there comes one problem: What about if a speaker utters a speech act when the commitments that it could have created about already hold? One way to deal with this is to say that the repeated act has no illocutionary effect at all, as it does not entail a change of commitments. Repeated speech acts may also increase the strength of the commitments; on saying, falsely, *I did not cheat on you* several times, the speaker would commit several lies, and hence face more serious consequences. A speech act may also increase the salience of existing commitments, just as it has been proposed for discourse referents. There are various ways for dealing with this situation; here I will simply assume that speech acts that would create commitments that obtain already have no effect.

2.3 *Model Frames of Interpretation, and the Semantic Interpretation Language*

I will assume a minimal model frame, for the purpose of a transparent exposition. We will assume four basic types, and one rule to build up complex types:

(5) Simple types:
e: entities (objects, kinds, events)
t: truth values (True and False)
s: indices (world-time-points)
c: contexts (specifying speaker c_s, addressee c_a, world-time index c_t).
Complex types:
If σ, τ are types, then $(\sigma)\tau$ is a type (functions from σ-entities to τ-entities); if σ is a simple type, parentheses will be omitted.

A model contains a set of entities **E**. Entities come in different sorts, like objects, kinds, or events. The world-time indices **I** are ordered with respect to a relation $\leq$ of precedence. This relation is transitive and reflexive, but not total (or linear). Rather, it is left-linear (if $i' \leq i$ and $i'' \leq i$, then either $i' \leq i''$ or $i'' \leq i'$), which captures the intuition that the past is fixed but the future allows for different developments. We take $<$ to be the corresponding irreflexive order: $i < i'$ iff $i \leq i'$ and $\neg[i' \leq i]$. We also assume for simplicity that $\leq$ is a discrete order, that is, for any $i < i'$ there is an i''

with $i < i'' \leq i'$ such that there is no i''' with $i < i''' < i''$; we will call i'' an immediate successor of i, and write $i \multimap i''$.

I will assume a semantic representation language with constants and variables of various semantic types, with the usual Boolean operators, with existential and universal quantifiers, and with lambda abstraction and lambda conversion.

The constants of this language will usually be specified in SMALLCAPS. The meaning of the constants of the semantic representation language is specified by an interpretation function F that assigns them to functions belonging to a type as specified in (5). For example, RUN may be a constant of type **set**, which means that F(RUN) is a meaning of this type, a function from indices to a function from entities to truth values. There are also variables of every type, whose meaning will be specified by variable assignments g, as usual. Complex expressions of this semantic representation language are assigned a meaning representation recursively. For example, if SUE is a constant of type **e**, and if i* is an index, then the expression RUN(i*)(SUE) is assigned a meaning relative to an interpretation function F and a variable assignment g, namely F(RUN)(g(i*))(F(SUE)), which informally should capture the truth value of the proposition that Sue runs, at the index i*.

Expressions of type **t** can be combined by Boolean operators $\wedge$, $\vee$, $\neg$, $\rightarrow$, $\leftrightarrow$, and variable assignments can be modulated by the quantifiers $\exists$, $\forall$ and by the lambda abstractor λ, in the usual way. The interpretation function F and the set of indices **I** as structured by $\leq$ should be considered as interdependent, in the following sense: If one index i succeeds another one i', then the two indices i, i' must differ in assigning meanings to certain expressions. That is, if $i' \multimap i$, then it must hold that for at least one constant α that $F(\alpha)(i') \neq F(\alpha)(i)$. The change from i' to i must manifest itself in at least one semantic change, otherwise the set of indices I and the relation $\leq$ would be too fine-grained.

The order relation $\leq$ is related to a temporal order, insofar as it holds that whenever i temporally precedes i' within one history, then it also holds that $i < i'$. But the other way does not hold; there may be indices i, i' such that $i < i'$, but i and i' indicate the same clock time. That is, clock time is a coarser relation than the relation $<$. For example, an illocutionary act may change an index i' to an immediately succeeding index i without elapsing time. In the following, I will use the symbol $\ll$ to express temporal precedence, e.g. $i \ll i'$.

2.4 Denotational Meanings: Propositional Relations

We can distinguish between denotational meanings, specifying the reference and truth conditions of expressions with respect to an index (individuals and propositions), and actional meanings, resulting in the change of indices and option spaces.

Denotational meanings are generally functions from contexts to functions from indices, where pure indexicals depend only on the context, and pure non-indexicals depend only on the index of interpretation. If two such meanings $[\![\alpha]\!]$, $[\![\beta]\!]$ are combined, with $[\![\alpha]\!]$ the functor category, the standard extensional meaning combination rule for non-intensional constructions is $\lambda c\lambda i[[\![\alpha]\!](c)(i)([\![\beta]\!](c))]$, that is, the context and the index of the combined meanings are passed down to the constituent meanings. For intensional operators $[\![\alpha]\!]$, the argument is the intension,. $[\![\beta]\!](c)$ We can give a general type-driven rule for functional applications, as follows:

(6) $[\![\alpha]\!]([\![\beta]\!]) = \lambda c\lambda i[[\![\alpha]\!](c)(i)([\![\beta]\!](c)(i))]$ or $\lambda c\lambda i[[\![\alpha]\!](c)(i)([\![\beta]\!](c))]$, whichever is well-formed.

Let me illustrate the construction of denotational meanings with a simple example, the sentence *I admired Sue*. I assume that the thematic roles of the verb *admire* are filled within a constituent vP, and that temporal information is located in a category TP, where the subject moves out of the vP to SpecTP, and the verb moves to T, where it combines with the tense operator:

(7) $[_{TP} I_1\ [_{T'}\ [_T\ admire_2\text{-PAST}\ [_{vP}\ t_1\ [_{vP}\ [_v\ t_2]\ Sue]]]]]$

This structure is interpreted compositionally, where the moved constituents are interpreted in their underlying positions.

(8) a. $[\![admire]\!] = \lambda c\ \lambda i\ \lambda y\ \lambda x[\text{ADMIRE}(i)(y)(x)]$, type **cseet**
b. $[\![Sue]\!] = \lambda c\ \lambda i[\text{SUE}]$, type **cse**
c. $[\![_{v'} admire\ Sue]\!] = \lambda c\ \lambda i\ \lambda x[\text{ADMIRE}(i)(\text{SUE})(x)]$, type **cset**
d. $[\![I]\!] = \lambda c\lambda i[c_s]$, type **cse**
e. $[\![[_{vP}\ I\ admire\ Sue]]\!] = \lambda c\lambda i[\text{ADMIRE}(i)(\text{SUE})(c_s)]$, type **cst**
f. $[\![\text{PAST}]\!] = \lambda c\lambda p\lambda i'\lambda i[i \ll i' \wedge p(i)]$, type **c(st)sst**
g. $[\![[_{TP}\ \text{PAST}\ [_{vP}\ I\ admire\ Sue]]]\!] = \lambda c\lambda i'\lambda i[i \ll i' \wedge \text{ADMIRE}(i)(\text{SUE})(c_s)]$, type **csst**

The result is a function from contexts c to a propositional relation between two indices i and i′, where i temporally precedes i′, and the speaker admires Sue at i. Notice that the index i′ is not set to the time c_t of the context yet, to accommodate the interpretation of tense in embedded clauses such as *I said that I admired Sue.* The identification of i′ with c_t in assertions like *I admired Sue* will happen when the semantic object in (8) is used to perform a speech act.

2.5 Actional Meanings: Assertions

So far, we have derived a regular semantic object, a propositional relation of type **csst**. To express assertional mood, we will make use of an assertion operator, ASSERT. It takes an index i, an addressee variable y, a proposition p and a speaker variable x, and yields the value True iff at i, x is liable for the truth of the proposition p to the addressee y.

(9) $\text{ASSERT}(i)(p)(y)(x)$

$\Longleftrightarrow$ at i, the speaker x is liable for the truth of p at the index i towards the addressee y.

Notice that ASSERT is a state predicate; it denotes the state of being liable for the truth of a proposition. The assertion of a proposition involves a change of state, namely a change from an index at which the state of having assertive commitments does not hold to one at which it does hold. The following definition introduces an operation that expresses such index changes. Here F[i] and G[i] are formulas denoting truth values depending on a free index variable i in it.

(10) $i' \multimap\!\bullet\, i\ [F[i]]$

$\Longleftrightarrow_{\text{def}}\ i' \multimap\!\bullet\, i \wedge \neg F[i'] \wedge F[i] \wedge$

for all formulas G such that F and G are logically independent:

$G[i'] \leftrightarrow G[i]$

The condition $i' \multimap\!\bullet\, i\ [F[i]]$ can be rendered as: i follows i′ immediately, and i′ differs from i only insofar as the condition F holds of i′. This expresses a minimal change from i′ to i, consisting in the change of truth value of the condition F[i]. – Performing an assertion now can be described as follows:

(11) $\lambda i \exists i'[i' \multimap\!\bullet\, i\ [\text{ASSERT}(i)(p)(y)(x)]]$

This proposition is true for all indices i that differ from the immediately preceding i′ only insofar as at i, x is liable towards y for the truth of the proposition p at i. This means that the transition from i′ to i consists in the illocutionary act of assertion of the proposition p by the speaker x to the addressee y. That is, at i′, the speaker x was not liable towards the addressee y for the truth of p at i, and at i, x is liable for it.

Several remarks are in order at this point. Notice, first, that this is a momentaneous transition, as i follows i′ immediately. This is similar to achievement verbs like *arrive*, which also denote momentaneous changes of states. This reflects the fact that the illocutionary act does not take up any time; a grammatical indication of this is that explicit performatives do not occur in the progressive tense (cf. **I am promising you hereby that I will come to your party*). This is different from the locutionary act, or utterance act, which may take up time, as it involves an utterance event, and to which we can refer with progressive tense – as in *I was just promising you that I will come to your party when I received an phone call from my boss.* In the current paper, I will not attempt to model utterance events, and so we can be silent about the precise causal and temporal relation between the locutionary act and the illocutionary act.

Second, as with index changes in general, the change in (11) spawns an event. This event may be referred to later in discourse – for example, it can be referred to by a pronoun, as in *That's a lie.*

Third, the definition (10) implies that the assertive commitments did not hold at i′ already. At the end of Sect. 2.2 I have discussed possibilities to construct a change even if the assertive commitments obtained before i – as a change of saliency of the commitment, or by referring to a corresponding utterance act. Here, I just assume that in typical cases where formula (11) appears, an implicature is generated that $i' \neq i$, as otherwise the simpler formula ASSERT(i)(p)(y)(x) would have been used.

The following diagram illustrates this index change in a discrete model frame. The left-hand side shows the possible courses of history at an index i. In particular, there is one course at which an event e occurs (recall that events are not explicitly represented in the current model, but implicitly as changes between states). The right-hand side shows the possible courses of history after that event e has occurred.

(12) Possible courses at an index i Possible courses after event e has occurred

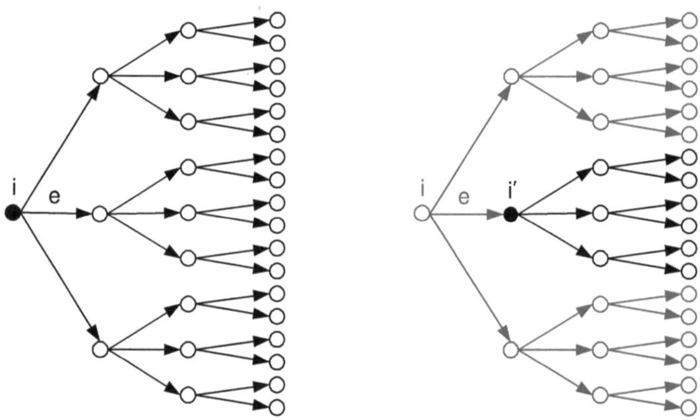

We now consider as an example the assertion of the propositional relation derived in Sect. 2.4. I assume with Rizzi (1997) that assertion has a syntactic reflex, a category ForceP. This is the category of root clauses discussed in Sect. 1. I assume that the ForceP is headed by a syntactic speech-act operator ASSERT and has a TP as its complement. The subject is moved to the specifier position of ForceP but is interpreted within the TP, and the tensed verb is moved to the ASSERT head but is interpreted at the head of TP. This leads to the following structure:

(13) $[_{\text{ForceP}}\ I_1\ [_{\text{Force}'}\ [_{\text{Force}}\ [\textit{admire}_2\text{-PAST}]_3\text{-ASSERT}]\ [_{\text{TP}}\ t_1\ [_{\text{T}'}\ t_3\ [_{\text{vP}}\ t_1$ $[_{\text{VP}}\ t_2\ \textit{Sue}]]]]]]$

The ASSERT operator is interpreted as an operator that takes a propositional relation and changes an input index i′ to an output index i so that at the output index, there are assertive commitments of the speaker c_s towards the addressee c_a with respect to the proposition $R(c_t)$. In addition, the input index i is the index of the context c at which the sentence is interpreted.

(14) $[\![\text{ASSERT}]\!] = \lambda c\lambda R\lambda i'\iota i[c_t = i' \wedge i' \multimap\!\bullet i\ [\text{ASSERT}(i)(\lambda i''\exists\, i'''\ [R(i'')(i''')])(c_a)(c_s)]]$, type **c(sst)ss**

After plugging in the propositional relation that stands for the TP meaning, e.g. the one derived in (8), we get the following interpretation:

(15) $\lambda c[[\![\text{ASSERT}]\!](c)([\![[_{TP}\ I_1\ [admire_2\text{-PAST}]\ [_{vP}\ t_1\ [t_2\ Sue]]]]\!](c))]$
$= \lambda c\lambda i'\iota i[ct = i' \wedge i' \multimap\!\bullet i\ [\text{ASSERT}(i)(\lambda i''\exists i'''[i''' \ll i'' \wedge \text{ADMIRE}(i''')(\text{SUE})(c_s)])(c_a)(c_s)]]$, type **css**

For a given context c, this can be applied to the index of the context i' and changes this minimally to the index i that differs from i' only insofar at i', the speaker c_s is liable towards the addressee c_a for the truth of the proposition $\lambda i''\exists i'''[i''' \ll i'' \wedge \text{ADMIRE}(i''')(\text{SUE})(c_s)]$ at i. And this proposition is true at the index i iff there was a temporally preceding index i''' before i such that at that index i''', Sue admired the current speaker.

It is worthwhile to have a closer look at the temporal relation between the index at which the assertive commitments arise, and the index at which the asserted proposition is said to be true. Recall that with the condition $i' \multimap\!\bullet i$, the change from i' to i is momentaneous, hence i' does not temporally precede i – it does not hold that $i' \ll i$. This means that the index i''' in (15) at which the speaker actually admires Sue must temporally precede i', the index representing the state of the world just before the assertive commitments arise.

One might wonder how this works out for the assertion of present tense propositions. Let us simply assume that present tense requires the identity of the two indices (cf. (16) in contrast to (8)):

(16) $[\![\text{PRES}]\!] = \lambda c\lambda P\lambda i'\lambda i[i = i' \wedge P(i)]$, type **c(st)sst**

We then get the following interpretation for the assertion *I admire Sue*:

(17) $\lambda c\lambda i'\iota i[ct = i' \wedge i' \multimap\!\bullet i\ [\text{ASSERT}(i)(\lambda i''[\text{ADMIRE}(i'')(\text{SUE})(c_s)])(c_a)(c_s)]]$, type **css**

In this case the assertive commitment that arises at i is that the proposition 'the speaker admires Sue' is true at i. Now, recall that according to (10) the only proposition in which i' and i differ is that in i, the speaker has the assertive commitment that the proposition that he admires Sue is true at i. The proposition that the speaker admires Sue is logically independent from that, and hence it has to be true already at i'.

There are other illocutionary operators besides ASSERT, e.g. for commands and for questions. They differ in the semantic type of function that they take as an argument. For example, the question operator can be analyzed as taking sets of propositions as arguments, the Hamblin meaning of questions (cf. Hamblin

1973). Hence, the sentence radical of a question is not a proposition, but a set of propositions. With a question, the speaker puts the addressee under the obligation to make an assertion that identifies one of the propositions in the meaning of the sentence radical. In spite of such differences, speech acts can, in general, be assumed to have the same semantic type, **css**, contextualized functions from indices to indices.

2.6 Speech Act Reports

We have seen how the ASSERT operator, which expresses assertive commitments, can be used to model the illocutionary operator of assertion. It also can be applied to express the report of assertions. We can assume that ASSERT is part of the meaning of speech-act verbs like *tell* in examples like *I told you that I admired Sue.* (18) gives the essential meaning component of *tell*, with three arguments, a proposition p, a direct object y for the addressee, and a subject x for the speaker of the reported assertion.

(18) $[\![\textit{tell}]\!] = \lambda c\lambda i\lambda p\lambda y\lambda x\exists i'[i' \multimap\!\!\bullet\, i\ [\text{ASSERT}(i)(p)(y)(x)]]$, type **cs(st)eet**

The proposition p is specified by a *that* clause, which turns a propositional relation like (8) into a proposition:

(19) a. $[\![\textit{that}]\!] = \lambda c\lambda R\lambda i\exists i'[R(i)(i')]$, type **c(sst)st**

b. $[\![[_{\text{CP}}\ \textit{that}\ [_{\text{TP}}\ \textit{I}_1\ [\textit{admire}_2\text{-PAST}]\ [_{\text{vP}}\ t_1\ [t_2\ \textit{Sue}]]]]]\!]$
$= \lambda c[[\![\textit{that}]\!](c)([\![[_{\text{TP}}\ \textit{I}_1\ [\textit{admire}_2\text{-PAST}]\ [_{\text{vP}}\ t_1\ [t_2\ \textit{Sue}]]]]\!](c))]$
$= \lambda c\lambda i\exists i'[i' \ll i \wedge \text{ADMIRE}(i')(\text{SUE})(c_s)]$, type **cst**

This contextualized proposition can fill the p argument of (18), as illustrated below:

(20) $[\![[_{\text{vP}}\ \textit{John}\ [_{\text{VP}}\ [_{\text{V}'}\ [_{\text{V}}\ \textit{tell}]\ \textit{Mary}]\ [_{\text{CP}}\ \textit{that}\ [_{\text{TP}}\ \textit{I admired Sue}]]]]]\!]$
$= \lambda c\lambda i[[\![\textit{tell}]\!](c)(i)([\![\textit{Mary}]\!](c)(i))([\![\textit{that I admired Sue}]\!](c))([\![\textit{John}]\!](c)(i))]$
$= \lambda c\lambda i\exists i'[i' \multimap\!\!\bullet\, i\ [\text{ASSERT}(i)(\lambda i''\exists i'''[i''' \ll i'' \wedge$
$\text{ADMIRE}(i''')(\text{SUE})(c_s)])(\text{MARY})(\text{JOHN})]$

The resulting proposition applies to indices i that originate from an immediately preceding index i′, where i differs from i′ insofar as John is liable towards Mary for the truth of the proposition that there is an index i‴ preceding i such that at i‴, the current speaker admires Sue. Notice that applying tense, like past tense, will shift the index i to the past of the context time for *John told Mary that I admired Sue*, and that the resulting propositional relation can itself be asserted.

The current analysis captures the reading under which John told Mary something like *He admired Sue*, referring with *he* to the speaker of c in (20). For the reading in which John said something like *He admires Sue*, past tense just indicates agreement with past tense in the main clause (cf. Ogihara 2007). This can be expressed by the past tense variant in (21) that forces the two indices i and i′ to be identical, and restricts them to being past the context index c_t. This results in (22) as the interpretation of (20).

(21) $[\![\text{PAST}_{\text{agr}}]\!] = \lambda c \lambda P \lambda i' \lambda i.i \ll c_t[i = i' \wedge P(i)]$, type **c(st)sst**

(22) $\lambda c \lambda i \exists i'[i' \multimap\!\bullet\, i\ [\text{ASSERT}(i)(\lambda i'' \exists i'''.i''' \ll c_t[i''' = i'' \wedge \text{ADMIRE}(i''')(\text{SUE})(c_s)])(\text{MARY})(\text{JOHN})]$

Now this proposition applies to i iff i results from an i′ by John expressing assertive commitments towards Mary that the proposition that the speaker admires Sue is true at the index i itself.

2.7 *Explicit Performative Speech Acts*

Speech-act verbs like *tell* can also be used in explicit performative speech acts, as in *I hereby promise to come*, or – to stick with one example – in (23):

(23) *I hereby tell you that I admired Sue.*

One analysis of such cases is to assume an illocutionary operator PERFORM as the operator of the ForceP that applies to a propositional relation R. Just like other operators, it changes the commitments of participants; the type of change is expressed by the proposition, here 'speaker tells addressee that speaker admired Sue'. We have the following structure:

(24) $[_{\text{ForceP}}$ *I*$_1$ $[_{\text{Force}'}$ $[_{\text{Force}}$ $[$*tell*$_3$-PRES$]_2$ PERFORM$]$ $[_{\text{TP}}$ t_1 $[_{\text{T}'}$ $[_{\text{T}}$ $t_2]$ $[_{\text{vP}}$ t_1 $[_{\text{VP}}$ $[_{\text{V}}$ $t_3]$ $[_{\text{DP}}$*you*$]$ $[_{\text{CP}}$ *that I admired Sue*$]]]]]]]]$

The PERFORM operator is similar to the ASSERT operator defined in (14), except that it does not specify the nature of the commitment of the speaker. Rather, this is given by the propositional relation R: It states that the input index i′ changes minimally to i such that the embedded propositional relation holds.

(25) $[\![\text{PERFORM}]\!] = \lambda c \lambda R \lambda i' \iota i[i' = c_t \wedge i' \multimap\!\bullet\, i\ [\exists i'' R(i)(i'')]]$

The propositional relation expressed by the TP *I tell you that I admired Sue* should be rendered as follows, where present tense is expressed by identity of the indices i and i′, as in (16):

(26) $\lambda c \lambda i \lambda i''[i = i'' \wedge \exists i''' [i''' \multimap\!\!\bullet i'' [\text{ASSERT}(i'')(\lambda i'''' \exists i''''' [i'''' \ll i''''' \wedge \text{ADMIRE}(i''''')(\text{SUE})(c_s)])(c_a)(c_s)]]]$

After application of the meaning of PERFORM to that propositional relation we get the following result, simplified due to the condition $i = i''$:

(27) $\lambda c \lambda i' \iota i[i' = c_t \wedge i' \multimap\!\!\bullet i\, [\text{ASSERT}(i)(\lambda i\, \exists i''''' [i \ll i''''' \wedge \text{ADMIRE}(i''''')(\text{SUE})(c_s)])(c_a)(c_s)]]]$

Notice that this is the same function as the simple assertion, (15). Hence, the explicit performative based on the assertive verb *tell* in (23) and the assertion lead to the same illocutionary act.

The PERFORM operator can be used for explicit performatives in general, for example in cases like *I promise you to come*, or *I ask you whether you were at home*, but also *The meeting is hereby opened.* A general precondition is that the context is such that the required index change can be performed by a simple utterance. To take up an example by Searle, this is not possible for *I hereby fry an egg.* A full account of explicit performatives would also have to make reference to the locutionary act as a cause of the index change, as expressed by *hereby.*

2.8 *Illocutionary Acts in Conversation*

The notion of an illocutionary act that we have derived so far should properly be called an "illocutionary act potential" – a function that, relative to a context, maps the indices of the context to an index. When such an illocutionary act potential of type **css** as in (15) is applied to, or performed at, a particular context, this results in a specific illocutionary act.

Contexts specify a speaker, an addressee, and a world-time index of the utterance. Hence, contexts c can be modeled as triples $\langle c_s, c_a, c_t \rangle$. A illocutionary act applied to such a context results in a new context in which the speaker and the addressee are the same (we do not model turn-taking here), but the world-time index is changed by the speech act. We write $c + A$ for the update of a context by the illocutionary act A:

(28) $c + A = \langle c_s, c_a, A(c)(c_t) \rangle$

For example, the performance of (15) at a context c will have the following result:

(29) $c + (15) = \langle c_s, c_a, (15)(c)(c_t) \rangle$

The resulting context differs from c insofar as its world/time index $(15)(c)(c_t)$ differs from c_t minimally insofar as c_s is liable towards c_a for the truth of the asserted proposition that c_s admired Sue at the index c_t (which means in this case that the proposition must have been true already immediately before c_t).

The suggested way of modeling does not capture that the asserted proposition becomes part of the common ground of the participants. One way to model common grounds is to assume, as context, a triple $\langle c_s, c_a, C_t \rangle$, where C_t is a set of world-time indices that are candidates for the actual world/time index at the current point of conversation. Update of a context by a speech act A then is pointwise update:

(30) $\langle c_s, c_a, C_t \rangle + A = \langle c_s, c_a, \{\langle c_s, c_a, i \rangle + A | i \in C_t\} \rangle$

In this way, the common ground C_t captures the liabilities for propositions, not the propositions themselves that may become part of the common ground if the addressee does not object against an assertion. As the focus of this paper is on embedded speech acts, I will not develop a model here to capture such aspects of common ground development.

This concludes the overview of a semantic account of speech acts. One important addition, argued for in Cohen and Krifka (2014), is to consider as common grounds not just the candidates for the actual index c_t, but also the possible future developments. This is because certain speech acts cannot be expressed as transitions from an index to another index, but rather as excluding possible transitions in the future. One example is denegation of speech acts, as in *I don't promise to come*, which is properly analyzed as an elimination of promises to come in the future. However, for the purpose of this paper we will disregard such types of speech acts, and concentrate on those that can be modeled by a simple change of commitments.

3 Embedding of Speech Acts

3.1 Preliminaries

In the preceding sections, we differentiated between describing situations in which speech acts occurred, and the performance of speech acts. In particular, we distinguished between context-dependent propositions, type **cst**, which are evaluated at an index i, yielding a truth value, and illocutinary act potentials, type **css**, which map an index to another index. In terms of dynamic semantics, propositions test an index whereas illocutionary acts change an index.

We can now see why illocutionary act potentials often resist syntactic recursion. Linguistic expressions typically denote functions that, when their arguments are supplied, refer to entities (then their type ends in **e**) or describe state of affairs (then their type ends in **t**). Only rarely do they change the world. If we disregard ancient magical spells or modern speech-driven user interfaces, the ability for utterances to change the world is limited to expressing commitments and obligations in the right circumstances. To take up Searle's example again, the speech act *I hereby fry an egg* does not work; what might work is to commit an addressee to fry an egg by *I hereby order you to fry an egg.*

Creating commitments can be seen as the ultimate goal of typical linguistic activity, and linguistic rules are typically used to define these commitments, with the help of expressions that are interpreted as functions on object-related or truth-related expressions. So, it is not astonishing that these ultimate goals, the creation of commitments, are typically not fed back into the rules to form even more complex expressions.

However, this does not exclude that linguistic expressions that create commitments are sometimes used to build up more complex expressions. And the model developed in the previous Sect. 2 allows us to explain how semantic operators can be applied to illocutionary acts. Crucially, illocutionary act potentials are meanings of a particular type, **css**, and it should be possible, in principle, that operators take such functions as arguments. In the current section, I will turn to such functions as illustrative examples to make this point. First, I will discuss the embedding of speech acts by speech act denoting predicates like *tell*. Secondly, we will discuss speech act adverbials like *frankly*. Finally, we will have a look at conditionals of the type *In case you are hungry, there are biscuits on the counter.*

3.2 Speech Acts as Arguments: Direct Speech

We can distinguish between different subtypes of embedding speech act related meanings. First, there is direct speech. Direct speech often is understood as a verbatim representation of what has been said, with a full shift of first and second person pronoun and other context-sensitive expressions from reference to the current speaker and addressee to the speaker and addressee of the reported speech act. However, verbatim representation is not required for direct quotes – for example, direct speech can be translated:

(31) a. John, to Mary: *Ich bewundere Sue.*
b. *John said to Mary "I admire Sue."*

This liberal use of direct speech is defended in Thucydides' *Peloponnesian War*: "... so my habit has been to make the speakers say what was in my opinion demanded of them by the various occasions, of course adhering as closely as possible to the general sense of what they really said." So, it seems that there are two conventions for direct speech, one literal, and one liberal. For the liberal use it is sufficient that the utterance cited would result in the same commitments as the original utterance. The embedding predicate may favor one or the other interpretation; for example, verbs of manner of speaking like *whisper* specify properties of the utterance act, hence their direct speech argument will be interpreted more literally.

Direct speech should be treated as embedding a locutionary act. In the current paper, I focus on illocutionary acts, and so I will not provide for a model of locutionary or utterance acts that are related to illocutionary acts. As a consequence, I can provide just for a rough sketch of how direct speech can be treated. I will concentrate here on the more liberal variety of direct speech; for the literal variety, we must simply assume that the linguistic form of the original utterance is reproduced.

Liberal direct speech can be understood in the following way. Assume that there is a type **u** of utterance types (as contrasted to utterance tokens). The objects in type **u** are related to illocutionary acts of type **css** by the rules of a language. For example, the utterance types *Ich bewundere Sue* and *I admire Sue* are related to the illocutionary act derived in (15) by the rules of German and English, respectively. If the assertion reported in (31)(a) happened, then John assumed liability with respect to Mary for the proposition that he, John, admires Sue. This assumption of liability can be reported by (31)(b), which has to be analyzed as follows: John performed a speech act towards Mary as the addressee which leads to the commitment that is conventionality associated by the utterance of the token *I admire Sue*, with John as a speaker and Mary as an addressee. So, the utterance type that is the complement of *said* is just a way to identify the illocutionary act that is denoted by that utterance type.

3.3 Speech Acts as Arguments: Indirect Speech

Putting direct speech aside, there are two further uses of *tell* to be considered. In one, *tell* subcategorizes for a *that*-clause, in the other, for a root clause:

(32) a. *John told Mary that he admired Sue.*
b. *John told Mary he admires Sue.*

In the case of (32)(a), *tell* simply embeds a proposition; we have dealt with this case in Sect. 2.6. We concentrate here on (b). It might be argued that this is nothing but a simplified form of (a) that lacks an overt complementizer. But the root property is evident in German, where we find verb-second syntax characteristic for root clauses (cf. Reis 1997)[5]:

(33) a. *John sagte zu Mary, dass er Sue bewunderte.*
b. *John sagte zu Mary, er bewundert Sue.*

I would like to propose that the embedded clause *he admires Sue*, or German *er bewundert Sue*, differs semantically from both direct speech, *"I admire Sue"* and

[5] In addition, there is a subjunctive form (Konjunktiv I) with special morphology: *Maria sagte, dass sie John bewundere* and *Maria sagte, sie bewundere John.* It generally indicates a speaker different from the speaker of the utterance context.

from a *that*-clause, *that he admired Sue.* It neither denotes an utterance type, nor a proposition, but an illocutionary act of type **css** that can be taken as an argument by *tell* (or *sagen*, in German).

Embedded illocutionary acts require a slight formal modification, as speaker and addressee of the embedded act are specified by way of the embedding clause, and not as the speaker and addressee of the complex sentence. For this reason we assume instead of (14) the meaning (34) for the assertion operator of embedded clauses, where x stands for the speaker and y for the addressee.

(34) $[\![\text{ASSERT}']\!] = \lambda c \lambda R \lambda y \lambda x \lambda i' \iota i[i' \multimap\!\bullet\, i\ [\textsc{assert}(i)(\lambda i'' \exists i'''[R(i'')(i''')])(y)(x)]]$, type **c(sst)eess**

When applied to the meaning of the TP *he admires Sue*, with an agreeing present tense operator, this results in the propositional relation in (35), and we get the interpretation in (36), where binding of the subject pronoun *he* to the speaker of the speech act x is assumed without specification of an explicit mechanism, for simplicity.

(35) $[\![[_{\text{TP}}\ he_1\ [_{\text{T}'} admire_2\text{-PRES}\ [_{\text{vP}}\ t_1\ [t_2\ Sue]]]]]\!] = \lambda c \lambda i''' \lambda i''[i'' = i''' \wedge \textsc{admire}(i'')(\textsc{sue})(x)]$, type **csst**

(36) $[\![[_{\text{ForceP}}\ he_1\ [_{\text{Force}'}\ [_{\text{Force}}\ [admire_2\text{-PRES}]_3\text{-ASSERT}']\ [_{\text{TP}}\ t_1\ [_{\text{T}'}\ t_3\ [_{\text{vP}}\ t_1\ [t_2\ Sue]]]]]]]\!]$
$= \lambda c[(34)(c)((35)(c))]$
$= \lambda c \lambda y \lambda x \lambda i' \iota i[i' \multimap\!\bullet\, i\ [\textsc{assert}(i)(\lambda i''[\textsc{admire}(i'')(\textsc{sue})(x)])(y)(x)]]$,
type **ceess**

The verb *tell*, in the version that takes an illocutionary act type as an argument, specifies the y and x argument as identical to its direct object and its subject. It does not do much else than that except that it restricts the embedded illocutionary act to assertions (forms like *John told Mary to come* and *John told Mary who came* do not embed illocutionary acts). The meaning of *tell* as taking an illocutionary act can be represented as in (37), where the sortal restriction to assertions is expressed as a restriction for the illocutionary act argument A. When combined with its arguments and a past tense operator, we get the propositional relation (38).

(37) $[\![tell]\!] = \lambda c \lambda i \lambda A{:}\text{Assertion}\ \lambda y \lambda x \exists i'[i = A(y)(x)(i')]$, type **cs(eess)eet**

(38) $[\![[_{\text{TP}}\ John_4\ [_{\text{T}'} tell_5\text{-PAST}]\ [_{\text{vP}}\ t_4\ [_{\text{VP}}\ t_5 Mary\ [_{\text{ForceP}}\ he\ admires\ Sue]]]]\!]$
$= \lambda c \lambda i \lambda i'''[i \ll i''' \wedge \exists i'[i = \iota i[i' \multimap\!\bullet\, i\ [\textsc{assert}(i)(\lambda i''[\textsc{admire}(i'')(\textsc{sue})$
$(\textsc{john})])(\textsc{mary})(\textsc{john})]]]]$

The index i (which temporally precedes the index i‴, due to the past operator) qualifies for this description if there is an index i′ such that i is that index that minimally differs from i′ insofar John takes on liability with respect to Mary that the proposition that he, John, admires Sue is true at the index i. That is, the index i results from an assertion of John to Mary that he, John, admires Sue.

The resulting meaning is similar to the meaning of the proposition-embedding clause *John told Mary that he admired Sue* derived in (20). In that use, the verb *tell* expresses that an illocutionary act of the type of assertion happens. In the speechact-embedding form (37), the verb *tell* does not denote such a speech act, but subcategorizes for this kind of speech act as its argument.

3.4 The Range of Embedding Predicates

Hooper and Thompson (1973) discuss five types of clause-embedding predicates, three of which allow for embedded root clauses. Meinunger (2006) lists four (or five) classes of verbs that allow for embedded V2 in German, and five classes that do not. It is not possible here to discuss the full range of these predicate types in the current paper.

One class that definitely allows for embedded root clauses are verbs of saying. Hooper & Thompson do not mention *tell*, but they discuss verbs like *say, announce, exclaim, vow* etc. They can be treated similar to *tell*, and differ insofar as they express subtypes of assertions with different restrictions on the asserted proposition, on the kind of commitments, and perhaps other aspects (cf. Searle and Vanderveken 1985 and Vanderveken 1990 for one theory on the dimensions in which verbs expressing speech acts can differ). These verbs can be used with the same meaning in explicit performatives, as in *I hereby vow that*...

We would now have to go through other clause-embedding predicates and check whether the idea that subcategorized root clauses denote speech acts makes sense. Not every case discussed in Hooper and Thompson (1973) will qualify for that. For example, they also list predicates like *it's true* or *it's obvious* in their "A" class, which do not report on speech acts. One could perhaps propose that *it's true* subcategorizes for an assertion A expressed by a root clause, with the meaning that the speaker considers A to be assertable because its sentence radical is obviously true. As another example, in German verbs like *glauben* 'believe' and *denken* 'think' allow for verb-second embedded clauses (and also for root-clause phenomena in English). As these verbs do not express speech acts but propositional attitudes, this is in conflict with the idea that root clauses always express speech acts. However, we may say that a propositional attitude can be characterized by a speech act that an agent would utter if the agent has that propositional attitude. If Mary believes that Bill is at school, then she is willing to assert that Bill is at school, and hence *believe* can subcategorize for such an assertion.

One interesting class are question-embedding predicates. In Krifka (1999), I have argued that they fall into two classes: Those that embed question sentence radicals, like *know* and also *tell*, and those that embed question speech acts, like *wonder*.[6]

[6]In Krifka (2001) I assumed that both types of verbs embed question acts, but that verbs like *know* type-shift this question act to the set of true answers. This was designed to handle certain

This distinction corresponds to the distinction of predicates that embed question extensions vs. intensions in Groenendijk and Stokhof (1984). McCloskey (2005) has pointed out that *wonder*, *ask* and certain other question-embedding verbs, but not verbs like *know* or *find out*, allow for main clause syntax, at least in some varieties of English, especially Irish English. As an example, consider the following quote of James Joyces *Dubliners*:

(39) *The baritone was asked what did he think of Mrs. Kearney's conduct.*

In German, verbs like *sich fragen* 'wonder' allow for root modal particles like *denn*:

(40) a. *John weiß, wen Maria (*denn) getroffen hat.*
b. *John fragt sich, wen Maria (denn) getroffen hat.*
'John knows/wonders whom Maria PART met'

This can be taken as evidence that predicates like *ask* and *wonder* embed interrogative illocutionary acts, just as *tell* can embed assertive illocutionary acts.

3.5 Other Uses of Indirect Speech: Proxy Speech Acts

The current proposal differs from Meinunger (2006), who assumes that in cases with root clauses as arguments, the embedding clause and the embedded clause are paratactically combined, and that this combination forms the sentence radical of an illocutionary operator. With this, Meinunger wants to express that in cases like (41) the proposition of the embedded clause, 'Laura is pregnant', is asserted to be part of Dirk's belief world, but is also asserted by the speaker.

(41) *Dirk meint, Laura ist schwanger.*
'Dirk claims, Laura is pregnant'
a. Asserted: Dirk made the claim that Laura is pregnant.
b. Asserted: Laura is pregnant.

With his interpretation, Meinunger captures the intuition of Hooper and Thompson (1973) that such sentences have a reading in which the embedded clause expresses the main assertion of the sentence, and the embedding clause is interpreted like a parenthetical clause. This interpretation is particularly obvious with first person subjects as in (42), which does not report a thought but rather an

phenomena relating to quantification into embedded questions. Now, and even in 2001, I see advantages of the proposal of Krifka (1999); cf. also McCloskey (2005).

assertion with a somewhat reduced commitment for the proposition, by specifying the kind of evidence for the truth of the asserted proposition.

(42) *I think it just started to rain.*

But this interpretation also obtains for sentences with third person subjects, as in (43):

(43) *The weather report said there will be rain.*

I would like to propose that the intuition that the embedded clause expresses the main assertion derives from a plausible pragmatic inference, and is not part of the semantic representation of such sentences. What (43) says can be expressed according to the lines developed for (38): *say* embeds an assertive speech act that is ascribed to the weather report, and the proposition that the weather report made this assertion is itself asserted by the current speaker. But the impact of this, if the subject of the sentence is a trusted source, is that the content of the embedded assertion becomes part of the common ground. Of course, the speaker assumes that he himself is a trusted source, and hence this effect obtains in particular with first person subjects. With third person subjects, it is as if the speaker invites another person to join the conversation; the speaker acts as a proxy for that other source. The reason for this move is that otherwise it would not be relevant to add the commitments of third person sources to the common ground.

We have seen that cases with embedded root clauses and cases with embedded *that* clauses end up having a similar meaning. As a consequence, we should expect such proxy speech act uses also for embedded *that* clauses. And indeed, it is possible to insert *that* in (42) and (43) without necessarily changing the conditions under which these sentences can be used.

3.6 Adverbials Modifying Illocutionary Acts

We now turn from illocutionary acts as arguments to illocutionary acts as targets of modifiers, like *frankly.* One of the main reasons for the performative analysis were adverbials that appear to attach to the performative prefix, as the following example and its paraphrase suggest (cf. Schreiber 1972; Davison 1973; Sadock 1974 for early literature on the phenomenon, and Mittwoch 1977 for an early critical view).

(44) *Your tie and shirt frankly don't go together.*
'I tell you frankly [that your tie and shirt don't go together].'

Such adverbs can be used in a descriptive way, as in the following example:

(45) a. *Mary told Bill frankly that his tie and shirt didn't go together.*
b. *Mary frankly told Bill that his tie and shirt didn't go together.*

The two possible positions in the descriptive use lead to meaning differences, the lower position expressing that the way the act was carried out was frank, and the higher position expressing that the choice to carry out the act was carried out was frank (cf. McConnell-Ginet 1982; Shaer 2003). These two positions have been captured by differentiating within the VP between an outer layer with a DO phrase that relates the agentive subject to an inner VP that expresses a property change. The structures in question then would be syntactically represented as follows:

(46) a. $[_{vP}$ *Mary* $[_{v'}$ DO $[_{VP}$ $[_{VP}$ *tell Bill*$]$ *frankly* ... $]]]]$
b. $[_{vP}$ *Mary* $[_{v'}$ *frankly* $[_{v'}$ DO $[_{VP}$ *tell Bill* ... $]]]]$

It is the second sense that parallels the speechact-related use. In the following, I will not assume a DO in semantic representations and assume that *frankly* is a modifier of VP even in the reading of (45)(b).

We start with an example of the descriptive use of *frankly*, which has the interpretation (47), where FRANK(i)(R)(x) states that to carry out the action denoted by R is a frank action by x. As this information is backgrounded, it appears as a restriction of the function.

(47) $[\![frankly]\!] = \lambda c\lambda R\lambda i\lambda x{:}\text{FRANK}(i)(R)(x)[R(i)(x)]$, type **c(set)set**

After applying this meaning to the meaning of the VP in (48) and the meaning of the subject DP, *John*, we arrive at the non-tensed proposition (49).

(48) $[\![[_{VP}tell\ Mary\ [_{CP}that\ I\ admired\ Sue]]]\!]$
$= \lambda c\lambda i\lambda x\exists i'[i' \circ\!\!-\!\!\bullet i\ [\text{ASSERT}(i)(\lambda i''\exists i'''[i''' \ll i'' \wedge \text{ADMIRE}(i''')(\text{SUE})(c_s)])(\text{MARY})(x)]$, type **cset**

(49) $[\![[_{vP}John\ [_{VP}frankly\ [_{VP}tell\ Mary\ [_{CP}that\ I\ admired\ Sue]]]]]\!]$
$= \lambda c\lambda i{:}\text{FRANK}(i)((48)(c))(\text{JOHN})[(48)(c)(i)(\text{JOHN})]$, type **cst**

This maps contexts c and indices i to truth values provided that the assertion by John to Mary that the speaker at c was frank at i – e.g., because it violates certain norms of secrecy or politeness. Notice that if this act is considered frank at i, then it also must already be considered frank at the index i′ immediately preceding i, the index at which the assertion is uttered. If defined, it maps c and i to truth iff i is the resulting index after John asserted to Mary that the speaker in C admired Sue.

The use of *frankly* as speech act adverbial can be derived from its descriptive use in (47). Recall that we derived for the assertion *I admired Sue* the following meaning, cf. (15):

(50) $[\![[_{\text{ForceP}} I_1\ [[\textit{admire}_2\text{-PAST}]_3\ \text{ASSERT}\ [_{\text{TP}}\ t_1\ [t_3\ [_{\text{vP}}\ t_1\ [t_2 \textit{Sue}]]]]]]\!]$
$= \lambda c \lambda i' \iota i[c_t = i' \wedge i' {\multimap\!\bullet}\, i\ [\text{ASSERT}(i)(\lambda i'' \exists i'''[i''' \ll i'' \wedge$
$\text{ADMIRE}(i''')(\text{SUE})(c_s)])(c_a)(c_s)]]$, type **css**

The meaning of the speech act adverbial *frankly* then can be specified as in (51). It restricts the input indices i′ of the illocutionary act A to those for which it is frank to perform A by the speaker of the context, c_s. When combined with the illocutionary act in (50) we get the result in (52), which contains the presupposition that it is frank for the speaker to assert that he admires Sue. Otherwise, the meaning of the illocutionary act (50) stays the same.

(51) $[\![\textit{frankly}']\!] = \lambda c \lambda A \lambda i' {:} \text{FRANK}'(i')(A)[A(i')]$, type **c(ss)ss**

(52) $[\![[_{\text{ForceP}} \textit{frankly}'\ [_{\text{ForceP}} \textit{I admire Sue}]]]\!]$
$= \lambda c[[\![\textit{frankly}']\!](c)([\![[_{\text{ForceP}} \textit{I admire Sue}]]\!](c))]$
$= \lambda c \lambda i' {:} \text{FRANK}'(i')((50)(c)))[(50)(c)(i')]$

Thus, the speech act adverbial *frankly* is essentially similar to the adverb *frankly* in its descriptive use. The descriptive adverbial takes a VP meaning **set** and yields another VP meaning; the speech act adverbial takes an illocutionary act meaning **ss** and yields another act meaning. In either case, the background meaning is expressed that the action performed was a frank one.

3.7 Conditional Speech Acts

We will finally look at illocutionary acts that are dependent on a conditional clause, a type of conditionals that Austin (1961) drew our attention to.

(53) *If you want biscuits, there are some on the side board.*

This type of conditional does not relate to truth conditions; the biscuits would be on the side board even if the addressee did not want them. Such examples have been analyzed as involving speech acts in the consequent of the conditional, e.g. by DeRose and Grandy (1999), and by Siegel (2006), who assumed a quantification over "potential" speech acts. The example can be taken to mean 'For all indices at which you have the desire for biscuits, the speaker asserts that there are some on the sideboard'. Conditional speech acts are not restricted to assertions, as the following examples with conditionalized questions, commands and explicit performatives show:

(54) a. *If I want biscuits, where can I find them?*
b. *If she wants biscuits, give her some.*
c. *If you want biscuits, I promise you that there are some on the side board.*

In the present format, Austinian conditionals can be treated as a straightforward combination of the semantics of *if* clauses and the semantics of illocutionary operators.

We assume a standard semantics of conditional clauses, as expressing a quantification over indices that are accessible via a particular accessibility relation. To be specific, the conditional clause expresses a modal quantification over indices, and the *if* clause specifies the restrictor of this quantifier. The essential steps in a truth-related conditional are given in (55): (a) specifies the meaning of the propositional relation *it is warm*, and (b) the constituent C′ that contains a non-overt universal modal quantifier (which can be expressed overtly, by *must*). AR stands for the accessibility relation, here an epistemic relation based on general meteorological knowledge. The next line, (c), gives the meaning of the *if* clause, a proposition, and (d) gives the full structure of a conditional CP, where the *if* clause expresses a condition for the indices i″ to be quantified over. The resulting propositional meaning can be asserted.

(55) a. $[\![[_{\mathrm{TP}}\ \textit{it is warm}]]\!] = \lambda c \lambda i' \lambda i[i' = i \wedge \textsc{warm}(i)]$
b. $[\![[_{\mathrm{C'}}\ \mathrm{MUST}\ [_{\mathrm{TP}}\textit{it is warm}]]]\!] = \lambda c \lambda p \lambda i' \lambda i \forall i'' \in \mathrm{AR}(i).p(i'')[\textsc{warm}(i'')]$
c. $[\![[_{\mathrm{CP}}\ \textit{if}\ [_{\mathrm{TP}}\textit{the sun shines}]]]\!] = \lambda c \lambda i[\textsc{sunshine}(i)]$
d. $[\![[_{\mathrm{CP}}\ [_{\mathrm{CP}}\ \textit{if the sun shines}]\ [_{\mathrm{C'}}\ \mathrm{MOD}\ [_{\mathrm{TP}}\textit{it is warm}]]]]\!]$
$= \lambda c \lambda i' \lambda i \forall i'' \in \mathrm{AR}(i).\textsc{sunshine}(i'')[\textsc{warm}(i'')]$

With Austinian conditionals, the conditionalization does not happen at the propositional level (TP and CP), but at the illocutionary level, FP. We assume here that Austinian conditionals do not express a modal quantification, but that the *if* clause just expresses a restriction on the input index.

(56) a. $[\![[_{\mathrm{FP}}\ \mathrm{ASSERT}\ [_{\mathrm{TP}}\ \textit{there are biscuits}]]]\!]$
$= \lambda c \lambda i' \iota i[i' ⊸\!\!\bullet\, i\ [\textsc{assert}(i)(\lambda i''[\textsc{there-are-biscuits}(i'')])(c_a)(c_s)]]]$
b. $[\![[_{\mathrm{F'}}\ [_{\mathrm{FP}}\ \mathrm{ASSERT}\ [_{\mathrm{TP}}\ \textit{there are biscuits}]]]]\!]$
$= \lambda c \lambda p \lambda i'.p(i')\ \iota i[i' ⊸\!\!\bullet\, i$
$[\textsc{assert}(i)(\lambda i''[\textsc{there-are-biscuits}(i'')])(c_a)(c_s)]]$
c. $[\![[_{\mathrm{CP}}\ \textit{if you are hungry}]]\!] = \lambda c \lambda i[\textsc{hungry}(i)(c_a)]$
d. $[\![[_{\mathrm{FP}}\ [_{\mathrm{CP}}\ \textit{if you are hungry}]\ [_{\mathrm{FP}}\ \mathrm{ASSERT}\ [_{\mathrm{TP}}\ \textit{there are biscuits}]]]]\!]$
$= \lambda c \lambda i'.[\textsc{hungry}(i')(c_a)]\ \iota i[i' ⊸\!\!\bullet\, i\ [\textsc{assert}(i)(\lambda i[\textsc{there-are-}$
$\textsc{biscuits}(i)])(c_a)(c_s)]]]$

The resulting illocutionary act can be applied only to those indices i′ at which the condition holds that the addressee c_a is hungry. If applicable, it changes that index to one in which the speaker c_S has assertive commitment towards the proposition that there are biscuits (which can be constructed as an indirect speech act that the addressee may eat them). If not applicable because the addressee happens to be not hungry, then nothing will change, that is, the assertion is not made. If, as discussed in Sect. 2.8, an illocutionary act is performed with respect to a set of indices, a common ground, then the assertion will only hold at those indices of the common ground at which the addressee is hungry. This is because at other indices, the assertion that there are biscuits would violate the pragmatic conditions of relevance.

4 Conclusion

To sum up: The goal of this paper was to develop a theory that allows for speech acts, and in particular illocutionary acts, to be acted upon by semantic operators, thus folding apparently pragmatic phenomena back into semantics. This is of considerable relevance for the topic of recursion, as it shows how an apparently semantically motivated restriction against recursion – the ban against embedding of speech acts – does not obtain. Once we have found proper semantic types for illocutionary acts, nothing prevents, in principle, the assumption of operators that take such types as arguments. This allows us to treat illocutionary acts as arguments of predicates, and as the target of modifiers. It also allows for illocutionary acts to be subject to operations like conjunction, disjunction and negation, but these operators were not in the focus of this paper (cf. Krifka 2001; Cohen and Krifka 2014).

The critical step in designing illocutionary acts as semantic objects was to get away from the static, propositional view in previous literature, e.g. Lewis (1970) and Vanderveken (1990). The dynamic view proposed in Szabolcsi (1982), in which they change indices, was essential to capture their potential as actions, and to make them accessible to semantic operators. To vary a quote that is still displayed, in golden letters, at a prominent place in my university: Semantic operators have hitherto been seen as evaluating the world; the point is that some of them are also able to change it.

References

Alston, W. P. (2000). *Illocutionary acts and sentence meanings*. Ithaca: Cornell University Press.
Austin, J. L. (1961). Ifs and cans. In G. J. Warnock & J. O. Urmson (Eds.), *Philosophical papers* (pp. 153–180). Oxford: Oxford University Press.
Bach, K., & Harnish, R. M. (1979). *Linguistic communication and speech acts*. Cambridge, MA: MIT Press.
Cohen, A., & Krifka, M. (2014). Superlative quantifiers as meta speech acts. *Linguistics and Philosophy, 37*, 41–90.
Davison, A. (1973). *Performatives, felicity conditions, and adverbs*. Doctoral dissertation, University of Chicago.
DeRose, K., & Grandy, R. E. (1999). Conditional assertions and "biscuit" conditionals. *Noûs, 33*, 405–420.

Emonds, J. E. (1969). *Roots and structure-preserving transformations*. Doctoral dissertation, MIT, Cambridge, MA.
Frankfurt, H. (1986). *On bullshit*. Princeton: Princeton University Press. (Republished 2005).
Frege, G. (1879). *Begriffsschrift, eine der arithmetischen nachgebildete Formelsprache des reinen Denkens*. Halle: L. Nebert.
Green, M. S. (2000). Illocutionary force and semantic content. *Linguistics and Philosophy, 23*, 435–473.
Groenendijk, J., & Stokhof, M. (1984). *Studies in the semantics of questions and the pragmatics of answers*. Doctoral dissertation, University of Amsterdam: Ms.
Groenendijk, J., & Stokhof, M. (1987). *Dynamic predicate logic. Towards a compositional, non-representational semantics of discourse*. Amsterdam: University of Amsterdam, ITLI.
Hamblin, C. L. (1973). Questions in Montague English. *Foundations of Language, 10*, 41–53.
Harnish, R. M. (2005). Commitments and speech acts. *Philosophica, 75*, 11–41.
Heim, I. (1982). *The semantics of definite and indefinite noun phrases*. University of Massachusetts at Amherst.
Heycock, C. (2006). Embedded root phenomena. In E. Martin & H. van Riemsdijk (Eds.), *The Blackwell companion to syntax* (Vol. II, pp. 174–209). Oxford: Blackwell.
Hooper, J., & Thompson, S. (1973). On the applicability of root transformations. *Linguistic Inquiry, 465–497.*
Kamp, H. (1981). A theory of truth and semantic representation. In J. A. G. Groenendijk, T. M. V. Janssen, & M. B. J. Stokhof (Eds.), *Formal methods in the study of language* (Vol. 135, pp. 277–322). Amsterdam: Mathematical Centre Tracts.
Kaplan, D. (1978). Dthat. In C. Peter (Ed.), *Syntax and semantics 9* (pp. 221–243). New York: Academic.
Karttunen, L. (1974). Presuppositions and linguistic context. *Theoretical Linguistics, 1*, 181–194.
Katz, J., & Postal, P. M. (1964). *An integrated theory of linguistic descriptions*. Cambridge, MA: MIT Press.
Krifka, M. (1999). Quantifying into question acts. In *Proceedings of semantics and linguistic theory 9*. Cornell: CLC Publications.
Krifka, M. (2001). Quantifying into question acts. *Natural Language Semantics, 9*, 1–40.
Lewis, D. (1970). General semantics. *Synthese, 22*, 18–67.
McCloskey, J. (2005). Questions and questioning in a local English. In Z. Raffaella, H. Campos, E. Herburger, & P. H. Portner (Eds.), *Crosslinguistic research in syntax and semantics* (pp. 87–126). Washington, DC: Georgetown University Press.
McConnell-Ginet, S. (1982). Adverbs and logical form. *Language, 58*, 144–184.
Meinunger, A. (2006). On the discourse impact of subordinate clauses. In V. Molnar & S.-s. Winkler (Eds.), *The architecture of focus* (pp. 459–488). Berlin: Mouton de Gruyter.
Mittwoch, A. (1977). How to refer to one's own words: Speech-act modifying adverbials and the performative analysis. *Journal of Linguistics, 13*, 153–368.
Moens, M., & Steedman, M. (1988). Temporal ontology in natural language. *Computational Linguistics, 14*, 15–28.
Montague, R. (1973). The proper treatment of quantification in ordinary English. In J. M. E. Moravcsik, P. Suppes, & K. J. J. Hintikka (Eds.), *Approaches to natural language* (pp. 221–242). Dordrecht: Reidel.
Nunberg, G. (1977). *The pragmatics of reference*. Doctoral dissertation, City University of New York.
Ogihara, T. (2007). Tense and aspect in truth-conditional semantics. *Lingua, 117*, 392–418.
Partee, B. (1987). Noun phrase interpretation and type-shifting principles. In J. Groenendijk (Ed.), *Studies in discourse representation theory and the theory of generalized quantifiers* (pp. 115–143). Dordrecht: Foris.
Pustejovsky, J. (1995). *The generative lexicon*. Cambridge, MA: MIT Press.
Recanati, F. (1987). *Meaning and force*. Cambridge: Cambridge University Press.

Reis, M. (1997). Zum syntaktischen Status unselbständiger Verbzweit-Sätze. In C. Dürscheid, K. H. Ramers, & M. Schwarz (Eds.), *Syntax im Fokus*. Festschrift für Heinz Vater (pp. 112–144). Tübingen: Niemeyer.
Rizzi, L. (1997). The fine structure of the left periphery. In H. Liliane (Ed.), *Elements of grammar* (pp. 281–337). Dordrecht: Kluwer.
Ross, J. R. (1970). On declarative sentences. In R. A. Jacobs & P. S. Rosenbaum (Eds.), *Readings in English transformational grammar* (pp. 222–272). Waltham, MA: Ginn & Co.
Sadock, J. M. (1974). *Towards a linguistic theory of speech acts*. New York: Academic.
Schreiber, P. A. (1972). Style disjuncts and the performative analysis. *Linguistic Inquiry, 3*, 3121–3357.
Searle, J. R., & Vanderveken, D. (1985). *Foundations of illocutionary logic*. Cambridge: Cambridge University Press.
Shaer, B. (2003). "Manner" adverbs and the association theory: Some problems and solutions. In L. Ewald, C. Maienborn, & C. Fabricius-Hansen (Eds.), *Modifying adjuncts* (pp. 211–260). Berlin: Mouton de Gruyter.
Siegel, M. E. A. (2006). Biscuit conditionals: Quantification over potential literal acts. *Linguistics and Philosophy, 29*, 167–203.
Smaby, R. (1979). Ambiguous coreference with quantifiers. In F. Guenthner & S. J. Schmidt (Eds.), *Formal semantics and pragmatics for natural languages* (pp. 37–75). Dordrecht: Reidel.
Stalnaker, R. (1974). Pragmatic presuppositions. In M. K. Munitz & K. Unger (Eds.), *Semantics and philosophy* (pp. 197–214). New York: New York University Press.
Staudacher, P. (1987). Zur Semantik indefiniter Nominalphrasen. In B. Asbach-Schnitker & J. Roggenhofer (Eds.), *Neuere Forschungen zur Wortbildung und Historiographie in der Linguistik*. Festgabe für Herbert E. Brekle zum 50. Geburtstag (pp. 239–258). Tübingen: Narr.
Stenius, E. (1967). Mood and language game. *Synthese, 17*, 254–274.
Szabolcsi, A. (1982). Model theoretic semantics of performatives. In F. Kiefer (Ed.), *Hungarian linguistics* (pp. 515–535). Amsterdam: John Benjamins.
Vanderveken, D. (1990). *Meaning and speech acts. Vol. I: Principles of language use. Vol. II: Formal semantics of success and satisfaction*. Cambridge: Cambridge University Press.
Vendler, Z. (1957). Verbs and times. *Philosophical Review, 66*, 143–160.
Wittgenstein, L. (1958). *Philosophische Untersuchungen*. London: Basil Blackwell.

Recursion, Legibility, Use

Peter Ludlow

Abstract Certain kinds of cognitive abilities are required for recursive structures to be legible. Once the basic abilities are in place and recursive structures become legible, our ability to construct and express complex ideas is unbounded. It is argued that the prerequisite cognitive abilities are rules of semantic composition, understood expressivistically – that is, they are higher-level operations on expressivist attitudes. These operations on attitudes place constraints on the grammar, since different operations make legible different kinds of recursive structures.

Keywords Recursion • Legibility • Semantics • Expressivism • Use theory • Faculty of language • Negation • Disjunction • Attitudes

What comes first, the recursive structure of natural language or our ability to have complex thoughts? As we will see, this question presupposes that there are only two options. In this chapter I argue that there is a third, more nuanced, option to be explored: Certain kinds of cognitive abilities are required for recursive structures

---

An earlier draft of this paper was presented at the University of Massachusetts conference on Recursion, held in May 2009. Thanks are due to Sean Ebels Dugan for comments on that draft. Special thanks are due to Angelika Kratzer for her commentary at the conference, to Noam Chomsky for a key suggestion involving the integration of the use theory, and to Jason Stanley an anonymous reviewer for comments on the antepenultimate draft of this paper. Discussion in a Rutgers reading group on Mark Schroeder's *Being For*, helped me to formulate the final version of the expressivist proposal, which I subsequently ran past Andy Egan's graduate seminar on the foundations of semantics. I am particularly endebted to Andy, Simon Goldstein, and Una Stonjnic for questions and comments. None of the above are yet convinced that any of this works.

P. Ludlow (✉)
Department of Philosophy, Northwestern University, Evanston, IL, USA
e-mail: peterjludlow@gmail.com

T. Roeper and M. Speas (eds.), *Recursion: Complexity in Cognition*, Studies in Theoretical Psycholinguistics 43, DOI 10.1007/978-3-319-05086-7_5,

to be legible, hence these basic cognitive abilities come first. However, once the basic abilities are in place and recursive structures become legible, it is, as it were, off the races for our ability to construct and express complex ideas. What are the prerequisite cognitive abilities? I will argue that they are rules of semantic composition, but rules understood expressivistically – that is, they are higher-level operations on expressivist attitudes. As we will also see, these operations on attitudes place constraints on the grammar, since different operations make legible different kinds of recursive structures.

Thinking of the rules of semantic composition in this way also opens up a tantalizing new way for us to think about the theory of language as use. If we take the primary use of language as being the expression of attitudes – for example, pro-attitudes (like *being-for*) – then it becomes possible to construct a formal semantics of natural language for a use-based theory of meaning. This is surprising, if true, because it is widely if not universally supposed that use theories of meaning resist formalization. To be sure, constructing a formal expressivist semantics is not trivial and there are a number of puzzles to be dealt with. I sketch what I consider to be a plausible solution path for these puzzles.

1 Background

To a first approximation, generative linguistics is the theory of the system, cognized by human agents, by virtue of which they are able to pair perceivable linguistic forms with their meanings. One way to develop this idea is to say that representations that interface with the human perceptual/articulatory (PA) system are paired with representations that interface with the human Conceptual/Intensional (C/I) system. Recent work in generative linguistics – in particular within the minimalist program initially articulated in Chomsky (1995) – has hypothesized that this is accomplished via a *minimal* system that satisfies *legibility* constraints imposed by both the P/A system and the C/I system. Following Hauser et al. (2002) we can call this minimal pairing system the *faculty of language faculty narrowly construed* (FLN). Let's call the level of representation that interfaces with the P/A system *PF* (for phonetic form) and let's call the level of representation that interfaces with the C/I system *LF* (for logical form). We thus have the following picture:

P/A <====> PF ←---→ LF <====> C/I

FLN

Hauser, Chomsky, and Fitch and subsequently others have hypothesized that the FLN is an optimally efficient wiring solution to the problem of linking these interfaces. One plausible working hypothesis about the FLN is that it uses very simple operations like *merging* two elements and moving elements as necessary (and only as necessary) to meet the interface legibility requirements. The hallmark of the resulting wiring solution is recursion, meaning that the process of merging elements in this way yields recursive structures (more on this below).

Of course, a simple merge operation and the resulting recursive structures do not, by themselves, put strong constraints on linguistic theory – nature is full of various kinds of simple processes generating recursive patterns after all (for example spiral patterns in shells and galaxies) – and some of the most interesting properties of natural language (subjacency for example or the basic principles of binding theory) don't seem to have anything interesting to do with recursion by itself. So the interesting properties of language appear to fall out from the fact that the system yielding these recursive structures must also meet some rather strict interface conditions. If there are no constraints imposed by the interfaces, then nothing forces the interesting properties of language into relief – we just get vanilla recursive structures, and not the quirky structures we find in natural language.

The question is: What kinds of constraints do the interfaces impose? For current purposes let's devote our attention to the constraints imposed by the C/I interface. (There is some reason to believe that most of the interesting linguistic constraints are imposed at the C/I interface). The problem is that the available descriptions of the C/I system have been amorphous at best, and the danger is that the C/I system becomes just the theory of everything, or at least a system that is "cognitive science complete" – you would have to solve every problem in cognitive science to have a theory of the system, and in the meantime talk of the legibility constraints becomes vacuous.

Is there any way to articulate a constrained picture of the C/I system? My goal is to offer a proposal – or rather a family of proposals – by which we can begin to construct concrete theories of the C/I interface, and begin exploring in detail the kinds of constraints placed on the grammar by this interface (including the constraints on kinds of recursion imposed by the interface). In particular, I will argue that we can think of semantic theory as being a theory of the C/I interface. But what kind of semantic theory?

One approach would be to take on standard truth-conditional semantics for this task. A number of philosophers and linguists have expressed worries about whether such an approach to semantics is naturalizable and thus embeddable in linguistic theory understood as a chapter of naturalistic inquiry. I am less pessimistic, but I think it is at least worth exploring other options.

A number of researchers (ranging from Chomsky 2000 and numerous other publications to Horwich 1998, 2003, 2005) have suggested that the C/I system might be likened to a linguistic use theory. Of course, a use theory could also lapse into a theory of everything, so without a constrained picture of the mechanisms of language use, a use theory puts few legibility constraints on the interface. Again, we need a concrete story here if talk of "legibility" is to be non-vacuous.

One problem in any attempt to resolve this situation is that there has been a great divide in Anglo-American philosophy over the past 50 years between use theorists on the one hand (Wittgenstein, Austin, Searle) and truth-conditional semanticists on the other (Montague, Kaplan, Kripke). The traditional view has been that use theories resist formalization. When use theories *are* formalized, as in Kaplan (2001), it is done by assigning truth-conditional content where we might be surprised to find it. As Kaplan frames the move space, on Wittgenstein's view 'I am in pain' means 'ouch', while on Kaplan's view 'ouch' means 'I am in pain'. The underlying assumption of Kaplan is that the raw use theory cannot be directly formalized; formalization is only possible if we posit truth-conditional content.

Horwich (a use theorist) has resisted this conclusion, going so far as to suggest that the formalization of the use theory is a trivial matter, basically only requiring that we understand the contribution of the components.

> ... one might wonder how it can be that the content of SEM(E) [the semantic object associated with an expression] is determined by the contents of its parts – i.e., why there is no possibility that the underlying property in virtue of which a complex mental expression has its meaning fails to square with the properties in virtue of which the lexical items have their meanings (given the way these items have been combined). But this can be explained trivially – in a way that has nothing to do with truth conditions. It suffices to suppose that the content-property of a complex mental expression, SEM, is *constituted by* – one might even say *identical to* – its property of being constructed as it is from LIs [lexical items] with certain meanings.
>
> ... compositionality is accommodated without making any assumptions whatsoever about what sort of property of a primitive is responsible for its embodying the concept it does. (2003: 174–175)

Chomsky, on the other hand, is more cautious:

> Similar uneasiness is elicited, at least for me, by Horwich's account of compositionality. We have to account for the fact that "the content of SEM(E) is determined by the contents of its parts ..." This "can be explained trivially," Horwich suggests, if we recognize that understanding an expression, "is, by definition, nothing over and above understanding its parts and knowing how they have been combined." It is true that we could understand the expressions (1)–(10) when we understand their parts and know how they are combined, but to work this out seems anything but trivial. One could also say that perception of a cube in motion should be explained in terms of firing of cells in the visual cortex and the way the effects are integrated, but there s nothing trivial about the task of explanation, even for perception of a straight line. It might be true that the task can be carried out for language without reference to truth conditions, as Horwich suggests, just as it might be true that it can be carried out with reference to truth conditions. But it has to be shown, which doesn't seem easy, in either case. (2003: 303).

As we will see, Chomsky is right to be cautious here. As semanticists know, understanding how meanings combine is far from trivial in the truth-conditional case (indeed, you could argue that trying to figure this out is precisely what semanticists do for a living). But as we will also see below, the move to a use theory of meaning does not necessarily preserve the hard won discoveries of truth-conditional semantics.

Nevertheless, as we will also see, the potential rewards of working out the details of a use theory are great, so it is worth pursuing the question here, if only to expose the impending obstacles for further exploration.

Accordingly, I will pursue a strategy for constructing a formal use theory that can serve as the locus of contact between the C/I system and the FLN. I will not construct the use theory out of whole cloth however; my aim will be to borrow heavily from recent work in semantic theory and show that with some manageable if non-trivial adjustments, components of the truth-conditional semantics can be retasked as a formalized use theory. To do this I will borrow conceptual resources from recent work in expressivism and extend those resources to a general theory of expressive content that applies to all discourse, not just normative discourse.

I will move through several steps, offering working hypotheses along the way. The hypotheses are the following.

1. **The Semantic Legibility Hypothesis:** Formal Semantics is the theory of the C/I Interface. Legibility means legibility to the interpretation rules of the semantics.
2. **The Semantic Precondition Hypothesis:** Certain kinds of semantic rules are preconditions for certain kinds of recursive syntactic structures to be legible.
3. **The Semantic Naturalization Hypothesis:** Semantics can be naturalized via an expressivist semantics (primitives are referential intentions and pro-attitudes rather than reference and truth). That is, semantics is a calculus of expressive attitudes.
4. **Expressivist Semantics as Use Theory:** A plausible way of understanding an expressivist semantics is along the lines of a use theory that details the contributions that linguistic structure makes to the expression of attitudes.

I'll argue that there are two potential payoffs from hypothesis 3 – first it gives us a way of understanding how human language can be continuous with animal languages (which are arguably expressivist) – human language becomes an expressivist animal language plus recursion. Second, when hypothesis 3 is combined with hypotheses 1 and 2 it predicts that second order attitudes are a precondition for recursive structures to be legible in natural language, and thus explains the relation between the emergence of recursion and false belief tests etc.[1] Finally, the introduction of hypothesis 4 will allow us to understand formal semantic theory as a component in a formalized use theory – in effect, a system that shows how the expression of complex attitudes can extrude from the addition of recursion to the capacity to express more basic attitudes.

Despite the attractiveness of this research program there are deep conceptual and technical problems that need to be solved for an expressivist semantics to be viable. I'll conclude with some cautiously optimistic remarks on these problems.

With this plan in the background, let's turn first to hypothesis 1.

1. The Semantic Legibility Hypothesis: *Formal Semantics is the theory of the C/I Interface. Legibility means legibility to the interpretation rules of the semantics.*

[1]If correct, this suggests an oversimplification in recent discussions over whether propositional attitudes come first or whether recursion comes first in human development (see deVilliers 2007; deVilliers and de Villiers 2003; de Viliers and Pyers 2002). Related to this is the issue of whether recursive linguistic forms are a precondition for thought or whether this goes in the opposite direction and thought comes first. As I argued above, the unsimplified approach to the question is that certain kinds of cognitive abilities are a pre-condition for recursion, but that in turn complex recursive structures are a precondition for our ability to express complex attitudes, and perhaps also for our even having complex attitudes.

Most work in semantics over the past 40 years has assumed a two-stage approach to the theory of meaning. The two-stage theory rests on the framework initially articulated by Paul Grice (1989) as an alternative to the then extant use theories of Wittgenstein and Austin. As Grice saw it, rather than have a theory that went straight from syntactic forms to use, we will find it fruitful to posit an intermediate level of interpretation. For now, let's call this intermediate level the *semantic* level. Applying our talk of interfaces to this idea, we have the following picture.

LF Syntax -- interfaces --- Semantics -- interfaces --- Use Theory

For the moment let's avoid controversy about what a semantic theory is to look like (eventually we'll attempt to convert it to a *stage 1 use theory*), let's just focus on the idea that meaning might come in two stages – what Grice characterized as the Proposition Expressed (we might say *the proposition literally expressed*) and the Proposition Meant (that is, what the speaker intended to convey).

Our ordinary use of the term 'meaning' is notoriously fluid (taking in "the meaning of life" at the extreme) but we can nevertheless clarify two different conceptions of meaning in part by spelling out the theories responsible for each component. I want to avoid talk of propositions here (for reasons that will become clear later), and introduce the following way of talking about these components. I'll say that semantics (or the stage 1 use theory) is concerned with "literal meaning" or if you prefer the "official meaning" and that the stage 2 use theory is concerned with "conveyed meaning".

LF Syntax -- interfaces --- stage 1 (semantics) -- interfaces --- Stage 2 Use Theory
(literal meaning) (meaning conveyed)

I make this distinction because we can use an expression having a particular literal meaning to communicate something rather different from that literal meaning. To use Grice's famous example, I might write a letter of recommendation for my student saying "He is very punctual and has excellent handwriting" and thereby use an expression that is literally about handwriting and punctuality to communicate something about hireability.

This division of labor is not driven by conceptual necessity, but it has 40 years of very successful science behind it. My sense is that the results and discoveries in semantics have been at least as impressive as those in syntax, and semantic theory has accomplished this in part by focusing on what I am here calling literal meanings and the progress has also been made possible by ensuring that semantics has been tightly constrained by the syntactic form of natural language constructions – i.e. a very direct connection is assumed.

Likewise, progress has been made on the stage 2 use theory (what we might call the theory of pragmatics), not just on the descriptive end but on the explanatory end as well. We have successfully reduced much of the pragmatic component to a set of Gricean maxims (or even a smaller set as proposed by Sperber and Wilson 1995) and subsequent work has shown that the origin of these maxims or principles can be explained by evolutionary game theory – i.e. these are not "just so" stories but stories for which we actually have working predictive models. (see Grim (2007), Asher et al. (2001), Parikh (2001), and van Rooy (2004).)

Notice that talk of the C/I system has temporarily dropped out, but we can reintroduce it if we like, since I take it that semantics is one component of the C/I system and it is the component of the system with which the language faculty (FLN) interacts. Semantics is *locus of contact* between the FLN and the C/I system.

PF -------- LF Syntax --interfaces--- Semantics ----- stage 2 Use Theory --- etc

FLN C/I System

The puzzle in this is that while semantics sits at the crossroads of two very naturalistic research programs – linguistics on the one side and the stage 2 Use Theory and the rest of the C/I system on the other – it *seems* to many linguists and philosophers to be an outlier in one respect. That is, we have a picture about how the FLN is grounded in low level bio-physical principles, and we have a picture about the origin of the basic principles of use theory (or at least the Gricean maxims) but what about semantics – what grounds *it* naturalistically?

PF -------- LF ----- Semantics ----- stage 2 Use Theory --- etc

FLN

I'll return to this question in the discussion of hypothesis 3. First, however, I want to turn to the question of the kinds of constraints that Semantic Theory can place on the FLN, and in particular on how certain axioms of the semantic theory are actually preconditions for the legibility of recursive structures.

2. ***The Semantic Precondition Hypothesis:*** *Certain kinds of semantic rules are preconditions for certain kinds of recursive syntactic structures to be legible.*

To make this discussion clear I want to be as concrete as possible about what I mean by recursion. Consider the following toy phrase structure grammar.

1) XP $\rightarrow$ XP ConjP (where X is I, D, V or P)
2) ConjP $\rightarrow$ Conj XP (where Conj is 'and', 'or')
3) S (IP) $\rightarrow$ DP I$'$/ not S
4) I$'$ $\rightarrow$ I VP
5) DP $\rightarrow$ det NP
6) NP $\rightarrow$ N$'$ (PP)
7) N$'$ $\rightarrow$ (adj) N$'$
8) N$'$ $\rightarrow$ N
9) NP $\rightarrow$ NP CP
10) CP $\rightarrow$ wh C$'$
11) C$'$ $\rightarrow$ C S
12) VP $\rightarrow$ V XP

As I am using the term 'recursion', rules 1, 7, 9 would be recursive rules because the rules "call themselves", and *combinations* of rules like 3-5-9-10-11 would be recursive, because the topmost rule in the derivation is called again lower in the derivation. I am calling the output of such rules (the trees generated by such rules) recursive *structures* because the structures have "descendent" nodes that are the same type as their ancestors. For example, I consider the following two structures to be recursive structures because a category S occurs within S in the first structure and N' occurs within N' in the second.

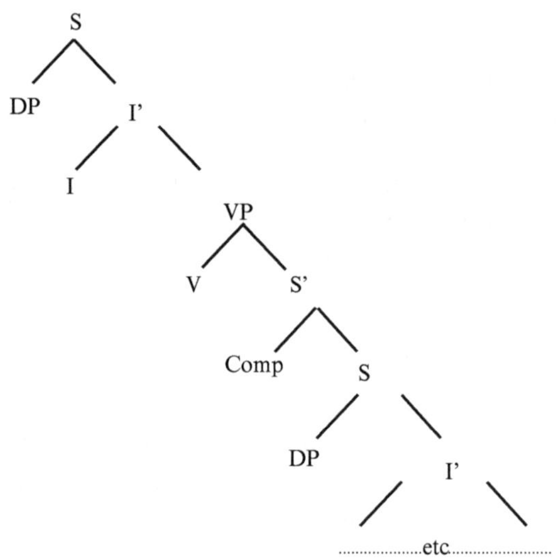

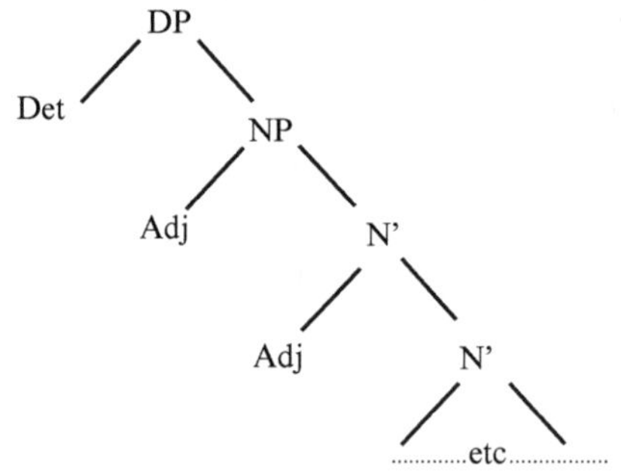

I assume that in point of fact linguistic structures are more complex than this and I also assume that they need not be the product of recursive rules like those I specified in the toy grammar above.

To be as concrete as possible in our discussion of semantics I am going to assume a standard approach to natural language semantics, and hence I'm going to adopt the notation and basic theoretical background articulated in the Heim and Kratzer introductory textbook. Obviously, it is an introductory textbook, and specific proposals are offered with a heavy emphasis on their pedagogical value, but the theory articulated there is rich enough to illustrate the kind of research program I am envisioning here.

For our purposes we assume the basic elements of a semantic theory are elements of type e and t, where we can think of elements of type e as being entities or referents and elements of type t as being truth values. Predicates of the language can be thought of as functions that map from elements of type e to elements of type t. So, for example, the predicate 'food' can be thought of as a function that maps from individual entities onto true or false. More complex elements like adjectives can be thought of as mapping from function to function etc. To a first approximation, avoiding more complex constructions like generalized quantifiers the picture is something like this.

Syntactic category	Semantic type (extensionalized)	Expressions
ProperN	<e>	names (*John*)
S	<t>	sentences
CN(P)	<e, t>	common noun phrases (*cat*)
NP	<e>	"referential" NPs (*John, the king*, he_i)
	<e, t>	NPs as predicates (*a man, the king*)
ADJ(P)	<e, t>	predicative adjectives (*carnivorous, happy*)
	<<e, t>, <e, t>>	adjectives as predicate modifiers (*skillful*)
REL	<e, t>	relative clauses (*who(m) Mary loves*)
VP, IV	<e, t>	verb phrases, intransitive verbs (*loves Mary, walks*)
TV	<e, <e, t>>	transitive verb (*loves*)
is	<<e, t>, <e, t>>	*is*

Lexical items specify the particular functions. To illustrate, let's use the notation [[α]] to indicate the semantic value of α, and consider a couple of examples of lexical rules by which the semantics specifies the meaning of an expression.

(1) [[**Schleiermacher**]] = Schleiermacher

(2) [[**ponders**]] = $\lambda x \in D_e$. true iff x ponders

Of interest to us a bit later will be the rules for the connectives, which we can give as follows:

(3) [[**and**]] = $\lambda p \in D_t$. [$\lambda q \in D_t$. true iff p = q = true]

(4) [[**or**]] = $\lambda p \in D_t$. [$\lambda q \in D_t$. true iff p = true or q = true]

(5) [[**not**]] = $\lambda p \in D_t$. true iff not p = true

Now for my money the most interesting feature of formal semantics is just discovering how much one can accomplish with such limited resources. You might think that given the infinite number of syntactic constructions you might encounter there must be an infinite number of semantic rules to interpret them, or at least an awful lot of rules. But once the lexical rules are in place we get by with interpreting all those structures just using a handful of very simple semantic rules. Following Heim and Kratzer (1998), let's characterize them as follows.

(i) *Terminal Nodes* (TN): If α is a terminal node, then $[[\alpha]]$ is specified in the lexicon.
(ii) *Non-Branching Nodes* (NN): If α is a non-branching node, and β is its daughter node, then $[[\alpha]] = [[\beta]]$.
(iii) *Functional Application* (FA): If α is a branching node, $\{\beta,\gamma\}$ is the set of α's daughters, and $[[\beta]]$ is a *function* whose domain contains $[[\gamma]]$, then $[[\alpha]] = [[\beta]]\ ([[\gamma]])$.
(iv) *Predicate Modification* (PM) : If α is a branching node, $\{\beta, \gamma\}$ is the set of α's daughters, and $[[\beta]]$ and $[[\gamma]]$ are both in $D_{<e,t>}$, then $[[\alpha]] = \lambda x \in D_e$. true iff $[[\beta]]\ (x) =$ true and $[[\gamma]]\ (x) =$ true.
(v) *Predicate Abstraction* (PA): If α is a branching node whose daughters are a relative pronoun and β, then $[[\alpha]] = \lambda x \in D_e.\ [[\beta]]^x$ (where, $[[\beta]]^x$ indicates the predicate formed by replacing the subject argument position of $[[\beta]]$ with the variable x.)

Clearly this is just a first pass and I'm leaving out detail, but it won't matter for the discussion that follows. The point is that these composition rules are rich enough to interpret all the structures that can be generated by the toy grammar I gave above.[2] And then some. And this, to me, is impressive.

But now we get to the central point of hypothesis 2. If we think of the semantics (or the semantics module) as being the locus of contact with the C/I interface (as in hypothesis 1), then these semantic rules determine the conditions for structures being legible. That is to say, if a structure is to be legible then it must be a structure that can serve as an input for these rules. It must be *visible* to the rules.

For example, only binary branching and nonbranching structures are visible to these rules – ternary branching structures are thus not legible. Furthermore, while these rules can "read" binary structures in which both daughters are nouns (as in 'foil ball') they cannot read structures in which there are pairs of referential expressions (expressions of type e). So these rules cannot read 'that that' or 'she she'.

But notice also that if some of these rules were missing then certain recursive structures would not be legible. For example:

If **Predicate Modification** was missing then recursive structures that involved noun noun recursion would be out, as would structures involving NPs with recursively iterated PP modifiers, like 'dog in a city in a county in Texas ...'

If **Predicate Abstraction** was missing, then recursive structures involving relative clauses would not be legible (at least on the Heim and Kratzer analysis of

[2]The semantic rules I gave would need to be supplemented to handle complex DPs but the composition rules need no additions.

relative clause structures), since predicate abstraction is a device that takes a CP and converts it into an expression of type <e, t>. Without PA the structure corresponding to 'this is the cat that ate the rat that ate the cheese that ...' would not be legible.

Of course, some recursive structures are made possible not by these basic semantic rules but by the lexical rules. Canonical examples of recursion involve the logical connectives.

So for example consider a lexical rule like the following, which is the 'and' used to conjoin predicates as in (sang and danced):

(6) $[[\text{and}]] = \lambda f \in D_{<e,t>}. [\lambda g \in D_{<e,t>}. [\lambda x \in D_{e}. \text{ true iff } f(x) = g(x) = \text{true}]]$

Without a lexical rule like this, then barring the introduction of conjunction reduction there would be no way to interpret structures with conjoined predicates – for example a construction like 'She sings and dances and acts and ...'

We can put the point this way. Certain kinds of lexical rules are preconditions for a large class of recursive structures – for example, those involving logical connectives. To a first approximation, these are lexical rules that involve mapping from a function of a particular type onto a function of the same type. That is, the lexical rules must specify functions that are...

(i) higher order
(ii) type reflexive

First order functions – for example those that are <e, t> are not sufficient by themselves for recursive structures like those generated by our toy grammar to be legible. Higher order functions that are not reflexive are also not sufficient (by themselves) for recursive functions to be legible. For example determiners are higher order, but not reflexive in the sense I have indicated:

You can get recursion without type reflexivity, but only if you have a semantic derivation that forms a type reflexive theorem. So for example, sentential complements appear to be of this form: [John believes that Mary believes that ...]. Successive application of rules can generate the type reflexive theorem. On the semantic end, what is permitting these recursive syntactic structures (like 'John believes that Mary believes that ...') to be legible is what we could call a **derived type reflexive theorem**. In this case, what is allowing the recursive structure is that once axioms are combined in a derivation, the resulting step or theorem is type reflexive.

Summing up, we can chart out the preconditions for the legibility of recursive structures as follows.

PM: required for noun noun compounds, and NP-PP recursion
PA: required for relative clause recursion
Type Reflexive lexical rules: required for [adj [adj [... [N]] recursion, logical connective recursion etc
Derived type reflexive theorems: required for sentential complement recursion

We can afford to be flexible about the size of the syntactic object that is handed off to the semantics. For example, it might be that an LF for the entire sentence must be constructed before it is handed off to the semantics, in which case legibility is determined by whether the semantics can successfully grind through the LF without choking – in other words, there must be a semantic rule that can interpret the semantic value of each node in the tree. Alternatively, it might be that the handoff occurs at the vP and CP level phases in the sense of Chomsky (2001) – in effect at nodes that correspond to propositions of some form, or the handoff could happen after every merge operation (as suggested in Epstein and Seeley (2006)). In the latter case after each merge operation there is an immediate handoff to the semantics. If the merged material is legible to one of the semantic rules outlined above, then the derivation continues.

Let's put all this on the back burner now. I now want to turn to a discussion of expressivism. We'll come back to this taxonomy in a bit.

3. ***The Semantic Naturalization Hypothesis:*** *Semantics can be naturalized via an expressivist semantics (primitives are referential intentions and pro-attitudes rather than reference and truth). That is, semantics is a calculus of expressive attitudes.*

Recent work in philosophy has explored the prospects of developing an expressivist semantics for natural language. Expressivism has its roots in attempts to naturalize normative discourse in ethics. For example, Stevenson (1944) proposed that when we make moral claims to the effect that 'x is wrong', we can be understood as expressing something without cognitive content (non-cognitive here means that there is no truth conditional content) – we are "emoting" in Stevenson's terminology. We could also say we are expressing an attitude of disapproval towards the contemplated action or (following Gibbard 2003) are expressing a commitment or plan to act in a particular manner.

Obviously there are a number of forms that expressivism could take and a number of ways we could describe the relevant attitudes, but let us use the term "Pro!" with an exclamation point "!" to signify the attitude that is expressed when approving a certain action and/or committing ourselves to a certain course of action (possibly a linguistic action). Similarly, we can use "Con!" with an exclamation point to signify the attitude that is expressed in disapproving a certain action or making a commitment not to act in a certain manner. It may be part of these attitudes that we are committed to "sticking to our guns" in holding the attitudes (this is a question outside the scope of this paper).

One outstanding problem with treating the varieties of normative discourse expressivistically is the so-called Frege-Geach problem (Ross (1939: 33–34), Geach (1958, 1960, 1965) and Searle (1962, 1969). The basic problem is this: if sentences expressing normative discourse don't have traditional truth values and only express attitudes, then how is it that we can embed those sentences within larger sentences that clearly *do* have truth values?

For example, consider an utterance of the following sentence:

(7) If eating foie gras is wrong, then Barry doesn't want to be right

It is one thing to say that one is merely emoting in the utterance of a sentence like 'eating foie gras is wrong', but the conditional is usually taken to have a definitive truth value. But this only makes sense if truth values can be assigned to both the antecedent and the consequent. What is one to do?

One strategy is to extend our expressivist account to conditionals as well (for example they might express conditional attitudes), but we might also suppose that the mistake was trying to apply expressivism piecemeal. Perhaps what we really need is a generalized expressivist semantics for natural language. In this way, there is nothing special about the semantic value of the antecedent of (7) or (7) itself. *Everything* has expressivist content.

Hypothesis 3 in effect calls us to explore the general expressivist strategy – i.e. to attempt the construction of an expressivist (or expressivist-friendly) semantical theory that is finitely axiomatizable, is in some interesting sense compositional, and which is robust enough to be able to tackle the formidable logical complexity of natural languages as explored by natural language semanticists. This may sound like the sort of project that could be executed in a brute strength manner, and is, at the end of the day, philosophically trivial (as noted above, this is a view that appears to be held by Horwich). As we will see, however, this is far from the case. But before we get into the problems, let's look at the positive side of this project.

2 An Expressivist Semantics for Hyper-opinionated Agents

The basic idea is the following. Assuming we are constructing an expressivist semantics for an agent that is hyper-opinionated (an agent that is has an opinion about every proposition), we can simply swap out our basic notions <e> (for entities) and <t> (for truth values) and replace them with expressivistically kosher basic elements.

For example, instead of truth values, we will introduce the attitudes Pro! and Con!. Instead of entities, we will introduce what I will call referential intentions. These are the attitudes we express when, for example, pointing.[3] The remaining semantic types will be built from these basic notions and will in effect be "functional attitudes", or "classificatory attitudes".

[3]The notion of referential intention can present problems if we aren't careful. I'm going to assume that referential intentions are individuated by real world objects, so that there is a one-one correspondence. Others may prefer to think of referential intentions as picking our "files" in the sense of file change semantics. I'm going with the external object route here simply to table issues that might arise about quantification over these elements.

To keep our two approaches to semantics separate then, let's use 'i' to indicate a referential intention and 'A' to indicate a Pro-attitude. Instead of indicating the semantic value of an element α thus: $[[\alpha]]$, we can use the following notation to indicate that we are talking about an attitude expressed: $!!\alpha!!$. Strictly speaking, we will be talking about an utterance of α by an agent *a*, in a context *c* etc, and we can annotate our notation accordingly – $!!\alpha!!_{a,c}$. but I will abstract from that in the discussion that follows.

We can then translate our basic semantic types as follows.

Syntactic category	Semantic type (extensionalized)	Expressions
ProperN	<i>	names (*John*)
S	<A>	sentences
CN(P)	<i, A>	common noun phrases (*cat*)
NP	<i>	"referential" NPs (*John, the king, he$_i$*)
	<i, A>	NPs as predicates (*a man, the king*)
ADJ(P)	<i, A>	predicative adjectives (*carnivorous, happy*)
	<<i, A>, <i, A>>	adjectives as predicate modifiers (*skillful*)
REL	<i, A>	relative clauses (*who(m) Mary loves*)
VP, IV	<i, A>	verb phrases, intransitive verbs (*loves Mary, is tall, walks*)
	<i, <i, A>>	transitive verb (*loves*)
is	<<i, A>, <i, A>>	*is*

The semantic rules, PM, PA, etc also remain unchanged except that, as I said before, we swap out the $[[\alpha]]$ notation for $!!\alpha!!$. So technically everything operates as it did before. There just remains the problem of interpreting what is happening.

Let's work our way into this by looking at the sentential connectives first. Rather than having truth values as our basic semantic values we have the attitudes Pro! and Con!.

(8) $!!\mathbf{and}!! = \lambda p \in D_A.\ [\lambda q \in D_A.\ \text{Pro! iff } p = q = \text{Pro!}]]$

(9) $!!\mathbf{or}!! = \lambda p \in D_A.\ [\lambda q \in D_A.\ \text{Pro! iff } p = \text{Pro! or } q = \text{Pro!}]]$

(10) $!!\mathbf{not}!! = \lambda p \in D_A.\ \text{Pro! iff not } p = \text{Pro!}$ (and because the agent is hyperdecided $p = \text{Con!}$)

At the subsentential level, a function like <i, A> indicates a functional/classificatory attitude. When you express this attitude you are expressing an attitude about mappings from referential intentions to pro-attitudes. For example, consider the noun 'food'. I am *for* applying this noun to some of my referential intentions, but not all of them. The attitude is a "functional attitude" in the sense that it maps from referential intentions onto Pro! When I say 'that is food' while gesturing at a carrot, I am expressing the attitude I am for (am Pro!) applying the term 'food' to that carrot.

Similarly, if I utter a sentence with a transitive verb, e.g. 'John saw Bill', I have referential intentions for John and Bill, and the verb 'saw' expresses an attitude that I have about applying the term 'saw' to certain ordered pairs. In this instance I am expressing the attitude that I am for (Pro!) applying the term to the ordered pair of referential intentions corresponding to John and Bill. Matters are a bit more complex than this because we are assuming a binary branching structure, so to get to the flattened expression we first need to combine the attitude expressed by the transitive verb with the referential intention for Bill. The attitude expressed by the verb, is a function that inputs referential intentions and then outputs functions of type <i, A>, so in effect when I say John saw Bill I am saying that if I input Bill to the function corresponding to 'saw', the result will be an attitude that maps from referential intentions to pro-attitudes, and I am for (Pro!) applying that resulting functional attitude to the referential intention for John. And so it goes with progressively more complex attitudes. Adjectives are higher order attitudes that map from functional attitudes onto functional attitudes and so on.

It might seem that this semantics has the consequence that our utterances are altogether metalinguistic. When I say 'that is a carrot' I'm not saying it's a carrot, I'm saying I'm for applying the term 'carrot' to that thing. And while it's true that our claim is metalinguistic, it isn't *merely* metalinguistic. Being for calling something 'carrot' has consequences – for example that barring allergies you are also for putting it in a salad. Similarly, the semantics for 'person' involves what referential intentions we are for applying the term 'person' to, but it isn't merely this. If we are for labeling someone or something 'person' we take on other important commitments as well – for example showing them respect. These linguistic pro-attitudes are not inert – they have consequences for our broader plans and actions. Similarly, we are often able to supply reasons for calling something 'person'. However, those reasons and consequences need not be encoded in the semantics. We can consider them extra-semantic.

Staying with the conceit that all agents are hyper-decided, let's revisit our taxonomy for what semantic rules are required for various recursive structures to be legible. Recall the taxonomy we came up with earlier.

PM: required for noun-noun recursion, and NP-PP recursion
PA: required for relative clause recursion
Reflexive lexical rules: required for [adj [adj [. . . [N]] recursion, logical connective recursion etc
Reflexive derived theorems: required for sentential complement recursion

Can we say more about the semantic precondition hypothesis if we assume an expressivist semantics? The answer appears to be yes. First, since as expressivists we are now thinking about semantic rules in terms of operations on attitudes we can see that certain kinds of attitudes are involved in these different semantic rules. Or better, the rules just describe attitudes and operations on attitudes.

For example, what I have called reflexive lexical rules involve the ability to map from attitudes of a particular type onto to attitudes of the same type – it involves our having second order expressivist attitudes.

PM involves what we might call a kind of attitude fusion. It is an operation that takes two functional expressivist attitudes and fuses them into a new expressivist functional attitude. Crucially then, PM not only involves our having attitudes about attitudes, but it introduces the ability to merge them.

PA involves the ability to take something for which we might have a pro-attitude (say a sentence) and convert it into a functional attitude. We might be inclined to say that it involves our ability to take basic attitudes of type A, and create more abstract attitudes of type <i, A>.

Finally, our derived type reflexive theorems involve our ability to not only operate on higher order attitudes, but to chain them together. So our new taxonomy is something like this.

PM: involves second order attitudes and attitude fusion.
PA: involves second order attitudes and abstraction on them.
Reflexive lexical rules: involve second order attitudes, and mapping from attitude to attitude.
Derived type reflexive theorems: involve second order attitudes, attitude mapping, and chaining.

Putting this all together we have the following correlation between kinds of recursive structures and cognitive abilities.

second order attitudes and attitude fusion: required for noun noun compounds, and NP-PP recursion.
second order attitudes and abstraction on them: required for relative clause recursion.
second order attitudes, and mapping from att to att: required for [adj [adj [... [N]] recursion, logical connective recursion etc.
second order attitudes, attitude mapping, and chaining: required for sentential complement recursion.

3 Some Predictions

If this story is correct, then several predictions fall out immediately. First and foremost: If these cognitive abilities are preconditions for linguistic recursion, then there should be a correlation between linguistic recursion and second order reasoning – for example the false belief tests. So far so good; children do pass the false belief test well before they are able to process recursive structures like relative clause recursion and complement clause recursion.

But there are also opportunities for probing some finer grained questions here. If there are certain kinds of aphasia that would target some of these cognitive abilities (for example the ability to abstract functional attitudes from basic attitudes), then the result would be not the loss of recursion altogether but rather the loss of relative clause recursion only. Similarly, the loss of the ability to fuse attitudes might wipe out certain kinds of N-PP recursion. Presumably if the maturation of these cognitive abilities happened at different rates, then we could predict the correlative onset of competence with different classes of recursive structures at different times (a prediction that appears to be supported by Roeper and Snyder (2005) among others).[4]

4. ***Expressivist Semantics as Use Theory:*** *A plausible way of understanding an expressivist semantics is along the lines of a use theory that details the contributions that linguistic structure makes to the expression of attitudes.*

One interesting feature of the program developed thus far is that it appears to provide us with the ingredients for a formalized use theory of meaning. Why? A standard conception of semantics since Frege has been the idea that linguistic expressions have referents and truth conditions. As Austin and others cautioned, however, it seems more apt to say that we *use* expressions to refer, etc., but that they do not in and of themselves *have* referents or truth conditions. By shifting our perspective to an expressivist semantics this admonition hardly seems necessary. It might seem plausible to think that words have referents, but the idea that words have referential intentions is borderline nonsensical. Likewise, while it might seem plausible to think that an utterance has truth conditions, it is hard to imagine that it (the expression) could, in isolation of us as speakers, express an attitude.

If we retask the tools of semantic theory as I have suggested, then what we get is a kind of tool kit for using language to recursively construct and express complex attitudes. This isn't the kind of use theory envisioned by Austin and Wittgenstein. For one thing, there is nothing social about this stage 1 use theory. The basic ingredients could be part of our biological endowment, and indeed the basic level components could be continuous with the systems that animals use to express attitudes (our system happens to have additional resources like higher order attitudes which in turn allow recursive structures and these in turn enable

[4]Less clear is the question of whether this has consequences for cross-linguistic variation. Examples in Roeper (2005) for example illustrate that the kinds of recursive structures allowed vary significantly across languages. French, for example, does not allow recursion on possessives, so that constructions like 'John's mother's dentist's cousin' are not happy. Is it to feasible to think that these variations are the result of variation in the kinds of cognitive operations employed? This is certainly doubtful, given that the semantic rules deployed in the interpretation of possessive recursion are active elsewhere in the semantics. This linguistic variation appears to have its source in parametric variation in the FLN that is not tied to the C/I interface. Options are limited here; presumably one looks in constraints driven by differences between the English and French perceptual/articulatory system.

the expression of attitudes of unlimited complexity). The stage 2 use theory, which introduces Grice's Cooperative Principles and the Conversational Maxims will be the point where social elements enter into the use theory.

Two concerns must now be addressed. First, given the broad range of attitudes that we can express, there is a question of how attitudes are to be individuated. What makes one attitude distinct from another one? The concern is that when we try to answer this question we will be tempted to say that the attitudes correspond to different states of affairs, and this suggests that the whole project rests on a truth conditional semantics at its foundations. This tracks a traditional worry about expressivism in general: what is going on when people are disagreeing about something? If the first party to the disagreement is simply saying "Yay! p", and the second party is simply saying "Boo! p" then its not clear what sense we can make of the disagreement (two fans can cheer for different teams, but we don't suppose they are disagreeing).

The traditional response to this worry is to say that when we disagree we disagree "in attitude," but this response seems a bit thin. What does it mean to disagree in attitude? And how are the underlying attitudes individuated? Gibbard (2003) has argued that we might think of a disagreement in attitude as akin to a disagreement in plans (obviously plans not individuated by truth conditions). Here is one way to think about the matter (not attempting to be faithful to Gibbard). Suppose we have a shared referential intention. You point and say 'that is food'. I point and say 'that is a predator'. On the theory I have sketched we are expressing different attitudes (clearly), and with Gibbard's proposal the difference in attitude will be manifest in the kinds of plans they give rise to. For example your attitude might be manifested in a plan to chase, and mine in a plan to flee.

I don't think we need to introduce plans to explain the difference in attitudes. Agent A may be for applying 'person' to corporations and agent B may be against it, in which case they clearly have distinct attitudes that can be expressed by diverse functions of the form <i, A>. With respect to applying 'person' to corporations, A is Pro! and B is Con!. This will no doubt also give rise to different plans (for example A believes they should be allowed to donate to political parties and B does not). The point is that a difference in attitude will have important consequences and it is easy to understand why people dig in and argue about these things.

There is another problem that I alluded to earlier – that of referential intentions. The problem is that one can have many different referential intentions to a particular object, and this raises concerns when we start thinking in terms of cardinal quantifiers. We don't want to be in the business of counting referential intentions. I have suggested that we individuate and type referential intentions by their denotations, but this won't be palatable to everyone. It also raises the question or why we should take the detour through referential intentions at all; why not stick with objects themselves?

In spite of these concerns, I think the conjunction of hypotheses 1–4 sounds like a winning program. Let's take stock of what it can do.

1. It gives us a tightly constrained picture of what the C/I interface conditions must be. LF structures (or the outputs of syntactic derivations) must be visible to the semantic theory – either merge-by-merge or phase-by-phase or as a whole. Whatever the chunk handed to the interface it must be legible to the semantic theory – construed as a formal semantic theory.
2. It gives us a way of situating semantic theory within the C/I system and thus provides a foothold on any attempt to construct a modular theory of the C/I system. It gives us a well defined component of that system.
3. The introduction of expressivism gives us a plausible strategy for naturalizing semantic theory.
4. Correlatively, it gives us an account that shows that natural language is continuous with animal languages on one level – it is fundamentally expressivist – but that human language may be distinctive in that complex attitudes are built by recursion.
5. It gives us a picture of the kinds of cognitive abilities that are pre-conditions for recursive structures to be legible.
6. In turn, this gives us a powerful tool for exploring correlations between specific recursive structures and diverse cognitive abilities.
7. If we think of the semantic theory as stating constraints on the expression of attitudes then we have the makings of a formal and naturalizable use based theory of meaning.

As I said, this sounds like a winning program, but there are difficulties with executing the details of any such effort.

4 The Problem with Expressivism

Thus far I have been operating under the fiction that agents are hyper-decided. Clearly, that hypothesis can't be maintained, and this is the fundamental problem with expressivism – it breaks down (or appears to break down) – with agents that are not hyperdecided.

To see this, consider the direct translations of our rules for sentential connectives into our expressivist semantics:

(11) !!**and**!! $= \lambda p \in D_A. [\lambda q \in D_A.\text{Pro! iff } p = q = \text{Pro!}]]$

(12) !!**or**!! $= \lambda p \in D_A. [\lambda q \in D_A. \text{Pro! iff } p = \text{Pro! or } q = \text{Pro!}]]$

(13) !!**not**!! $= \lambda p \in D_A. \text{Pro! iff not } p = \text{Pro!}$

Initially the rule for 'and' seems unproblematic, but the rules for 'or' and 'not' seem to fail.

Consider the rule for 'or' first. I might approve of 'S1 or S2' but it does not follow that I either approve of S1 or approve of S2. To make this vivid, suppose that S2 is simply the negated form of S1 and suppose that S1 is some complex logical proposition. Then it is entirely reasonable that I approve 'S1 or S2', even though I am not ready to approve either S1 or S2.

Problems also arise for negation. Suppose again that S is some claim that I am decidedly opposed to and I assert 'not S'. The problem is that not being Pro! doesn't seem to be strong enough. In this case I am not merely not Pro! I am Con!

To a first approximation it looks like we need to introduce additional basic attitudes into the mix. For example it looks like we need a negation rule like the following.

(14) !!**not**!! = $\lambda p \in D_A$. Pro! iff p = Con!]]

But if we make this move then now we have a problem with conjunction, for what are we to say in a case where we are Con! a particular conjunction? I might be Con! the sentence 'S and ~S' but it doesn't follow that I am Con! both (or even one) of the conjuncts. Not surprisingly, we here revisit the problem we had with disjunction.

Expressivists themselves have identified another problem with expressivism and negation – that it seems to lack the ability to express certain kinds of negation. Schroeder (2008), following earlier work by Blackburn (1984, 1988) and Unwin (1999, 2001), gives us a nice explication of the problem. Consider the following sentence w, and the three different ways in which it could be negated.

w John thinks that murdering is wrong
n1 John does not think that murdering is wrong
n2 John thinks that murdering is not wrong
n3 John thinks that not murdering is wrong

The worry is, that the second form of negation, which really seems like the natural one, does not seem to be available for w*.

w* John disapproves of murdering
n1* John does not disapprove of murdering
n2* ???
n3* John disapproves of not murdering

We can say John doesn't disapprove of murdering, and we can say he disapproves of not murdering, but there does not seem to be a form of negation analogous to n2 – something that expresses him thinking people shouldn't be sanctioned for murder. Now it may well be that are ways to generate this understanding via pragmatics (perhaps a scalar implicature), but my concern here is not really with correctly diagnosing this problem so much as using it as a vehicle for introducing Schroeder's proposal (which I will then proceed to borrow in a modified form).

Schroeder (2008: 59) thinks that the problem here is in part structural (we need another landing site for the negation) and he offers that the trick is to introduce a new matrix attitude – "being for." The result is extra structure, and hence a landing site for the negation, allowing us to generate the missing reading with the structure in n2*.

w* John is for blaming for murdering
n1* John is not for blaming for murdering
n2* John is for not blaming for murdering
n3* John is for blaming for not murdering

There are plenty of questions to be asked here. One question is the mechanism by which 'disapproval' gets exploded into 'not blaming for', and then 'being for not blaming for'. I personally don't have problems with lexical decomposition, but this is an extraordinarily robust form of decomposition. Another question is whether the being-for attitude make sense. Am I really **for** blaming for murdering? Why can I not be against murdering yet thinking blaming-for is a waste of time and energy. I might be against it and think we should eliminate the conditions that give rise to murder. I don't think the problem here is that Schroeder hasn't chosen the right attitude – the problem is that any sort of attitude might accompany the judgment that something is wrong.

Driven in part by these concerns I would like to put forward a simpler proposal that absorbs, I think, the critical insight of Schroeder's proposal – the introduction of another layer of structure. My proposal is that if structure is the issue, we really don't need to add anything more than a matrix predicate with a bleached out content like "expressing an attitude." That is I would propose something like the following:

w* John expresses the attitude of disapproving of murdering
n1* John does not express the attitude of disapproving of murdering
n2* John expresses the attitude of not disapproving of murdering
n3* John expresses the attitude of disapproving of not murdering

So you express your having an attitude. Like you express your having a headache not headache simpliciter.

The interesting thing about this proposal is that once pressed into service it gives us a way to resolve the difficulties we saw with the connectives for non-hyperdecided agents. Consider again, the case of disjunction. The problem we encountered was for such agents, it might be possible to be Pro! a disjunction but not be Pro! either of the disjuncts (for example when I believe I have either an even or odd number of coins in my pocket).

Introducing the extra layer of structure allows that I *express the attitude* that I am Pro! I have an even number of coins in my pocket or Pro! I have an odd number of coins in my pocket. I construct an either-or attitude, or if you prefer, a disjunctive attitude. The resulting theory is presumably not compositional in that the semantic value of the whole is not built out of the semantic values of the parts, but it is nevertheless recursive. It is a kind of calculus of attitudes.

The story for negation is similar. We considered a case in which I was decidedly against a proposition, and it seemed that not being Pro! was insufficient to capture that. But now we gloss the account as me *expressing the attitude* of not being Pro! the proposition. It is baked into my attitude that I am against this, something that is quite different from it not being the case that I have the attitude of being Pro! the proposition.

Of course, I've been advocating a much broader application of expressivism than just the handling of ethical and normative discourse – I've been suggesting that the semantics of natural language itself can be given an expressivist account. To this end, it doesn't seem fruitful to unpack 'murder is wrong' as 'has the attitude of disapproving of murder' – the expressivist part comes in at the point where we decide what is to count as being in the extension of 'wrong' – are we Pro! putting murder in the extension or range of the predicate 'wrong'. There is nothing in the semantics about disapproving or blaming.

There is still plenty of disapproval and blaming to be had – it just isn't delivered in the semantics proper. If we agree (in attitude) that an act belongs in the extension of 'wrong', presumably that means we are ready to disapprove of the act and probably will blame someone who commits the act – this is what we do when we think something is wrong. Similarly, there are farreaching consequences if we agree to classify something as in the extension of 'person' rather than 'fetus', and whether we allow marital rape within the extension of 'rape'. Disputes over meaning are never *merely* disputes over meaning because the choices we make have consequences for a vast array of legal and social institutions. Word choice is important.

Word choice is important, so one might think that an expressivist account just doesn't do it justice. This may be – an expressivist semantics doesn't require that you be an expressivist about all normative questions. You might even be a moral realist about whether someone *should* be Pro! classifying marital rape as being in the extension of 'rape'. The semantics is neutral on this; the only real support it gives to the expressivist about moral discourse is its demonstration of the possibility of articulating expressivism without fear of logical incoherence.

Even if we can work out an expressivist semantics for the basic conditionals there is the question of whether we can extend it to modals, quantification, etc. For the record, Schroeder (2008) thinks the worries just surveyed are surmountable but that modals and certain quantifiers can't be given a viable expressivist treatment. Perhaps, but it is hard to see how this could be so given the resources typically available to the semanticist. For example, if an expressivist account of basic relational predicates is possible then it is difficult to see why a version of generalized quantifier theory couldn't be treated expressivistically, since, after all, generalized quantifiers are simply relations between sets. To the extent that modality can be treated as a species of quantification, the same point holds for modality. I'm not suggesting that there is no problem – I'm merely suggesting that these constructions don't *appear* to present any problems more complicated than those we have already examined.

5 Conclusion

I'm not pretending that I have a solution to all the problems faced by expressivist semantics. For the record, I also don't claim to have a solution to the problems with truth conditional semantics. We don't expect quick solutions in this domain. It may be that the expressivist program fails and takes the formal use theory down with it, and it may be that we need to pursue the idea that a truth-conditional semantics can be a part of naturalistic inquiry.

However, I think that the fact that the problems are hard is a good sign here. If this was just a matter of changing e's to i's and t's to A's it would seem that we hadn't really accomplished much. But clearly the move to taking basic attitudes as the primitives of our semantics has profound consequences for the shape of our semantic theory.

It is also important to keep in mind that while the expressivist research program is very difficult, there is much to recommend it. In effect, the payoff would be a novel way of naturalizing semantic theory, and this in turn provides a plausible candidate for the locus of C/I interface conditions. Once it is installed as the interface locus – then we immediately gain a powerful explanation for why recursive structures should be legible, and we develop a number of predictions about correlations between kinds of cognitive abilities and the kinds of recursive structures we encounter. And last but not least, we gain a working hypothesis about the connection between human languages and animal languages. To wit: at the base, human language is grounded in expressing referential intentions, approval, and disapproval. What separates human language is that we have ways of taking these basic expressions of attitudes and combining them recursively into expressions of unbounded richness and complexity.

References

Asher, N., Sher, I., & Williams, M. (2001). Game theoretical foundations for Gricean constraints. In R. van Rooy & M. Stokhof (Eds.), *Proceedings of the 13th Amsterdam Colloquium* (pp. 31–37). ILLC Publications.

Blackburn, S. (1984). *Spreading the word*. Oxford: Oxford University Press.

Blackburn, S. (1988). Attitudes and contents. *Ethics, 98*, 501–517.

Chomsky, N. (1995). *The minimalist program*. Cambridge: MIT Press.

Chomsky, N. (2000). *New horizons in the study of language and mind*. Cambridge: Cambridge University Press.

Chomsky, N. (2001). Derivation by phase. In M. Kenstowicz (Ed.), *Ken hale: A life in language* (pp. 1–52). Cambridge: MIT Press.

Chomsky, N. (2003). Reply to Horwich. In N. Hornstein & L. Anthony (Eds.), *Chomsky and his critics*. Oxford: Blackwell.

de Villiers, J. (2007). The interface of language and theory of mind. *Lingua, 117*, 1858–1878.

de Villiers, J. G., & de Villiers, P. A. (2003). Language for thought: Coming to understand false beliefs. In D. Gentner & S. Goldin-Meadow (Eds.), *Language in mind: Advances in the study of language and cognition*. Cambridge, MA: MIT Press.

de Villiers, J. G., & Pyers, J. E. (2002). Complements to cognition: A longitudinal study of the relationship between complex syntax and false-belief-understanding. *Cognitive Development, 17*, 1037–1060.
Epstein, S., & Sealy, D. (2006). *Derivations in minimalism*. Cambridge: Cambridge University Press.
Geach, P. (1958). Imperative and deontic logic. *Analysis, 18*, 49–56.
Geach, P. (1960). Ascriptivism. *The Philosophical Review, 69*, 221–225.
Geach, P. (1965). Assertion. *The Philosophical Review, 74*, 449–465.
Gibbard, A. (2003). *Thinking how to live*. Cambridge, MA: Harvard University Press.
Grice, P. (1989). *Studies in the way of words*. Cambridge, MA: Harvard University Press.
Grim, P. (2007, November). *The emergence of Gricean constraints: Lessons from simulation*. Talk given at Games and Decisions in Pragmatics III, the Center for General Linguistics, Berlin.
Hauser, M. D., Chomsky, N., & Fitch, W. T. (2002). The faculty of language: What is it, who has it, and how did it evolve? *Science, 298*, 1569–1579.
Heim, I., & Kratzer, A. (1998). *Semantics in generative grammar*. Oxford: Blackwells.
Horwich, P. (1998). *Meaning*. Oxford: Oxford University Press.
Horwich, P. (2003). Meaning and its place in the language faculty. In N. Hornstein & L. Anthony (Eds.), *Chomsky and his critics*. Oxford: Blackwell.
Horwich, P. (2005). *Reflections on meaning*. Oxford: Oxford University Press.
Kaplan, D. (2001). The meaning of Ouch and Oops (Explorations in the theory of *Meaning as Use*), Draft #3, ms., UCLA.
Parikh, P. (2001). *The use of language*. Stanford: CSLI Publications.
Roeper, T. (2005). *The prism of language: How child language illuminates humanism*. Cambridge: MIT Press.
Roeper, T., & Snyder, W. (2005). Language learnability and the forms of recursion. In A.-M. Di Sciullo (Ed.), *UG and external systems: Language, brain, and computation*. Amsterdam: John Benjamins.
Ross, D. (1939). *Foundations of ethics*. Oxford: Oxford University Press.
Schroeder, M. (2008). *Being for*. Oxford: Oxford University Press.
Searle, J. (1962). Meaning and speech acts. *The Philosophical Review, 71*, 423–432.
Searle, J. (1969). *Speech acts, an essay in the philosophy of language*. Cambridge: Cambridge University Press.
Sperber, D., & Wilson, D. (1995). *Relevance: Communication and cognition* (2nd ed.). Oxford: Blackwell.
Stevenson, C. (1944). *Ethics and language*. New Haven: Yale University Press.
Unwin, N. (1999). *Quasi*-realism, negation and the frege-geach problem. *The Philosophical Quarterly, 49*(196), 337–352.
Unwin, N. (2001). Norms and negation: A problem for Gibbard's logic. *The Philosophical Quarterly, 51*(202), 60–75.
van Rooy, R. (2004). Signalling games select horn strategies. *Linguistics and Philosophy, 27*, 493–527.

Recursion and Truth

Wolfram Hinzen

Abstract The paper describes reasons for departing from a traditional 'T-model' architecture of grammar and moving towards the 'no interface' view, on which grammar, inherently, organizes certain kinds of semantic information which are hypothesized to not exist in the absence of grammatical organization. In this sense, these are forms of *grammatical meaning*, and the claim is that there is no such meaning in some pre-linguistic 'semantic component', located on the other side of some 'interface'. The paper particularly articulates this view with regards to reference and truth as inherent aspects of grammatical semantics, and argues that apparent constraints on recursion fall out from the way in which grammar organizes the forms of intent/sional reference that we universally find in human language.

Keywords Recursion • Truth • Reference • Grammatical semantics • X-within-X • Intensionality • Interfaces

1 The Origins of Recursion

On a common understanding, human language is an instance of a large range of systems that exhibit a discrete infinity of hierarchically organized objects. If we model each of these objects mathematically as a binary (two-membered) set and the operation of creating such a set is recursive, the sets will contain other such sets, until we hit lexical items, at the bottom of this system. These are its atoms: elements exhibiting no hierarchical syntactic complexity at all. Regarding the origins of atoms in hominin mental evolution – some mechanism for the encapsulation of

W. Hinzen (✉)
Catalan Institute for Advanced Studies and Research (ICREA), Barcelona, Spain

Department of Linguistics, Universitat de Barcelona, Barcelona, Spain
e-mail: hinzen@ub.edu

T. Roeper and M. Speas (eds.), *Recursion: Complexity in Cognition*, Studies in Theoretical Psycholinguistics 43, DOI 10.1007/978-3-319-05086-7_6,

complex conceptual structures in a non-structured primitive element – little appears to be known (see Uriagereka 2008 for some discussion). Regarding the origin of recursion, nothing much appears to be known either. Recent research has rather raised it to the level of the one significant mystery that distinguishes human language and mentality from animal systems of thought and communication (Hauser et al. 2002). Often the thought has been that the basic operation defined to generate the above structure (the operation Merge of recent minimalist theorizing: see e.g. Chomsky 2008a), comes 'for free', in the sense that it is minimally required for *any* system with the basic features of language. Put differently, it is 'virtually conceptually necessary'. But this is not an explanation, any more than Kant's claim that, in order for scientific knowledge to be possible, our minds need to have certain 'categories', provides an explanation for how or why our mind has that organization.[1]

Modelled as above, moreover, recursion tells us no more about human language than about any other hierarchically organized, discretely infinite system – from organic growth (Prusinkiewicz and Lindenmeyer 1996) to music (Katz and Pesetsky 2009) to DNA (Searls 2002), if not to a large range of other human and non-human systems of planning, foraging, and navigation. Whenever a system has discrete infinity, Merge can be employed to apply to whatever its atoms are and to form sets from them that contain other sets. And unless we block the 'internal' application of Merge (its application to sets and lexical items contained in other sets), the system will have 'transformations' as well. So, linguistic specificity will not come from Merge. It is also not clear why much specificity would come from the use of some of its atoms as 'labels' of the resulting sets (Hornstein 2009), if indeed labels are required for the workings of the linguistic system at all (see Chomsky 2008b).[2]

Nowadays, the standard answer to the question of what leads from Merge to the kind of recursions specific to human language is interface conditions, leaving Merge itself to apply completely freely (see e.g. Boeckx, this volume). That is, how recursion is constrained follows from constraints externally imposed on Merge by language-external systems, particularly the conceptual-intentional (C-I) system. This picture is broadly in line with (though not tied to) a traditional view on which syntax is 'autonomous', 'formal', and 'modular', whereas semantics is external to it and independently motivated. An alternative is that there is no distinctive syntax-module at all, any more than a distinctive semantics module – and hence

[1]Chomsky (2010) also argues that the question of precursors for Merge is mute – hence it evolved all at once: 'There are many proposals involving precursors with Merge bounded: an operation to form two-word expressions from single words to reduce memory load for the lexicon, then another operation to form three-word expressions, etc. There is no evidence for this, and no obvious rationale either, since it is still necessary to assume that at some point unbounded Merge appears. Hence the assumption of earlier stages seems superfluous.' While this may well be plausible, it doesn't make the problem easier.

[2]Thus, to construct a purely hypothetical example, one might propose that the Merging of a Path with a Goal in insect navigation is still a Path, namely a bounded one: the Path 'projects' in this sense and becomes the 'label'.

there is no C-I interface between them either, though there will be relevant systems of discourse processing that process the outputs of the language faculty further (Hinzen 2006; an idea that Chomsky 2008a, describes as a 'radical' one).[3] There is only one system and it generates all the distinctions we need. In this case, constraints on recursion need to follow from the workings of this one system itself: they can't be externally imposed.

The present paper pursues this 'radical' model, building on Arsenijevic and Hinzen (2013). The latter paper observes apparently universal and somewhat surprising *restrictions* on recursion in human language. Specifically, (i) Universal Grammar appears to ban the occurrence of any syntactic category immediately in itself, i.e. the configuration X[X], where X is any category: typically, the two occurrences of X are kept distinct through an interleaving category $Y \neq X$.[4] Moreover, (ii), where a category X embeds in itself indirectly (via interleaving of other categories), the embedded XP behaves differently from the embedding one, hence strictly there is no recursion (in the linguistic sense of *self*-embedding) in this case either.[5] I return to this view here, tracing its consequences for semantics and compositionality more specifically. In the Arsenijevic and Hinzen model, the unit of recursion is the cycle: it is the primitive on which recursion in human language is based. As viewed in the phase-based model of syntax (Chomsky 2008a, b), the cycle is inherently also a unit of semantic evaluation. Syntax thus revolves around semantic distinctions that correspond to different cycles. In this sense it is organized around semantic principles, and recursion is a consequence of the way in which syntactic derivations subserve semantic distinctions (specifically, semantic evaluations at the syntax-discourse interface).[6] This has consequences, I aim to show here, for the naturalization of truth, the fundamental primitive of semantic theory.

There is no dispute about the fact that speakers evaluate some linguistic expressions as true or false – a contingent aspect of human semantics which a recursive, systematic, and compositional form of communication or thought does not need to exhibit (Carstairs-McCarthy 1999). It therefore needs to be explained: we need to understand why the language system exhibits this specific form of organization.

---

[3]This architecture is in stark contrast to a system such as Jackendoff's (2002), say, where the autonomy of syntax remains axiomatic. See e.g. p. 124: 'meaning has an inventory of basic units and of means to combine them that is as distinct from syntax as syntax is from phonology . . . we ought to attempt to evacuate all semantic content from syntactic structure, just as we removed phonological content'. The present view is not in conflict with the old and valid insight that, as Lasnik (2000: 186) puts it, 'there is no rule of *red*-extraction': a syntactic rule constrained to apply only when a lexical item with the content RED is inserted.

[4]See also Roeper and Hollebrandse (2008) on recursion being typically 'indirect'.

[5]None of this is in conflict with language being recursive in the sense of computable (though I don't regard the computability of language as an empirically established fact either).

[6]This is not in contradiction with the above view that there is no C-I interface as on the standard minimalist view C-I systems are richly (in particular propositionally) structured, and they impose systematic forms of organization on the syntax. On the view developed here, there is a discourse representation, but it is 'passive' and consists of a set of discourse referents (which are then updated by syntactic structures) only.

The core idea explored in this paper is that the answer to this question is actually linked to the question about the origin and nature of recursion. The naturalization of recursion and of truth should proceed together.

2 A Correlation Between Truth and Recursion

I begin by observing some correlations between truth (as a specific form of semantic evaluation at particular moments of a derivation) and recursion. As I argue in this and the next section, there is recursion (in the sense of hypotactic embedding: coordination and recursion in adjunction will not be discussed here) only and as long as expressions are not evaluated for truth. Semantic evaluation for truth constrains syntactic recursions both so as to generate only objects of a certain kind and so as not to let these objects, once they are constructed, recur in themselves.

Note, then, that the paradigmatically recursive structures in language are the truth-evaluable ones, but that the range of truth-evaluable structures in language is in fact highly restricted.[7] To start with, it is clear and interesting that *words*, even though, crucially, they can be hierarchically, thematically, and compositionally structured, are not evaluable for truth. As di Sciullo (2005) discusses, words can encode some thematic (i.e., event) structure, as when incorporating an agent or theme (1a, b), even if not both (1c):

(1a) tut-or
(1b) tut-ee
(1c) *or-tut-ee

Compound words can resemble sentences and phrases in a similar way, as when 'doll house' denotes a house for a doll, a 'doll house shop' a shop whose purpose is the selling of houses for dolls, and a 'doll house shop seller' a person in the habit or profession of selling doll house shops. Words also can encode an ontological object-event distinction, as in (2), where the same lexical root, $\sqrt{}$GROW, is morpho-syntactically specified so as to denote a temporally extended activity located as simultaneous with the point of speech (2a), or an objectified process that cannot incorporate Agentivity (2b) (cf. **John's growth of tomatoes*), or an objectified process that can (2c) (cf. *John's growing of tomatoes*):

(2a) $\sqrt{}$GROW-s
(2b) $\sqrt{}$GROW-th
(2c) $\sqrt{}$GROW-ing

[7] In 'type-driven' approaches to semantic compositionality this is often not observed. Thus, it is common to find that VPs that contain a subject position are said to be 'semantically of type t' (e.g. Heim and Kratzer 1998: 217). While this is justified to the extent that it makes the recursions run smoothly, VPs don't e.g. contain temporal information, which any evaluation for truth in natural language requires.

Finally, as di Sciullo also notes, words have, again like sentences, a systematically compositional semantic structure, as the difference between (3a) and (3b) suggests:

(3a) tavol-in-etto 'funny little table'
(3b) tavo-lett-ino 'little funny table'

Against all of this semantic potential of words stand their intrinsic semantic limitations. Thus no English word can refer to truth-values: 'John' e.g., cannot denote one:

(4) *John is true

Relatedly, a word, no matter how complex and even if configuring an event or thematic structure, cannot anchor that event in time: (5a) is incapable of expressing what (5b) does, in which the (compound) word is replaced by a phrase (di Sciullo 2005: 51):

(5a) John is a <u>bank robber</u>
(5b) John is robbing a bank

Nor is (5c) possible, showing that however complex the semantic structure of a word becomes, it retains a curious generic character:

(5c) *At six, Mary is a <u>writer</u>

Relatedly, while words can expand productively, their expandability is inherently bounded in a way that prevents us from talking about a concrete instance of what abstract kind they denote (example from di Sciullo 2005: 36):

(6) constitute-ion, constitute-ion-al; constitute-ion-al-ity;
anti-constitute-ion-al-ity; *th-anti-constitute-ion-al-ity

Thus, words cannot encode predications, and for the same reason, no compound word can say, of some person, that he is in the habit of selling doll house shops.

These limitations even show in the use of proper names. Thus, the name 'John' is not only generic in the sense that millions of people have it, but it is also generic in the sense that when used to denote one particular individual familiar to the interlocutors, it cannot denote John at this or that moment, in this or that time of his life, in this or that state of his mind. That is 'John' denotes John generically, hence in a more abstract way, disregarding the vagaries of his actions, mental state, time, and place. Similarly, Russell referred to simply as 'Russell' is not the early or the late Russell, Russell as a logician or as a lover, he is just Russell. Nor is a name, in and of itself, even capable of fixing what it is that it refers to. Thus, 'John' in 'John stands here' might denote his car; 'Johns' will denote persons having this name; and 'John was all over the floor' might denote his flesh, after a massacre (Hinzen 2007: ch. 4). Generally speaking, names need a phrase or sentence before their reference is fixed and fully specific. Thus, (7), where the name 'Nixon' occurs in a compound,

(7a) John is a [Nixon-admirer] in every respect except that he does not admire Nixon
(7b) ?John admires Nixon in every respect except that he does not admire Nixon

paradoxically shows that a 'Nixon-admirer' is not necessarily someone who admires Nixon (di Sciullo 2005: 52):

Neither can names that are parts of compounds be targeted for extraction (8), unlike in the case of phrases (9):

(8a) John is a [Nixon-admirer]
(8b) *who is John an [t admirer]?
(9a) John [admires Nixon]
(9b) who does John [admire t]?

The right conclusion therefore appears to be that far from being the paradigms of referentiality, words as well as even names only attain referential specificity and definiteness, in short full referential potential, *after* entering phrases and sentences, i.e. syntax. Syntax mediates semantic reference. At the level of a lexical root such as √HOST, where the syntactic derivation has not yet begun to operate, it is not even clear yet whether we will use this root to refer to an object or an act (as in *the √HOST* versus *√HOST-ing some guests*); and although this particular semantic distinction is attainable at the level of compounds, we still cannot yet relate the abstract concepts of our minds as encoded in such compounds to specific moments in time and discourse, as we have seen, let alone put forward a proposition as true or false.

The widely assumed axiom in the philosophy of language and formal semantics, that semantic interpretation begins from the assignment of referents to *words* and the meaning of a sentence is built up 'compositionally' from there, thus needs to be qualified. Referential semantic interpretation is delayed at least until the phrase, if not the phase, as we will argue later. As we have just seen, the compositional aspects of words and sentences are also logically *independent* of their referential aspects (cf. Uriagereka 2008): we could have the former in the absence of the latter (or at least the absence of referential specificity), and there could be a language where all expressions are semantically compositional (as words and compounds are), yet no specific object could ever be referred to with an expression of this language. In human language, reference only *gradually* arises as the syntactic derivation proceeds and relevant levels of categorial complexity are reached. This is true even for names, as we have noted, which go far beyond the 'sticking of labels to objects' and only in fact function as names when the syntax cooperates. Arguably, namehood presupposes full DP-syntax, explaining why naming is persistently absent in non-human animal calling and apparently more taxing in human cognitive processing than descriptions are (Semenza 2009).

As it turns out, in fact, the potential for an evaluation of truth arises only towards the very *end* of the syntactic process. Thus, no Noun Phrase, no matter how complex, can grammatically be said to be true or false, owing primarily to their lack of Tense, e.g. (10):

(10a) Mary's smile
(10b) Caesar's destruction/destroying of Syracuse

Neither need clausal structures exhibiting Tense have what it takes to be true or false:

(11a) [$_{IP}$ to be an idiot]
(11b) [$_{CP}$ for John to be an idiot]
(11c) [$_{CP}$ if he is handsome]
(11d) [$_{CP}$ that endorses the proposal]

Clauses used as root phrases need not qualify either:

(12a) You idiot!
(12b) Good cook, your mother!

Thus, (12a) is bound to an expressive rather than propositional meaning (Potts and Roeper 2007), and (12b) is evaluative as well, in a way that a proposition need not be. Neither of them can be hypotactically embedded in the way propositions can be, hence give rise to recursive structures. The same is true of small clauses, as (13) shows:

(13a) *[$_{CP}$ John considered [$_{SC}$ Joan see [$_{SC}$ her fall]]]
(13b) *[$_{CP}$ John thought [$_{SC}$ Joan happy [$_{SC}$ Mary sad]]

In both cases, this is a consequence of their structural impoverishment. Again, Tenses are needed to generate recursions productively at the clausal level (14a), recursions that start running smoothly and productively only when these Tenses are finite ones in the scope of a subordinating conjunction:

(14a) John [believed [Joan to consider [Mary to be happy]]]
(14b) John [believed [that Joan thought [that Mary was happy]]]

In short, *both* truth and recursion come into their own once we have built structures that are, in some sense, 'complete' enough: specifically, we need a fully sequenced hierarchy of functional projections up to the lower regions of the C-field as described in Rizzi (1997), which is where we have a truth-evaluable syntactic object. Before we are there, productive embedding at the clausal level won't happen. That in itself points to our conclusion that truth and recursion are inherently correlated in human language design, and that recursion is not 'free'. What recurs in language, in paradigmatically recursive structures usually used to illustrate recursivity (e.g. *Sam believed John thought Bill knew Mary was pregnant*), are *truth*-evaluable structures. The other paradigmatic case of recursion (in hypotactic contexts) is recursion at the level of referentially evaluable objects, as in (15), a case to which I return:

(15) the vase on the table in the room in the country house in France

Syntax therefore not only mediates reference. Truth and reference also mediate syntactic recursion: here again, the units of recursion are objects which are semantically sufficiently complete, so as to be usable for purposes of reference. Rather than being free, recursion waits until that completeness is reached (or sufficiently approximated). Syntax is in this sense constructed around semantic principles of organization.

3 More Correlation

Let us now note that C by itself, e.g., is simply not recursive: C does not embed in C, directly, but only after the embedding has unrolled sufficiently, so that a whole sequence of projections intrinsic to the C-domain is in place (cf. Hollebrandse and Roeper 2008). Thus, e.g., (16) is ruled out, what we get instead is *cyclic* recursion, as in (17a) or (17b):

(16) [C[C[C]]] . . .]
(17a) [C-T-V [C-T-V . . . [C-T-V]]]
(17b) [C-T-V-D [C-T-V-D . . . [C-T-V-D . . .]]]

To be specific now, what productively recurs appears to need to be at least as complex as IP plus perhaps the lower regions of CP as described in Rizzi (1997):

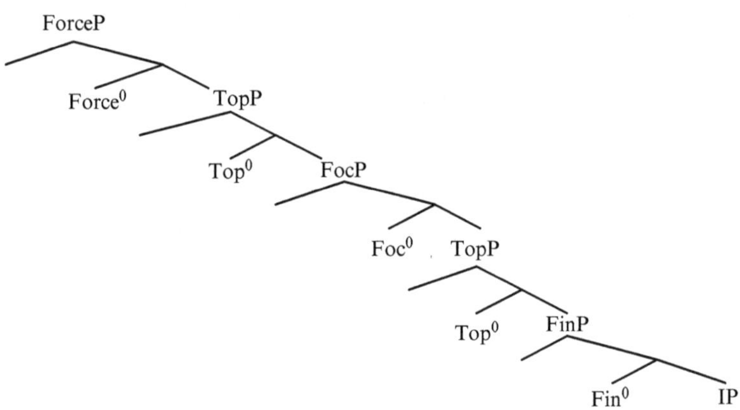

(18)

What comes above FinP here is elements governing the embedding of a propositional unit configured at the level of IP/Fin into the discourse. But recursion not only happens with units that are sufficiently complex, it also happens only as long as these units are not *too* complex, and embedding into the discourse has not proceeded too far. As soon as propositions have assertoric Force, in particular, they don't embed: embedded propositions are never asserted (although their truth values can be assumed in discourse, or be presupposed by the speaker, as in factives).[8]

[8] Note that in adjunct structures clauses with distinct assertoric forces *can* occur embedded: cf. *John said that the dog, which is really big, is small*. Here *that the dog is small* is assigned no truth value by the speaker, but *which is really big* is. Thanks to Tom Roeper, p.c., for the example.

If we terminologically distinguish between truth-evaluability and truth-evaluation (or assertion: the online assignment of a truth-value at a moment of speech), and associate the latter with ForceP, a constraint in human grammar design thus seems to be that the configuration in (19) is not possible:

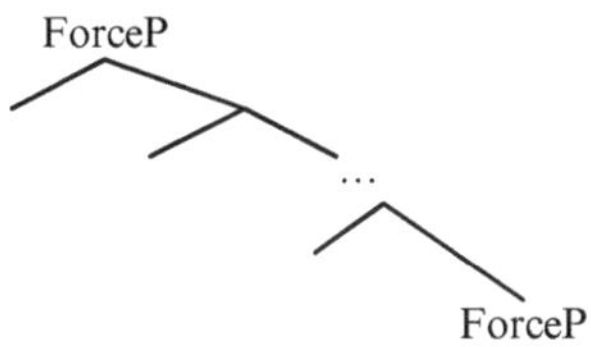

(19)

That is, a truth-evaluated object can never recur in itself.[9] The empirical observation is therefore that upon reaching appropriate complexity levels which correlate with evaluability for truth, the generated syntactic object can recur; if it goes beyond this level of complexity, recursion stops: the object cannot recur. Not only Force is not recursive: neither are evidentiality markers that can be added at this point, or tag-questions in languages like English or discourse particles in a language like German with which the truth assigned to a given proposition can be negotiated with the hearer. The 'purpose' of a syntactic derivation, then, is to compute a truth-value. Once this is done, the recursion is halted. And compositionality is halted too: the truth-conditions are fixed by the time the FinP is computed; evidentiality markers, tags, etc., don't *add* to the truth-theoretic content, on a common assumption. The same observation may also explain why the 'it's that' construction, as Andrea Moro points out (p.c.) does not iterate[10]:

(20) It's that John is sad
(21) *It's that it's that John is sad

In sum, the moral is that recursion in human language is highly restricted, that these restrictions are of a semantic nature, and that the notion of truth has to be brought in to account for them. Linguistic recursions are not only bound to the

[9] One might note that, for probably related reasons, the referential Force specified at the left edge of the DP cannot recur either (cf. remarks in Higginbotham 1985: 560–3).

[10] On Moro's analysis, the structure of the latter sentence is this (cf. Moro 1997: 193):

$[_{IP}$ it$_i$ is $[_{SC}$ $[_{CP}$ that John is sad] t$_i$]]

Here, *it* is base-generated in a Small Clause (SC) complement to the copula 'is', where it functions as a propredicative element that attaches as a predicate to the small-clausal subject, the clause 'that John is sad'. When this element is raised (leaving the trace *t*), the result is interpreted as involving the assignment of a sentential predicate to the embedded proposition. Basically, this is the truth-predicate: (20) essentially means *It's that it's true that John is sad* (or *It's the fact that John is sad*). It doesn't mean anything else that it might logically speaking have meant, e.g. *It's that it's probable that John is sad* or *It's that it's highly certain that John is sad*. This suggests that truth, and no other form of semantic evaluation, is indeed what inherently (or by default) 'closes off' a given proposition. If truth evaluation is not recursive, or functions as a recursion-stopper, we predict the oddness or ungrammaticality of (21).

cycle, they also stop altogether when semantic complexity has reached the level of the truth-value. Syntactic derivations are mappings from lexical roots, where referentiality is maximally unspecified or minimal, to a fully projected clause, where it is maximal. A derivation is the computation of an extension from a given intension, which in turn is a pure concept not evaluated for reference.

There is no moment in the derivation on this model where syntax is 'autonomous', or language is arbitrarily 'recursive'. Everything is bounded. It is only when we abstract from the inherent *uses* of specific syntactic configurations, that language appears to be generated by an unbounded recursive operation. Within the cycle, surely, everything is rigidly hierarchical and asymmetric: nothing ever recurs in it. We don't get structures like (22) or (23), say:

(22) C-T-T-T . . . *v*-V

(23) C-T-C-T-C-T . . . *v*-V

The reason again is plainly semantic. The same is true within smaller cycles, like *v*-V itself. Where ASPect phrases embed one another, for example, the relevant aspectual projections are of different kinds, and the same ones *don't* seem to immediately recur (cf. Den Dikken 2009). Where VPs recur, as in serial verbs, a closer look reveals that the constructions in questions are either asymmetric in their supposedly directly recursive constituents (e.g. one being an argument, the other an adjunct), they involve different categories (as they certainly would on a 'cartographic' approach), or they are clearly templatic (Pawley 2006). *v*Ps can systematically embed other *v*Ps, as in complex causative constructions (cf. *cause to die*), but this process is not only highly limited, but the embedded *v*P is also never quite as complex or projected as the embedding one (semantically an event embeds a *state*, say, not another event of the same sort). Moreover, if an event-encoding *v*P contains sub-events, these are semantically necessarily *parts* of the larger events, not independent events: a complex *v*P will always denote a *single* event. NPs recur productively in themselves, but also only with other material interleaving (cf. **the mother the bride*). Even nominal root compounding, perhaps the next candidate for 'direct' recursion ('X-in-X'), are plausibly analyzed as involving lexical access in between two Ns, if indeed they are not reduced structures with more material intervening. We also know that argument structure is highly restricted rather than unboundedly recursive; and that no sentence can have more than one subject (see Arsenijevic and Hinzen (2013) for more discussion of all of these examples). In short, the ban on recursion in the above sense above appears to be quite robust.

Nothing happening within the cycle is recursive or unboundedly productive: starting a phase from a nominal head, we know we can project up to, say, D, and then recursion stops: a new cycle has to be begun. Similarly, as we have seen, when starting with whatever is the initiating head of the C-phase, we can't significantly project beyond C. So operations building the cycle are essentially templatic: they are predictable given the choice of the initial head and its inherent potential. If we let an unbounded operation generate the cycle, we would therefore have to restrict it. If we begin from the restricted mental template, or *Gestalt*, by contrast, there is no unbounded operation to be restricted. If the cycle was recursive, or there would be an infinite sequence of projections contrary to the finite one that Rizzi (1997) or Cinque (1999) depicted, there wouldn't in fact *be* the specific recursions that we find in human languages. Instead, like in the natural numbers, we would be looking at an infinite sequence of uniform objects, all of the same nature, with no cycles of the sort we find in language ever forming (or, perhaps, the only cycle there would be would be the single step from one number to the next) (cf. Hinzen 2009). The source of recursion in human language, therefore, is in fact the inherent *limitations* of language, or its *boundedness* – which is what the cycle reflects.

Evaluation for truth is thus woven into the dynamics of structure-building in language. A consequence of this is that truth, inherently, is a *structural* notion, which cannot be explained by anything like general-purpose 'relations to the world' or notions of 'correspondence', which seem non-explanatory and in fact to trivialize the problem. The problem of the origin of truth is not solved by positing a 'truth-relation' to the world in which certain abstract entities such as 'propositions' stand. Rather, syntax – and specifically the cycle – is an essential part of the answer to the question of the origin of truth (though certainly not a complete one: this is *not* being claimed here). An alternative answer to this question has been to reject it. Thus, one might maintain that truth is a paradigmatically externalist notion that falls strictly outside the domain of 'internalist' inquiry into the mind/brain as generative grammar has conceived it (cf. Chomsky 2000). If truth is primarily a relation to the world, this might indeed follow. But truth may not be a relation, and arguably, Tarski (1933) and more recent 'deflationist' theorizing in the metaphysics of truth has shown us how it might not be (e.g. Field 1994). Even if it is a relation, however, our question was not the nature of any particular truth, but the origin of truth. In that respect my argument has been that, quite clearly, evaluation for truth is woven into the fabric of syntax and it is an aspect of how grammar is recursively organized.[11]

4 Compositionality in Modern Semantics

The principle of compositionality is written into the modern conception of semantics. As discussed in Szabo (2009), a sensibly disambiguated version of the principle is that when we have (i) fixed the meanings of the syntactic constituents of an expression, and (ii) fixed its syntactic structure, its overall meaning is fixed as well: it won't depend on anything else. The first formalization of this idea appears to be due to Tarski (1933) (cf. Hodges 2001). Defining syntax precisely and without any reference to meaning, Tarski showed, for each formula φ of a formal language L,

[11]A reference-theoretic approach to semantics, I have argued earlier (Hinzen 2007), confuses the problem with the answer to the problem: reference in an explanatory semantic theory is the problem, not the answer (and language is part of the answer). How we refer to the world in the way we do is exactly what we want to know. This problem is not answered by assuming that there is a 'reference-relation' between language and world; nor do causal relations seem to turn the trick, in the human case (cf. Hauser 2000).

how to define, by recursion on the syntactic complexity of φ, the class $\mu(\varphi)$ of those assignments (of objects to the free variables of φ) which satisfy φ. We see here precisely what we questioned above for the case of human language (with which Tarski was not directly concerned): recursion is defined without regard to semantic considerations. The question now arises how this meaningless syntactic structure is mapped to a semantic structure. Compositionality was (and is) the answer. The meaning-function $\mu(\varphi)$ is given by a function $\chi(\varphi) = (\mu(\delta), \mu(\theta))$, where δ and θ are immediate constituents of φ and χ only depends on the syntactic operations generating φ. Note that every constituent is assigned an independent meaning, which never depends on what constituent it is embedded in.

As used in the Montagovian tradition and specifically by Montague himself, semantics in this sense is essentially neutral as to what exactly meanings are. Married with philosophical assumptions on the nature of meaning, on the other hand, the idea by and large has been that language is *externally controlled* somehow (it equates with denotation). The base case of meaning is thus where φ is simply a name and $\mu(\varphi)$ is an object, its bearer. With this assumption, contradicting conclusions reached above, standard introductions to the philosophy of language set out (e.g., Lycan 2008), and Szabo (2009) formulates a broad consensus when he remarks that it is 'very plausible that the meaning of proper names is nothing but their referent (for the point of having proper names in a language does not seem to go beyond the labelling of objects).'

So much was naming (thus understood) the base case, that all *other* denotations that a compositional semantics will have to assign to syntactic constituents have been defined *in terms* of the denotations of names, i.e. individual objects: e.g., as sets of such objects (in the case of adjectives and common nouns), or as sets of such sets (as in the case of generalized quantifiers). A predicate denotation thus is a mapping from an object to an object (type <e, t>, or a set of objects characterized by that function). Basically, then, every constituent is assigned a referent, and where these are sets they are construed as unary functions applying to objects in their domain of definition, until a single individual referent results, a truth-value. In a sense, then, the ontology of semantics never goes beyond the base case, the ontology of individuals. To put this differently, the ontology of semantics, as standardly conceived, is never *relational*: the only thing there is is individuals and set-theoretic constructions from them.[12] It is as if meaning could never arise from anything other than some sort of relation between a symbol and a thing. In short, although Tarski did formal work and that work is neutral on the nature of meaning, in actual practice it has been linked with a conception of meaning as reference, and every syntactic constituent is evaluated for reference (though some of these are functions/sets and most 'referents' are theory-internal constructs). The referents of a complex constituent are then recursively evaluated in terms of the referents of their immediate parts.

This conception, I will now argue, leaves a venerable Ancient problem unsolved. Suppose we wish to understand why (24) means what it does:

---

[12]This is a *nominalist* ontology (reflecting Occamist premises), as described in Leiss (2009).

(24) Theaetetus sits.

The standard referential conception in semantics begins my mapping 'Theaetetus' to Theaetetus – so far so good.[13] And then 'is wise' is mapped to another referent, let us call it a 'property'. So we have two referents: an object and a property. But the meaning, a proposition, is neither of these two. And it is not of the same kind as either of them. So it is something third. But then we have three objects or referents, and we are not closer to what makes for the unity of a proposition. So let us go back to the two first objects. What provides for their *link*? Plato posited a relation of 'instantiation' between them: the object 'partakes' in the property, which somehow embodies generality (though God only knows how). But this relation is again something third. We would have again three entities, and would again need something fourth that relates them, creating a regress leaving us with the same problem as before.

Plato's proposal was critically reviewed for two millennia. It kept intriguing the greatest of minds in all the centuries that passed. The problem has remained: what makes for the transition between a name and what is not a name and neither a sequence of names, namely a sentence? Somehow, *saying* something does not appear to be the same thing as *referring* to something. Semantics is *more* than reference, and is not resolved by positing a new external object ('properties', 'propositions', etc.) whenever we need one for some non-name like constituent to refer to: ontology doesn't save semantic theory.

How could an approach solve this problem in which anything that is relational is ultimately reduced to something that is not, namely names and set-theoretic constructions from them, as we have seen above? Set-theory is not the answer, even if we believed in the existence of sets, and could define the element-of relation consistently, and believed, unlike Pietroski (2003), that predicate extensions *are* sets. If compositionality is assumed, the problem aggravates: each constituent then independently refers, and before referents are composed, a sentence meaning is simply a set of referents. How does that set become a proposition? Looking at any referent assigned to a predicative expression won't answer that question: all it tells us is that a certain object (say, Theaetetus) is mapped to another object (say, truth). But all this says is that 'sits' does what we know predicates do. How do predicates do that?[14] Again, positing 'truth' as another object in one's ontology for sentences to refer to does precisely not explain what makes for the unity of a proposition, of

---

[13]Notwithstanding the fact no one quite knows what kind of object Theaetetus might be. Thus he is clearly dead, and dead objects are not somehow available still as objects of reference, like things stored in a warehouse. So is this referent a physical object? An object that once was a physical object, but is individuated irrespective of its physical instantiation? A physical object, when it still existed, at a certain mature time of its life? A person that as such is not a physical object?

[14]See Davidson (2005: 157–8) for discussion. The predicate 'being a member of the seated things' does not explain the original role of the predicate 'sits': it is a new predicate, for which the same problem arises. See also Higginbotham (2009: 149, for an illuminating discussion). The naïve idea that the extension of the common noun 'dog' is the set of individual dogs also ignores the fact that, in English, it is the word 'dogs' that refers to such a set, not the word 'dog'. Given compositionality

which, as Davidson notes, the truth value is the sign: 'only whole sentences have a truth value' (Davidson 2005: 87).

Davidson's brilliant late (2005) book is an admirable attempt to trace the history of this problem, which, in many ways, can be seen *as* the problem of the origin of truth. Davidson reads that history as a history of failed attempts, from Plato to Aristotle to Frege to Russell. Even Tarski, on Davidson's account, did not solve it, despite the fact that he defined truth. Tarski's achievement, on this reading, is to have told us how to set up a 'T-theory' for a relevant fragment of a given language, where T is a predicate applying to (structural descriptions of) sentences. T is defined such that, whenever a sentence *S* such as 'snow is white' is in its extension, snow is indeed white (and vice versa). More precisely, the theory is constrained so as to generate an infinity of sentences of this form ('T-sentences'):

(25) *S* is T if and only if *p*.

Here for '*S*' we may substitute a structural description of any sentence of the language for which we provide the theory (under certain restrictions), and for '*p*' we substitute that very sentence (if our metalanguage or the language of the theory contains the language in question, as we may assume for present purposes). So, intuitively, T appears as a *truth*-predicate: for that is the predicate for which correct T-sentences would indeed intuitively hold (it is indeed so in English that, if 'snow is white' is true, snow is white). Moreover, and crucially, the definition of T tells us how the truth-conditions of sentences systematically depend on the meanings of their parts, hence it is *informative* in regards to how truth conditions are computed. It is our general concept of truth, however, the one against which we can test whether any given T-theory (or any given T-sentence) does indeed apply to a given language, which is *not* defined by Tarski. If, e.g., a T-sentence unlike (25) said that, in English, 'snow is white' is T iff grass is green, it would likely be a bad theory. Tarski presupposes our prior grasp of this concept. All Tarski told us, then, is how our knowledge of the conditions under which sentences are true systematically exploits our knowledge of the structure of these sentences. But all this is on the *basis* of our knowledge of truth and ability to apply this notion to utterances.

Tarski thus didn't tell us how truth arises; and nor did he solve the problem of the unity of the proposition, or explain why sentences only have one truth value: this is what we will aim to improve on below. Put differently, it is not so that, when deprived of a notion of truth, Tarski's recursive and compositional assignment of semantic values to parts of sentences would make us understand what truth is. On the contrary, as Davidson (2005: ch. 7) stresses, the assignment of a compositional structure to sentences is in the service of validating a given formal T-theory for a given language L, in the process of testing it against our truth-theoretic intuitions about L. The truth-conditions we assign to T are what we start from. They are our basic data. Once we have them, we work out how the truth values of wholes depend

we cannot specify the meaning of a morphologically less complex word in terms of that of a more complex one.

on their parts. Tarski's merit, on the other hand, is to have shown us that we cannot advance in the problem of truth on an ontological or reference-theoretic path: it does not help to posit an entity that a true sentence denotes; or to postulate truth values as objects for sentences to refer to; or to posit various referents for sub-sentential expressions as a basis for understanding the notion of truth. That would neither explain why we ever map a word to a referent, nor how a truth value arises from a set of referents: Tarski's semantics is not a reference-theoretic one, and I take that to speak in its favour. It is from here that what I have to say about the unity of the proposition departs. Tarski wrote long before modern linguistics got going, and philosophers started taking linguistic form seriously, as something that might explain us why it is mapped to meaning in the way it is. Here the previous section may provide a lead: recursions as visible in linguistic form are not arbitrary; and as I will now argue, they are not compositional in the above sense either.

5 Truth and Recursion, Again

Let us return to Davidson's talk of 'the truth-value of a sentence as a sign of the unity of a sentence: only whole sentences have a truth value' (Davidson 2005: 87). What this means is no more and no less than that there is no recurrence on truth: if a sentence is assigned a truth value, no part of the sentence, not even if it contains many other sentences, is assigned a truth-value too. Non-recursion and the problem of the unity of the proposition are thus closely related. The fact is that there is always only one truth value – no matter how many sub-sentences a sentence embeds, and no matter whether these sub-sentences could perfectly well be evaluated for truth when occurring as roots. Why should this be so?

From a traditional, modern semantic view as described above there is no reason at all for why this should hold, and nothing in semantics predicts it. Indeed, it is so puzzling that *intensionality*, which simply *is* the observed impossibility of assigning normal semantic values such as truth values as the denotations of embedded sentences, has been one of the most persistent and recalcitrant problems in semantics, seemingly invalidating the basic assumptions of semantic theory and the compositionality principle (see e.g. Higginbotham 2009: 143–4). From the viewpoint of compositionality, the meaning of a whole is a function of the meanings of the parts and their syntactic combination, but the meaning of the parts is *not* a function of the meaning of any whole that contains these parts. So a sentence that is an embedded constituent should contribute the meaning that it also has otherwise, in different structures, or when it occurs non-embedded.[15] If that meaning is a truth value, as it is on Frege's traditional view, we predict that the truth value of a whole is to be computed from (or determined by) the truth values of its parts. That, exactly,

[15] Frege's contention that in embedded contexts, sentences refer to senses rather than truth values, is in conflict with this principle.

is what we never find – no human language is like that. The truth value of a whole never depends on whether any of its parts are true or false. This is well known to be what takes children about 4 years to understand: before this time, in a situation where (26a) holds but Mom in fact bought oranges, they answer 'oranges' to the question in (26b) (de Villiers 2007):

(26a) Mom said she bought apples.
(26b) What did Mom say she bought?

As the child approximates adult grammar, it learns that the truth value of a sentence like (26a) never depends in any systematic way on that of its embedded part. I interpret this as entailing three things: (i) only adult grammar analyzes complex sentences as cyclic, part-whole structures, in which (ii) the parts are never fully referential but always remain intensional; and thus (iii) referential force remains undivided and reserved to the structure as a whole. Put differently, the truth value bearing syntactic object is the ultimate non-recursive unit. It doesn't multiply and it doesn't occur in itself. Other things can occur in it, e.g. the denotations of embedded CPs, NPs, and *v*Ps, but semantically these are always weaker than a truth value: they can never determine one. Once the truth value is assigned at the root, the structure of these embedded parts gets fixated: compositionality and recursion stop. Whatever occurs embedded is thus to some degree intensionally interpreted: it is not referentially complete. This is obvious in the case of NPs: lacking even tense, they cannot serve to fully locate the denotation of the head noun of the NP in the world: the world is temporal as well as spatial, and no physically located object is without a temporal dimension. It is also clear in the case of *v*Ps: lacking Tense as well (though having Aspect), *v*Ps can only configure an event – but they cannot locate one in time and space, or in relation to the time and place of speech, for which the functional projections in INFL and higher up are needed.

None of this necessarily means that the principle of compositionality is wrong. As long as we factor into our account of it that whatever occurs embedded is destined to remain intensional, compositionality might be retained. However, assigning non-standard, intensional denotations would not only not explain intensionality. It would also not solve the problem of unity, and it would also not be how compositionality has standardly been understood (it was conceived in an extensional, reference-theoretic paradigm). The whole point of the compositionality principle is that embedded units have whatever meaning they have context-independently.[16] Only in this way can the meaning of a whole be composed from them. That embedded units should be intensional as a matter of principle makes no sense from this viewpoint. Why should hypotactic embedding and intensionality correlate?[17]

The answer seems to be that syntax matters: the meaning of a whole *cannot* be broken into parts: it is one single unit, which, like a *Gestalt*, has parts, but these

[16]See e.g. Fodor and Lepore (2002: 37): 'the *compositionality* of complex concepts is of a piece with the *context-independence* of the contents of their constituents' (emphasis in original).

[17]That, too, is an issue that Larson's (2011) interesting approach fails to answer.

are, inherently, parts of the whole that contains them. This solution would solve the problem of the unity of the proposition, of which the truth value is the sign. It would predict why evaluation for truth is not recursive: it is no more recursive than a perceptual *Gestalt*, which also doesn't recursively occur within itself. The same solution would solve Frege's Puzzle about the intensionality of propositional attitude contexts. For the relevant reasoning here is this: In (27), the embedded sentence denotes a truth value; in (28), it also does, indeed in any possible world. But the intuitive truth values of (27) and (28) do not at all have to be the same:

(27) Lois believes that Superman flies.
(28) Lois believes that Clark Kent flies.

The answer now is: the assumption that embedded sentences denote truth values and that the truth values of their matrix sentences is computed from them is simply wrong: extensions are *never* composed in Universal Grammar. Only intensions are, until an extension is reached, in which case recursion stops.

Again we could say that *intensions* are composed, even if not extensions. So, even though 'Superman flies' might be more descriptive than referential, that description (or its semantic value) is still what the truth value of (27) is composed from. However, this would be to *concede* that the matrix CP and the embedded CP are not assigned the same semantic value: one is assigned a truth value, the other an intensional description. In a contemporary minimalist architecture, where functional projections govern the way syntactic objects function at the interface and are embedded in discourse, this would mean that they are not the same objects syntactically, and therefore again there wouldn't be recursion in cases like this either, if this entails the occurrence of the same syntactic object in itself, even if there would be compositionality. Moreover it would be puzzling why the two clauses behave so differently. We could also let the two clauses be syntactically the same and assign the same semantic object to them: a function from possible worlds to truth values in both cases. The function denoted by the embedded clause could then be composed with the denotation of the matrix verb, which would be defined so as to take the function as an argument. Then again however the two recursive clausal units would not be evaluated in the same fashion and we wouldn't know why this is the case: why should 'Superman flies' be a function from a possible world that is compatible with Lois' beliefs to a truth value, whereas the matrix clause is a function from the actual world to a truth value in that world? Reference would still shift when moving from the matrix to the embedded clause, and the assignment of the same semantic value to both would make it mysterious why this should happen.

One might try arguing that the reason that reference shifts is the specific nature of the matrix verb. Thus, could we solve the problem by assigning this verb a special semantic value, namely that of a relation between a person and a set of possible worlds compatible with what that person believes in the actual world, so that, e.g., (27) would be true just in case Superman flies in all worlds that are compatible with Lois' beliefs (see e.g., Heim and Kratzer 1998: 306)? Yet, that the matrix verb is indexed to the actual world whereas the embedded one is indexed to a merely possible one is exactly the fact we set out to explain: the solution now considered restates that fact in formal terms. That a special principle, Intensional Functional Composition (Heim and Kratzer 1998: 308), needs to be invoked in these cases, seems to formalize the facts we needed to explain. Note, moreover, that the lexical semantics of the matrix belief-type verbs is as such entirely *consistent* with reference *not* being shifted in the above way. Semantically and compositionally, nothing speaks against the following interpretive rule:

(29) (26a) is true if and only if Mom bought apples and Mom said so.

According to this rule, a speaker uttering (26a) would commit himself to two things: that the embedded clause is true and that the matrix one is true too. What he says would be true if he was right on both counts. The meaning of the word ‘says’ or ‘believes’ is as such compatible with such an interpretive rule. We shouldn’t blame the matrix verb for the intensionality effects if its semantics is entirely consistent with the absence of these effects. It seems more likely therefore that the problem cannot be solved in semantic terms and must lie in syntax: somehow it must be the case that as the syntax creates a part, it creates the intensionality effect as well. That would contradict compositionality as customarily understood, for the embedded unit would then, upon being embedded, precisely *not* any more mean what it would normally mean, and contribute that normal meaning to the compositional process. It would also contradict recursivity, insofar as an embedded part would never function as the whole that embeds it insofar as we take into account how syntactic constituents function at the interface. This would also make sense of the fact that the intensionality effects in hypotactic embedding are more general: arguably, it has nothing specifically to do with belief-type verbs and their lexical semantics. Thus, (30) clearly doesn’t mean the same as (31), not even if, in the context where (30) is uttered, the referent of ‘You’ is John Smith and the speaker knows that:

(30) You are John Smith.
(31) John Smith is John Smith.

If the semantic value of ‘You’ and ‘John Smith’ was an extension, and it was the same object (as in the context above), the two sentences should have the same compositional meaning. But this is clearly not so. It makes for a difference whether we refer to a person grammatically as second-person or as third-person. Intensionality arises for grammatical reasons here, not for semantic ones. Similarly, suppose we refer to a particular object under the description in (32):

(32) the vase on the table

In this case even though the reference of the embedded constituent ‘the table’ will need to be worked out in order to determine the reference of the whole expression, the embedded NP does not at all function on a par with the embedding NP. Certainly a speaker uttering (32) does not refer to two objects, as he would in (33), a paratactic context:

(33) this vase, this table

(32) is different from (33) in that 'the table' in (32) plays a quasi-thematic role (a location): it specifies a surface on which the vase is located. So it specifies the referent of 'the vase' further, and does not function as an independently referential unit. Indeed, if functioning as an independently referential unit, in a non-embedded context, it would precisely *not* denote what it denotes in (32): It would denote an individual, whole object, not a surface, as in (32), which is not the same as a table, but a particular aspect of it.[18] Moreover, for predicational purposes, the embedded NP in (32) behaves as if it wasn't there: in *the vase on the table is beautiful*, say, nothing at all is said about the table. The contrast with (34) makes clear that we have a single referent in (32), but multiple independent referents in (34):

(34) This vase, this table *is/are beautiful.

The reason must be syntactic. Hypotactic, recursive embedding as opposed to paratactic conjunction creates an intensionality effect, entirely independently of any attitudinal verbs or modal contexts.[19] Semantics is innocent; syntax is to blame. Syntax is the reality on which the principle of compositionality is silent.[20] In the remaining Section, I will turn to how syntax might create this effect.

6 Intensionality from Syntax

In its earliest formulation, in the mid-1960s, the 'transformational cycle' was a general principle on rule application: if there was a particular linear sequence of transformations acting on a single phrase marker, it would need to apply to the most deeply embedded sentential structure of the phrase marker first, before it could proceed to the next higher one. In the guise of the Strict Cycle Condition (SSC) of Chomsky (1973: 243), the cyclicity principle maintains that given an object as in (35) with two cyclic domains A and B, one embedded in the other, no rule can apply to such an object by solely targeting B:

(35) $[_A \ldots\ldots [_B \ldots\ldots]]$

Freidin (1999), tracing the SSC's history and writing from the perspective of early Minimalism, argues that it is a natural assumption that once a category has

[18]Or, as the case might also be, the floor under it, as in *the vase under the table*.

[19]Similarly, events embedded within others encoded in a complex *v*P need to be interpreted as parts of the larger events, not as independent events, as noted earlier, again an intensionality effect, given the intensionality of the notion of 'part'.

[20]A principle that, both in Montague's case and that of his contemporary followers never depended on any particular assumptions about syntax, or indeed whether there was such a thing in any real sense (e.g. Jacobson 2009).

become a constituent of another category (like B in (35)), it cannot project any more (Freidin 1999: 120). Its syntactic life is exhausted, as it were. Thus, in particular, no constituent within B can be targeted and moved to the left edge of B once (35) has been built. Things can move out of B only when B is 'live', and this is only as long as B is not embedded. One way of putting this is that the computation always only 'sees' a single cycle, the one in which it is currently working. By the time it applies to the object A, the cycle B has been completed and is not modifiable any more: Whatever plays a role in the further derivation must, by that time, have been moved out of B into its left edge so as still to be 'visible'. All the derivation 'sees' when it goes to construct A, then, is the head of B and its left edge.

A cycle, then, is naturally conceived as a window of computational opportunity as well as memory, which is how it is conceived today (Chomsky 2008b): the derivation never looks back further than the left edge of the previous cycle, and when the cycle it is operating on is complete it is transferred out of the derivational workspace, with only its head and left periphery remaining, which then belong to the new cycle. In this model, the operation 'Transfer' replaces the old operation 'Spell-Out'. Whereas, in the older model, Spell-Out is the point where phonetically interpretable features are stripped off the derivation and handed over to the sensory-motor interface, and narrow syntax then continues with a covert cycle to LF, there is now only a single cycle with Transfer operations to both the sensory-motor systems and the semantic systems at the same periodic points in the derivation, i.e. each cyclic boundary:

(36a)

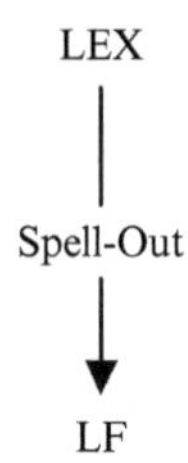

(36b)

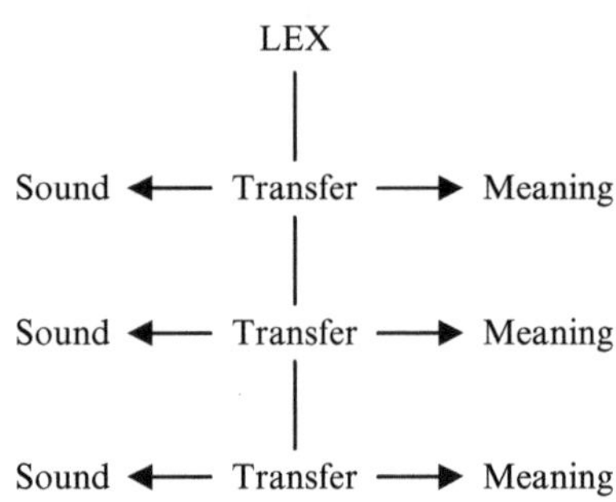

As a derivational 'phase', a cycle exhausts the projective potential of a head with which a given cycle sets out: if a verb, it will carry the cycle through to *v* (Aspect/Voice); if a nominal, it will carry it through to, say, D; if whatever is the head of the C-phase, perhaps T or Fin, it will be C, correlating with the computation of illocutionary Force, i.e. a full proposition put forward in discourse. Transfer thus creates a complete and then encapsulated unit of both sound and meaning. The only structure, then, that is ever computed in syntax, is of the form (37), and no syntactic object ever looks different:

(37)

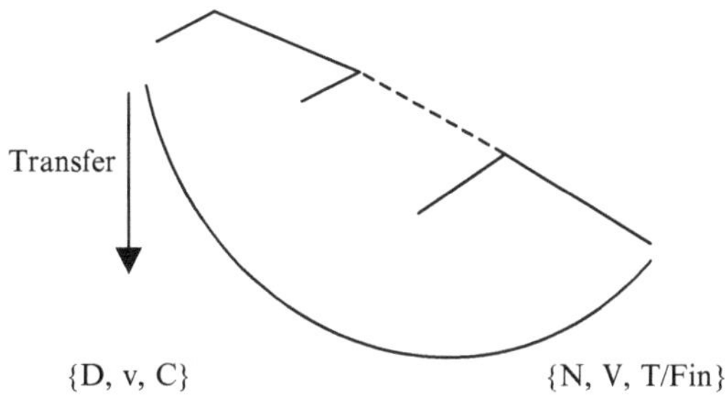

Each cycle has a left periphery, with indications that access to the syntax-discourse interface takes place after the completion of each of them (Aboh 2004; Jayaseelan 2008). What this means is that a particular discourse referent is delivered there: an object, event, or proposition, as the case may be, given that these are the intuitive ontological correlates of what I assume are the three phases. After it is delivered, the referent becomes part of the discourse representation – the structure that is updated by a given utterance – and the derivation continues with a new head that will take the head of the old phase as a complement. That old head is by that time fully referentially evaluated – in accordance with the referential potential of its internal structure – and indexed to the discourse referent it has determined. The complement of the old head has been transferred; it is gone from the derivation.

As noted, anything that goes on within a cycle is essentially bounded. The cycle is a rigid mental template: it consists in a fixed hierarchical sequence of projections, pre-determined in its basic outline with the initial choice of head: starting with N, one couldn't, say, end with T; and we couldn't reverse the pre-determined sequence, so as to have a determiner following a numeral, say (e.g. **three the cats*). So, again, there is in this sense no recursion within a cycle, of a sort that would yield unboundedness; there would be, only if we *abstracted*, gratuitously, from its inherent bounds. But there is no compositionality either, in a sense now described. Each cycle has to be looked at holistically or globally: as one single unit of computation as well as semantic interpretation. Before the phase boundary is reached, no interpretation is computed, and the discourse interface is not accessed. Consider (38), where 'unit' acts as a Classifier as in Chinese-type languages, 'individualizing' the mass-denotation of *cat* (cf. Hinzen 2007: 214–8):

(38) [these [three [units of [cat]]]]

A speaker using this expression in discourse would only be referring to three individual whole cats, not also or instead to what (38) would refer to, i.e. a set of sets consisting of three whole cats each:

(39) [three [units of [cat]]]

Neither would he refer to what (40) refers to, namely a set of whole cats:

(40) [units of [cat]]

Finally, he would surely not be referring simply to what (41) refers to, an unstructured and raw cat space:

(41) [cat]

In short, embedded constituents within a phase do not have full referential force: they do not determine objects of reference in the way the phase is used. They are part of an intensional description that helps to nail down a referent reached at the root. Reference is determined only at the phase level: before that, interpretation remains intensional. By the principle of compositionality, by contrast, we would be assigning independent referents to each of the constituents in (38), and within these constituents to each word. But this wouldn't explain why we only refer to one of these constituents when we use (38): why the referents of the embedded constituents are blanked out the moment we reach the root. It is a phase, therefore, which creates a unit of referential interpretation. The referent is there once the phase is complete. If it is broken off before it is maximally complete (given the projective potential of its initiating head), reference will be accordingly weaker: it may be reference to an abstract kind (cathood), a cat-space partitioned into chunks (a mass), or a cat-space partitioned into individuals: denotational specificity grows as the syntax proceeds.

Again one could, if one wanted, assign interpretations to the embedded parts of (38). The point would remain that from the viewpoint of speaker reference, their reference remains opaque. It is the phase that, by definition, determines a unit of compositional interpretation. If there were units of interpretation within a phase, there would be phases within a phase, which makes no sense: a phase is a unit of interpretation, so there are no units of interpretation within it. Compositionality in this sense fails within the phase. It makes no sense assigning independent referents to parts of phases, as compositionality does.

If phases embedded, though, compositionality might apply to their values. But strictly, as we have seen, a phase never is embedded: the complement of its head is 'transferred' the moment it is complete, and becomes inaccessible to operations then. The computation thus never sees one whole cycle embedded in another, and in this sense it never operates over a recursive structure (Arsenijevic and Hinzen 2013). Indeed, again, it makes no sense that a unit of computation embeds another unit of computation. So what gets embedded is ever only an edge, never a whole cycle. The edge corresponds to a referent, but the referent as such is an element in the discourse representation with which the syntax interfaces. It is not in the syntax. The only way the derivation can continue after a D-phase, therefore, is not by integrating the referent, John, but by integrating a *thematic role* that John *plays* with regard to an event encoded by *v*. Arguments, empirically, always play thematic roles, and are never just referential expressions. The syntax never combines referents: it creates a description to this effect. This explains why the meaning of *kill Bill* is not two referential acts juxtaposed (which it might semantically have been), one to an event of killing, one to Bill, but an event of killing of which Bill is the Theme.

Phases, then, strictly don't compose either: If the semantic value of a phase is an object, that referent does not compose or embed. What embeds is not John as such but the role of *the Agent of* a given event, e, say, and that description is not the same as *the Patient of* another event, e', say. If embedded meanings are not referents, two names in different embedded positions are typically not substitutable and extensionality fails. In a similar way, embedded propositions, too, will not denote truth values. They are descriptions of possible thought contents. They play a thematic role with regard to a mental state encoded in the verb taking the proposition as an argument. This is why compositionality fails. An embedded syntactic argument, *because* it is an argument, never functions as it would if it occurred alone or embedded elsewhere. It doesn't take its context-independent semantic value into the whole that contains it. It cannot take its truth value into this whole, since a truth value is a referent, and referents cannot compose. Anything that is embedded is ipso facto not maximally referentially complete. It remains intensional. If it was complete, the derivation would be cancelled: the assignment of the truth value is the stop to compositionality and recursion. As truth is the stop to compositionality, making it explicit as in any instance of (25), makes no difference to the truth-conditional content (we now understand *why* any instance of (25) holds, if it does, a fact that Tarski merely exploited). Anything that is embedded, therefore, falls short of reference to some extent; and whatever referent is determined at a phase boundary will be such that its description cannot be recovered from the derivation: by the time the referent is fixed the description is gone from the derivation.

7 Summary

The meaning of a sentence is not composed of independently referential parts and the way these referents are combined. By creating a part, the syntax creates an intensional description of the referent of the phasal head of which this part is a complement. It creates a whole that is *Gestalt*-like and of which the parts cannot be detached. The meaning of a part is what it is *because* the syntax has made it a part: an encapsulated unit transferred for interpretation, with only its head retained, which now becomes a description whose reference is opaque. The problem of the unity of the proposition arises only if we assume a multiplicity of independently referential parts. As for truth values, they must be unique, if syntax is bounded in the way described, and if whatever is embedded is never fully extensional. The truth value of a whole therefore never distributes over its parts, any more than the referent of a complex nominal or *v*P. Reference is only fixed at the left periphery of a phase. Semantic interpretation appears to wait until this happens.

Once upon a time, the phrase structure component of grammar was not recursive. A recursive phrase structure component was only introduced in the 1960s (Freidin 1999), but phrase structure has been abandoned since. Nowadays, there is only the generic operation Merge. But to understand the specific ways in which recursive structures in language arise, it is the cycle, as a mental template, that we need to understand. Recursion is mediated by phase boundaries (it is 'indirect'), and it occurs in the way it does because of these. Since a derivational phase has a semantic identity as much as a syntactic identity, recursion is semantically mediated or conditioned. Phases however don't as such embed. Derivations consist of cycles following other cycles, not cycles embedding cycles. This coheres with the fact that the objects they compute don't embed in themselves: necessarily, they become opaque, as described. It also explains why even indirect recursion is not strict: semantic values assigned to wholes are necessarily not ones we can assign to their parts; and what parts mean when they are parts is not what they would mean if they were not parts. Syntax is for real. It affects the semantic interpretation of the parts that it connects. The idea that syntax is merely set-theoretic Merge, and that semantics consist in the composition of the referents associated to all the sets and lexical items that are merged irrespectively of where they occur, does not model this fact.

References

Aboh, E. (2004). Topic and focus within D. *Linguistics in The Netherlands, 21*, 1–12.

Arsenijević, B., & Hinzen, W. (2013). On the absence of X-within-X recursion in human grammar. *Linguistic Inquiry, 43*(3), 423–440.

Carstairs-McCarthy, A. (1999). *The origins of complex language*. Oxford: Oxford University Press.

Chomsky, N. (1973). Conditions on transformations. In S. A. Anderson & P. Kiparsky (Eds.), *A Festschrift for Morris Halle* (pp. 232–286). New York: Holt.

Chomsky, N. (2000). *New horizons in the study of language and mind*. Cambridge: Cambridge University Press.

Chomsky, N. (2008a). Approaching UG from below. In U. Sauerland & H.-M. Gärtner (Eds.), *Interfaces + recursion = language? Chomsky's minimalism and the view from syntax-semantics* (pp. 1–29). Berlin/New York: Mouton de Gruyter.

Chomsky, N. (2008b). On phases. In C. Otero (Ed.), *Foundational issues in linguistic theory*. Cambridge, MA: MIT Press.

Chomsky, N. (2010). Some simple evo devo theses: How true might they before language? In R. K. Larson, V. Déprez, & H. Yamakido (Eds.), *The evolution of human language: Biolinguistic perspectives* (pp. 45–62). Cambridge: Cambridge University Press.

Cinque, G. (1999). *Adverbs and functional heads: A cross-linguistic perspective*. New York: Oxford University Press.

Davidson, D. (2005). *Truth and predication*. Cambridge: Harvard University Press.

De Villiers, J. (2007). The interface of language and theory of mind. *Lingua, 117*(11), 1858–1878.

den Dikken, M. (2009). On the functional structure of locative and directional PPs. In G. Cinque & W. Schweikert (Eds.), *The cartography of syntactic structure* (Vol. 6). Oxford: Oxford University Press.

Di Sciullo, A. M. (2005). *Asymmetry in morphology*. Cambridge, MA: The MIT Press.

Field, H. (1994). Deflationist views of meaning and content. *Mind, 103*(411), 249–84.

Fodor, J. A., & Lepore, E. (2002). *The compositionality papers*. Oxford: Oxford University Press.

Freidin, R. (1999). Cyclicity and minimalism. In S. Epstein & N. Hornstein (Eds.), *Working minimalism* (pp. 95–126). Cambridge: The MIT Press.

Hauser, M. D. (2000). *Wild minds*. New York: Holt.

Hauser, M. D., Chomsky, N., & Fitch, T. W. (2002). The faculty of language: What is it, who has it, and how did it evolve? *Science, 298*, 1569–79.

Heim, I., & Kratzer, A. (1998). *Semantics in generative grammar*. Oxford: Blackwell.

Higginbotham, J. (1985). On semantics. *Linguistic Inquiry, 16*, 547–931.

Higginbotham, J. (2009). Two interfaces. In M. Piattelli-Palmarini (Ed.), *Of minds and language* (pp. 142–154). Oxford: Oxford University Press.

Hinzen, W. (2006). *Mind design and minimal syntax*. Oxford: Oxford University Press.
Hinzen, W. (2007). *An essay on names and truth*. Oxford: Oxford University Press.
Hinzen, W. (2009). Hierarchy, merge, and truth. In M. Piattelli-Palmarini et al. (Eds.), *Of minds and language* (pp. 123–141). Oxford: Oxford University Press.
Hodges, W. (2001). Formal features of compositionality. *Journal for Logic Language and Computation, 10*(1), 7–28.
Hollebrandse, B., & Roeper, T. (2008). *Recursion as a restrictive semantic interface*. Handout from a presentation at the conference *Biosemantics*. Leiden.
Hornstein, N. (2009). *A theory of syntax*. Cambridge: Cambridge University Press.
Jackendoff, R. (2002). *Foundations of language*. Oxford: Oxford University Press.
Jacobson, P. (2009). Direct compositionality. In M. Werning, W. Hinzen, & E. Machery (Eds.), *The Oxford handbook of compositionality*. Oxford: Oxford University Press.
Jayaseelan, K. A. (2008). Topic, focus, and adverb positions in clause structure. *Nanzan Linguistics, 4*, 43–68.
Katz, J., & Pesetsky, D. (2009). *The recursive syntax and prosody of tonal music*. Ms., MIT.
Larson, R. K. (2011). Clauses, propositions and phases. In A.-M. DiSciullo & C. Boeckx (Eds.), *The biolinguistic enterprise: New perspectives on the evolution and nature of the human language faculty* (pp. 366–391). Oxford: Oxford University Press.
Lasnik, H. (2000). *Syntactic structures revisited*. Cambridge: The MIT Press.
Leiss, E. (2009). *Sprachphilosophie*. Berlin: De Gruyter.
Lycan, W. G. (2008). *Philosophy of language: A contemporary introduction*. New York: Routledge.
Moro, A. (1997). *The raising of predicates: Predicative noun phrases and the theory of clause structure*. Cambridge: Cambridge University Press.
Pawley, A. (2006). *On the argument structure of phrasal predicates in Kalam, a language of the Trans New Guinea family*. The 32nd annual meeting of the Berkeley Linguistics Society, Berkeley, CA.
Pietroski, P. (2003). Quantification and second-order monadicity. *Philosophical Perspectives, 17*, 259–298.
Potts, C., & Roeper, T. (2007). The narrowing acquisition path. In L. Progovac et al. (Eds.), *The syntax of nonsententials*. Amsterdam: John Benjamins.
Prusinkiewicz, P., & Lindenmeyer, A. (1996). *The algorithmic beauty of plants* (Reprint). New York: Springer.
Rizzi, L. (1997). The fine-structure of the left periphery. In L. Haegemann (Ed.), *Elements of grammar* (pp. 281–337). Dordrecht: Kluwer.
Searls, D. B. (2002, November 14). The language of genes. *Nature,* 420, 211–217.
Semenza, C. (2009). The neuropsychology of proper names. *Mind & Language, 24*(4), 347–369.
Szabo, Z. (2009). The case for compositionality. In M. Werning, W. Hinzen, & E. Machery (Eds.), *The Oxford handbook of compositionality*. Oxford: Oxford University Press.
Tarski, A. (1933). Der Wahrheitsbegriff in formalisierten Sprachen. *Studia Philosophica, 1*, 261–405.
Uriagereka, J. (2008). *Syntactic anchors*. Cambridge: Cambridge University Press.

Recursion in Language: Is It Indirectly Constrained?

Aravind K. Joshi

Abstract Recursion is a key aspect of formal theories of language. In this chapter we have presented a new perspective on recursion in language, claiming that recursion in language is *indirectly constrained*. After discussing some aspects of the relation between linguistic theory and mathematical linguistics, we show that this perspective arises from the properties of grammars that are beyond context-free grammars with context-sensitivity that is mild in a sense and that the derivation proceeds from structures to structures. Viewing the generation process inductively, we then show that for each index of recursion, there is a form (we call this a canonical form) that can be generated. However, corresponding to the canonical form, there are variants of the form, called here as non-canonical forms that exhibit gaps in the sense that for a given index of recursion, beyond two levels of embedding, not all non-canonical forms can be generated. This property emerges from the formal architecture of the system and not from any particular linguistic or processing constraint. We call this property Indirectly Constrained Recursion (ICR), suggesting a new perspective on recursion in language. ICR is a formal structural constraint and not a linguistic or processing constraint. In linguistic theories constraints on recursion are usually expressed either in terms of some linguistic constraints or in terms of processing constraints. ICR arises from the properties of the kinds of formal systems considered here. Linguistic constraints and processing constraints are then to be considered as additional constraints.

A.K. Joshi (✉)
Department of Computer and Information Science and Institute for Research in Cognitive Science, University of Pennsylvania, Philadelphia, PA, USA
e-mail: joshi@seas.upenn.edu

T. Roeper and M. Speas (eds.), *Recursion: Complexity in Cognition*, Studies in Theoretical Psycholinguistics 43, DOI 10.1007/978-3-319-05086-7_7,

The linguistic phenomena referred to in this chapter are well-known. Hence, the main contribution of this chapter is the presentation of a new perspective on recursion as described above.[1]

Keywords Dependencies:nested • Crossed • Grammars:context-free • Context-free • Context-sensitive • Phrase structure • Transformational • Tree-adjoining • Multi-component tree-adjoining • Languages context-free • Mildly context-sensitive • Extraposition from NP • Canonical forms • Non-canonical forms • Scrambling or liberation

1 Introduction

Recursion is a key aspect of formal theories of language, and the role of formal/mathematical linguistics in linguistic theory has always been one of the topics of investigation. It is well known that many of the formal models that go beyond context-free grammars provide novel ways of packaging context, allowing not only appropriate descriptions of linguistic phenomena but also extending the generative power (both weak and strong) beyond context-free grammars in a minimal (mild) manner. It turns out that grammars in this class of so-called mildly context-sensitive grammars (languages), MCSL, constrain recursion in an indirect manner. This indirectly controlled recursion (ICR) is a formal property and it is not a processing constraint.[2] Thus ICR provides a new perspective on recursion in language. Bringing out such a perspective is the main goal of this chapter.

1.1 Role of Mathematics in Linguistic Theory

We begin with some surprisingly little known comments by Noam Chomsky concerning the role of formal/mathematical linguistics in linguistic theory

> ... This seems to me what one would expect from applied mathematics, to see if you can find systems that capture some of the properties of the complex system that you are working with, and to ask whether those systems have any intrinsic mathematical interest, and whether they are worth studying in abstraction. And that has happened exactly at one level, the level of context-free grammar. At any other level it has not happened. The systems that capture other properties of language, for example, of transformational grammar, hold no interest for mathematics. But I do not think that is a necessary truth. It could turn out that there would be richer and more appropriate mathematical ideas that would capture other, maybe deeper properties of language than context free grammars do. In that case you

[1] In this chapter, we have presented ICR with respect to a particular framework. However, it is expected that similar results will carry over to some of the other systems, for example, in the class of Mildly Context-Sensitive Languages and perhaps, stronger versions of ICR can be found.

[2] Processing constraints associated with MCSL have been studied in the context of automaton models for MCSL; see, for example, Joshi (1990).

have another branch of applied mathematics which might have linguistic consequences. That would be exciting. Chomsky replying to a couple of questions concerning the role of automata theory (applied mathematics in general) in linguistics and the significance of that work for linguistics –In: Noam Chomsky on The Generative Enterprise: A Discussion with Piny Huybregts and Henk van Riemsdijk, Foris Publications, Dordrecht-Holland 1982, p. 15.[3]

Although Context-free Grammars, CFG's were enormously successful both mathematically and linguistically as the quotation above indicates, serious linguistic issues could not be addressed. Context-sensitivity was needed in the mathematical machinery to deal with some of these issues. However, adopting the context-sensitive rewriting rules was not the right direction, primarily because the context that needs to be specified has to be *structural,* i.e., one has to be able to specify structural context for a node (what is to the left or right of that node and/or what is above or below that node) and not just the left or right string context. Much of the work on transformational grammar since 1957 to late 1970s can be viewed in this way. However, any attempts at formal/mathematical characterization of these transformational grammars have led to full Turing computability (equivalent to the so-called type 0 grammars in the Chomsky hierarchy); see Peters and Ritchie (1973), for example. The net result was that this direction of work did not give any suggestions for introducing structural contexts in a limited (controlled) way such that it provided linguistically adequate machinery on the one hand and on the other, allowed a precise formal/mathematical characterization comparable to that achieved for CFG's, as described in the quotation at the beginning of this chapter.

1.2 Mildly Context-Sensitive Grammars (Languages)

One precise formal proposal consistent with the approach described above emerged during the period from 1975 to 1985 out of the work on Lexicalized Tree Adjoining Grammar (LTAG, also referred to as TAG for short) (Joshi 1985, 2004). The main ideas in TAG are as follows: There is a finite collection of elementary structures, which can be thought of as *minimal structural context packages.* These packages are assembled by two (universal, i.e., language independent) operations—*substitution* and *adjoining*.[4] These main ideas in TAG have roots in the mathematical theory of Context-free Grammars (CFG), especially in the so-called

[3]This book has been reissued in 2004, with a preface and an Introduction by Naoki Fukui and Mihoko Zushi, Mouton de Gruyter, Berlin, New York. See also "Formal Grammars in Linguistics and Psycholinguistics" William J. M. Levelt, John Benjamin Publishing Company, Amsterdam/Philadelphia, 2008 (reissued as a new edition of the original 1974 edition with an added Postscript summarizing some developments since 1974). This Postscript also refers to the quotation from Chomsky above.

[4]A derivation in a TAG proceeds from a sentence form to another sentence form via the operation of adjunction. This is reminiscent of the idea of a transformation as a relation between sentence forms as suggested by Harris (1957). However, in our case, the relation is between structured sentence forms and is formulated as a recursive system.

pumping lemma for context-free languages and linguistically, in the early work on transformational grammars, especially, generalized transformations, beginning with Chomsky (1957), see also Chomsky (2007). This class of grammars is such that it carries over the properties of CFG's with appropriate modifications, including the existence of an associated automaton.[5] TAG's thus represent a proper extension of CFG's in the spirit of the quotation in Sect. 1 (see Joshi 1985 and Joshi and Schabes 1997).[6]

Along with the development of TAGs, a characterization of the so-called Mildly Context Sensitive Grammars, (Languages), was proposed in terms of the following properties: (a) MCSL's (languages of mildly context-sensitive grammars) properly contain CFL's, (b) they are polynomially parsable, (c) they can describe a limited set of non-nested dependencies, and (d) they have the so-called constant growth property, i.e., the lengths of strings in the language, if ordered in increasing length grow only linearly (cf. semi-linear property of CFL's).

Subsequently it was shown that several other classes of grammars, in particular, some generalizations of TAGs, such as Multicomponent TAG, some formalizations of minimalist grammars (Frank 2002; Gärtner and Michaelis 2007), and combinatory categorical grammars (CCG), also belong to this class. Having identified the class of MCSG's, with TAG as the key example, in the spirit of the quotation in Sect. 1.1, we will now describe some aspects of TAG that are directly relevant to the issue of indirectly controlled recursion (ICR) in language.

1.3 Derivations in CFG and TAG

In a CFG a derivation proceeds from the start symbol to the terminal string. In a TAG the derivation proceeds from an elementary structure to another (derived) structure and then to another structure etc. In each step, a new structure enters the derivation via the operations of substitution or adjoining. A derivation in TAG need not proceed in one particular direction.

There are two special things to note about this process of derivation:

(a) Recursion enters the derivation via adjoining an auxiliary structure.[7] There is, of course, no recursion inside an elementary structure by definition.

[5]An extension of the pushdown automaton, called an embedded pushdown automaton, EPDA, equivalent to the language of a LTAG (in the same way as a pushdown automaton is equivalent to a CFG) exists and has been studied in the context of processing crossed and nested dependencies (Joshi 1990).

[6]In Appendix A, a rather simple example of an LTAG derivation is given for readers who may not be familiar with mildly context sensitive grammars and languages. For more details, including multicomponent LTAGs, see Joshi and Schabes (1997), Joshi (2004), and Vijay-Shanker (1988).

[7]In the formal treatment of LTAG recursion enters only via adjoining. In principle, recursion can also enter via substitution. In fact, one way of looking at adjunction is to describe it as a pair of substitutions in the following sense. If a node X in a tree t is targeted for adjoining by an auxiliary tree, say t', then the adjoining operation can be thought of as a pair of substitutions such that the tree t is pulled apart at the node X, creating two X nodes (top and bottom), and then t' is spliced

(b) Derivation proceeds from a sentence structure to another sentence structure and so on. *This property suggests an inductive definition of sentence forms, i.e., from sentence forms to sentence forms.* This way of viewing recursion is reminiscent of the 'transformational' approach of the period between the late 1950s and early 1960s. (See Footnote 4)

As soon as recursion appears in a grammar we have the possibility of deriving a potentially infinite set of sentence forms or structures. This potential infinity of sentence forms has been a topic of discussion for a long time. Center embedding of relative clauses as in (1) and (2) below is one of the standard examples.

(1) The gardener who the woman was calling all day finally came.
(2) The gardener who the woman who had lost all her keys was calling all day finally came.

In principle, this kind of embedding can be carried out up to any depth. Certainly human processing constraints (whatever they are) tend to limit recursion in practice. However, placing an arbitrary limit on the recursion in a formal grammar is not well motivated, first, because this constraint does not follow from the grammar formalism itself, and secondly, it hides an important aspect of language, namely the potential infinity of sentence forms and structures.

Given this centrality of recursion in language, are there ways of constraining this recursion without losing the potential infinity of sentence forms or structures? *Our answer is YES and this is the main theme of the chapter.*

It turns out that for grammars in the class of Mildly Context Sensitive Grammars, although recursion can lead to an unbounded number of sentence forms or structures, there are gaps in the space of all these forms, which arise at three or more levels of embedding, suggesting that recursion is indirectly constrained. In the rest of the chapter, we will illustrate these points by means of three examples in Sect. 2.

1.4 Indirectly Constrained Recursion (ICR)

The main idea is as follows: There is a canonical mode of recursion for sentence forms where the recursion is unbounded, i.e., for each index of recursion there is a canonical form. However, in the corresponding non-canonical forms there are gaps beyond a certain level, say, k, of recursion, i.e., beyond k, for each level of recursion not all forms are generated. Thus there are gaps. These gaps are not due to any specific linguistic or processing constraints. They are the properties of the formal system(s) under consideration. This is the reason for calling this property of recursion arising from the formal system itself as Indirectly Constrained recursion (ICR),

into the tree t by making a pair of substitutions. This way of looking at standard adjoining makes the standard TAG as a special case of Tree-Local Multi-Component TAG (TL-MCTAG), which we have discussed later,.

Examples (1) and (2) from Sect. 1.2, reproduced below as (3) and (4) illustrate the canonical forms.

(3) The gardener who the woman was calling all day finally came.
(4) The gardener who the woman who had lost all her keys was calling all day finally came.

In a context-free phrase structure grammar these forms are generated by rules that introduce recursion via a single rule or a set of rules. Clearly recursion introduced in this way is unbounded. We call these forms *canonical*[8] as these forms remain unbounded in all grammars in the class of Mildly Context Sensitive Grammars. However, corresponding to these canonical forms there are *non-canonical* sentence forms (variants of the canonical forms, transforms, in a sense) that can also be unbounded but there are 'gaps' in this unbounded set of non-canonical forms. In this sense recursion in language is *indirectly constrained.*

2 Three Illustrative Examples

Example 1 The following example is rather special in the sense that the non-canonical form breaks down already at Step 2. However, it is very useful for illustrating our main point in this chapter.

Canonical Forms	Non-canonical Forms
(1) Mary will win	(1') Mary will win
(2) John thinks Mary will win	(2') Mary, John thinks, will win
(3) Harry believes John thinks Mary will win	*(3') Harry believes Mary, John thinks, will win

(2') is the non-canonical form (2).[9] However, (3'), the non-canonical form corresponding to (3) breaks down. This is a very special case, as it happens right at the third step. This particular case is well-known under the heading "root transformation". (2') is a root transform of (2) and therefore resists being embedded under "Harry believes".

[8]We use the term canonical just as a convenient way of referring to the unbounded set of sentence forms, without gaps, that can be derived in a CFG. Nothing further should be read into this terminology.

[9]We can also have
*(3") Mary, Harry believes John thinks, will win.
*(3"') Harry, John thinks, believes Mary will win.
Judgments for (3') and (3") were provided by Tony Kroch and Mark Johnson, respectively.

Example 2 Let us consider again center embedding of relative clauses in English.

(1) The gardener who the woman was calling all day finally came
(2) The gardener who the woman who had lost all the keys was calling all day finally came

As we have seen before, this canonical mode of recursion can be continued, in principle, indefinitely. Now consider (1') where we have extraposed the relative clause from an NP (the so-called extraposition from NP transform).

(1') The gardener finally came who the woman was calling all day

We can also extrapose the first relative clause together with the clause embedded in it as in

(2') The gardener finally came who the woman who had lost all the keys was calling all day

However, it is not possible to extrapose just the second level relative clause as in

* (2") The gardener who the woman was calling all day finally came who had lost all the keys

This is perhaps due to the fact that in (2") both the relative clauses appeared to be 'stacked' on 'The gardener' and the extraposed clause cannot be interpreted as modifying "the woman".[10]

Extending this kind of extraposition to more than two levels of embedding does not seem to be possible. It is easily seen that in this case we have no gaps in the canonical mode. TAGs can model this indirectly constrained recursion (see Kroch and Joshi 1986, for some early work). The key point here is that there is a TAG for (2') and all other canonical variants, but there is no TAG that can generate (2").

---

[10]However, consider the following:

The gardener I told you about showed up again who had done such a bad job.

Here the extraposed relative clause is meant to modify the subject NP and it sounds much better than (2"). Judgment provided by Tony Kroch.

It is however possible to extrapose modifiers at the second level in certain cases, as in

(3) Supply troubles were on the minds of Treasury investors yesterday who worried about the flood of new government securities. (From WSJ, Penn Treebank Corpus).

Example 3 was provided by Tonia Bleam and Fei Xia. Stephan Muller has also pointed out similar examples in German, see http://www.cl.uni-bremen/~stefan/Pub/subjacency,html. The relative clause in (3) is non-restrictive. The conditions under which the relative clause can be discontinuous with its head are different for restrictives and non-restrictives.

Thanks to Tony Kroch for helping me sort out some of these issues.

In Appendix A we have shown a derivation of (1'), in a pictorial form, in the so-called tree-local MCTAG (Multi-Component LTAG), a variant of LTAG, which is weakly equivalent to LTAG but not strongly equivalent to LTAG, i.e., MCTAG gives structural descriptions not obtainable by LTAG.

Thus we have once again a situation where unbounded recursion is possible in the canonical mode but there are gaps in the non-canonical mode. In fact, in this case, it appears that the non-canonical mode has gaps beyond two levels of embedding, thus providing another example of indirectly constrained recursion.

Example 3 Examples 1 and 2 are rather special cases of the non-uniformity of recursion as the breakdown happens at level 2. Although these cases are illustrative of the property of Indirectly Controlled Recursion (ICR), one may argue, and quite convincingly, that this behavior could be easily demonstrated by a phrase structure grammar (PSG) and the mechanism of feature passing, and therefore, these examples perhaps do not serve as convincing examples illustrating ICR.[11] They do however help to set the scene.

Example 3, however, is of a different character. In this example, the non-uniformity of recursion kicks in at all levels of recursion, greater than 3.

In this example, we will consider center embedding of complement clauses in verb final languages (for example, in German). Thus we have center embedding of complement clauses as in (1) below.[12]

(1)	... dass	Hans	Peter	Marie	schwimmen	lassen	sah
		Hans	Peter	Marie	swim	let	saw
		NP1	NP2	NP3	V3	V2	V1

Each Vi takes Vi + 1 as its complement argument, i = 1, 2. V3 does not take a V-type argument. Each Vi also takes NPi, i = 1, 2, 3, as another argument. We have taken here the simplest case where each V has only one NP argument.

In (1) we have two levels of center embedding of complement clauses. As in the case of center embedding of relative clauses in English, we will have a potentially infinite number of sentence forms such as (1). These are canonical sentence forms as we have discussed before. A CFG can obviously generate such canonical forms.

Now, it is possible to generate variants of (1) by "scrambling" the NP's and V's in various ways. We will call these sentence forms non-canonical forms. For the present, we will consider the particular case where the V's remain fixed in the order they appear in (1), i.e., V3 (the verb that does not need a complement), then V2, and

[11] As pointed out by Mark Johnson.

[12] German is verb-final only in the subordinate clauses; so the 'that' ('dass') may not be left out. In our subsequent discussion we will ignore 'dass'.

We should really use full NP's in these examples. We have used proper names only for convenience.

finally V1 the top level tensed verb.[13] In (1) we have two levels of center embedding of complements.

Returning to (1), the NP arguments of (1) can however be scrambled (permuted). Thus, for example, we can have (2) below as a variant of (1).

(2) NP3 NP2 NP1 V3 V2 V1

In this case, corresponding to the canonical form (1) we have 5 different non-canonical forms corresponding to the different permutations of the three NP arguments, corresponding to the three levels of embedding, i.e., for k = 4, there are 24 possible permutations of which 23 are non-canonical, and so on.

What about TAG (and grammars in MCSG, in general)[14]? It turns out that the potentially infinite set of non-canonical sentence forms can be generated by TAG, except that there are gaps in this potentially infinite set. More specifically, for each k, k > 3 (i.e., with three or more embeddings), some permutations cannot be generated. For k = 4, there are 24 possible permutations, only 22 permutations can be generated, thus creating a gap of 2 sentence forms.[15] The two sequences below cannot be generated.[16]

(3) NP3 NP1 NP4 NP2 V4 V3 V2 V1
(4) NP3 NP2 NP4 NP1 V4 V3 V2 V1

Throughout of discussion about TL-MCTAG, it is not requited that the elementary trees must be binary. If we require that the elementary trccs must be binarized then it can be shown that (3) and (4) above can now be generated. This is due to the fact that binarization provides more nodes for adjoining. Thus with binarization the gap of 2 sentence forms at k = 4 disappears. However, with k = 5, binarization

---

[13]There are, of course, many other patterns of scrambling due to the verb sequence itself being scrambled, possibly along with the NP sequence, for example, as in NP1 NP2 NP3 V1 V3 V2. Further, a verb may have more than one NP argument, etc. These can be generated but we do not consider these sequences here.

[14]Actually, we use the so-called Tree-Local Multi-Component TAGs (Tree Local MCTAG), which is a generalization of TAGs, allowing more flexibility in composition without increasing the (weak) generative capacity (see Appendix A). We will continue to use the term TAG to include both TAGs and Tree Local MCTAGs.

[15]It is possible to add flexible composition (delayed composition) and multiple adjoining under certain circumstances without increasing the weak generative capacity of TL-MCTAG (see Chiang and Scheffler 2008 and Chen-Main and Joshi 2008). The particular derivable and non-derivable sequences mentioned in this paper have been discussed in detail in Chen-Main and Joshi (2008).

[16]Our concern here is about the existence of gaps and not about the acceptability judgments necessarily. The acceptability judgments for these sentences are hard in any case. Some are judged as clearly unacceptable and others as marginally acceptable, at best.

Further, there is no claim here about the gaps being directly related to processing constraints. Processing involves incrementality and that is not clearly an issue here.

does not help. Out of the possible 120 sequences, three sequences[17] cannot be generated, thus creating a gap of 3 sentence forms. Thus, recursion in the non-canonical sentence forms is indirectly constrained.[18]

For some detailed derivations, see Appendix B. The elementary trees for the verbs and the composition of these trees are shown for each figure in Appendix B. An elementary tree for a verb can be a single tree or it may be in two components, where the scrambled NP argument appears in a separate tree. The derivation then is described, for simplicity, in a pictorial form. If two trees (say, t1 and t2) adjoin to the same node in a host tree then this situation is interpreted as representing the two cases: (1) t1 is adjoined before t2 and (2) t2 is adjoined before t1.

It appears that recursion in language, in general, has the property that it is indirectly constrained; there is a canonical mode where there are no gaps in the recursion, however, in the non-canonical modes there are gaps beyond two levels of embedding (with the elementary trees not required to be binary) and beyond three levels of embedding even if the elementary trees are required to be binary.[19] *The indirect constraint on recursion is not due to any processing considerations.*[20] *It is due to the structure of TAG and the way recursion and structural context-sensitivity enters in the derivation, in contrast to strictly phrase structure based systems. More specifically, it is the operation of adjoining combined with substitution that leads to these gaps. Substitution alone cannot lead to such gaps.*

The situation in Example 3 (in contrast to Examples 1 and 2) cannot be modeled in a feature based PSG without recourse to some very general operations such as "scrambling" or "liberation", which involve, essentially freeing up the word order and then constraining it via some other rules. The formal power of such systems is not known. Example 3 thus serves to give a clear illustration of how an ICR like constraint on recursion in language can be formally characterized. It would be interesting to explore ICR like constraints in the class of grammars corresponding to the class of Mildly Context Sensitive Languages (MCSL).

---

[17]These three sequences are (1) NP4 NP2 NP3 NP5 NP1 V5 V4 V3 V2 V1 (2) NP4 NP2 NP5 NP3 NP1 V5 V$ V3 V2 V1 and (3) NP4 NP3 NP2 NP5 NP1 V5 V4 V3 V2 V.

For further details see Chen-Main and Joshi (2008).

[18]There is also no claim here that all the sentences outside the gap are equally acceptable.

[19]It is interesting to note that even if we consider discourse variants of a canonical sentence form with recursion we find that non-canonical discourse variants break down beyond a certain level. Thus for example, consider the canonical sentence form with tail recursion *Harry believes John thinks Mary will win the race* and the discourse variant *Mary will win the race. John thinks so. Harry believes so*. The discourse variant breaks down as the *believes* clause appears to be taking scope over the first sentence rather than over the *think* clause. See also Hollebrandse and Roeper (2007) for a discussion about recursion in discourse.

[20]There is an automaton model, called an embedded pushdown automaton (EPDA) that corresponds to a TAG in the same way as a pushdown automaton (PDA) is equivalent to a CFG. EPDA have been used to model the processing strategies for crossed and nested dependencies (see Joshi 1990) based on the so-called Principle of Incremental Interpretation as proposed in (Bach et al. 1986). Whether ICR as described above can be related to processing via EPDA is an open question at this time.

3 Summary

We have presented a perspective on recursion in language, claiming that recursion in language is indirectly constrained. This perspective arises from the properties of grammars that are beyond context-free grammars with context-sensitivity that is mild in a sense. TAGs and their variants belong to this class. Many other types of formal grammars for language also belong to this class (called Mildly Context Sensitive). Using TAGs we can view the generation of sentence forms inductively, i.e., from structures to structures, starting from initial structures. We then show by examples that there is a canonical mode for generating these forms such that for each index of recursion there is a form that can be generated. However, corresponding to the canonical form there are variants of the canonical form, called here as the non-canonical forms which exhibit gaps, in the sense that for a given index of recursion beyond two levels of recursion not all non-canonical forms can be generated. We have characterized this property as Indirectly Constrained Recursion (ICR) and suggested that this is a new perspective on recursion in language. The linguistic phenomena referred to in the chapter are well-known. The main contribution of this chapter is the presentation of a new perspective on recursion as described above. ICR is a formal structural constraint. This perspective arises from the architecture of the formal system and not from any particular linguistic or processing constraint. It would be interesting to carry out similar investigations in other frameworks, especially those corresponding to the class of mildly context- sensitive languages (MCSL), perhaps giving further insights into the nature of ICR like constraints and their relationship to linguistic structure. Further, although we have said repeatedly that ICR is not a processing constraint, there might be interesting relationships between ICR and processing constraints.

Acknowledgement I want to thank Rajesh Bhatt, Joan Chen-Main, Robert Frank, Julia Hockenmaier, Steve Isard, Mark Johnson, Anthony Kroch, Uwe Mönich, and Mark Steedman for very valuable discussions concerning some of the issues discussed in this chapter.

Appendices

Appendix A

LTAGs and MC-LTAGs

We will give a brief introduction to Lexicalized Tree Adjoining Grammars (LTAGs) and tree-local Multi-Component LTAGs (MC-LTAGs), which will be adequate to follow the derivations in Sects. 1, 2, and Appendix B.

An LTAG consists of a finite set of elementary structures (trees, directed acyclic graphs in general) where each elementary structure is associated with a lexical

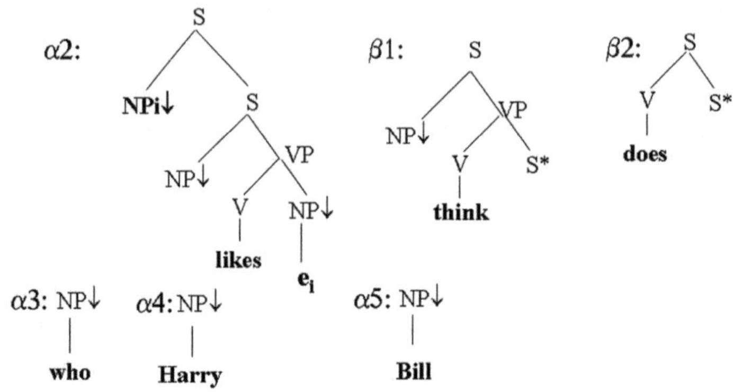

Fig. A.1 An LTAG derivation

anchor. Each elementary structure encapsulates the syntactic/semantic arguments and dependencies (corresponding to the lexical anchor associated with the structure) such as agreement, subcategorization, filler-gap dependencies, among others. These structures are large enough to localize all these dependencies and therefore can be larger than one level trees. This property, called *the extended domain of locality*, is in contrast to the case of CFG's where the dependencies are distributed over a set of rules.

The elementary trees are either initial trees or auxiliary trees. There is no recursion inside an elementary structure. Of course, an elementary structure may have the same label for the root node and the foot node in an auxiliary tree and can thus participate in a recursive structure. Since there is no recursion internal to an elementary structure, in a sense, in LTAGs recursion is factored away from the domain of dependencies. This is in contrast to CFG's where recursion is intertwined with the domain of locality, as both may be spread over several rules.

There are two composition operations, substitution and adjoining: These operations are universal in the sense that they are the same for all LTAGs. There are no rules in the sense of the rewriting rules in a CFG.

Since all dependencies of a lexical anchor are localized we may have several elementary structures corresponding to a lexical item, depending on the structural positions each argument can occupy in different constructions associated with the lexical item. There can be many such trees but the total number is finite. One can regard the different trees associated with a lexical item as extended projections associated with that item.

Figures A.1, A.2, and A.3 show an LTAG derivation. Figure A.1 shows the collection of trees[21] participating in the derivation. Figure A.2 shows the composition of

[21]We have not shown feature structures associated with the nodes of the elementary trees. These feature structures participate in the derivation via unification. We have left out these feature structures for simplicity.

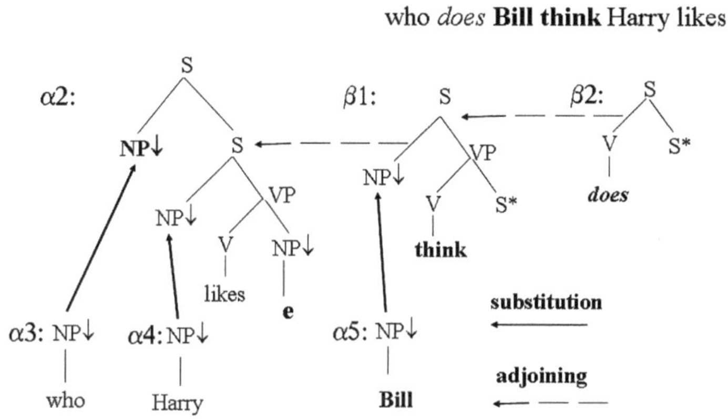

Fig. A.2 An LTAG derivation

Fig. A.3 A derivation tree

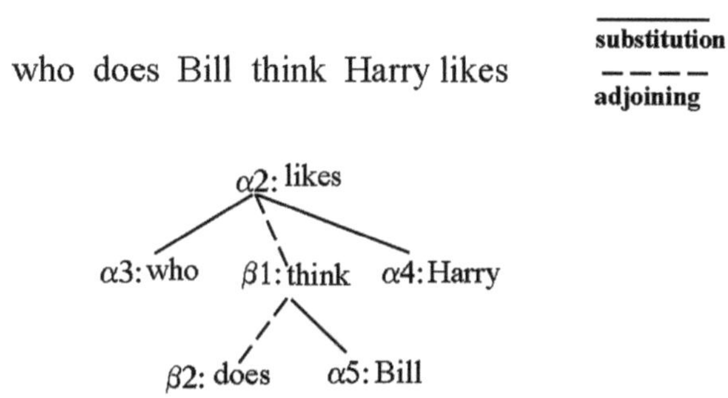

the elementary trees in Fig. A.1 by means of the operations of substitution (shown by a solid line) and adjoining (shown by a dotted line). Figure A.3 shows the derivation tree. The nodes are labeled by the names of trees as well as by the lexical anchors associated with the elementary trees. Solid lines represent substitution and the dotted lines represent adjoining. The lines should also be labeled by the addresses of the nodes in the host trees where substitutions and adjoinings take place. These have been omitted in order to simplify the diagram. Finally, Fig. A.4 shows the derived tree, which is the result of carrying out the compositions shown in Fig. A.3.

Note that the dependency between *who* and *e* has become long distance. However, it was local in the elementary tree associated with *likes* in Figs. A.1 and A.2.

The derivation tree is a record of the history of composition of the elementary structures by substitution or adjoining. The derived tree is the result of carrying out all the composition operations as indicated in the derivation tree. In a sense, the derivation tree has all the information and, in fact, compositional semantics can be defined on the derivation tree (Kallmeyer and Joshi 2003).

The order of derivation in the derivation tree in Fig. A.3 is shown bottom up in a sense. This is the normal convention that is followed in a TAG derivation. However,

Fig. A.4 A derived tree

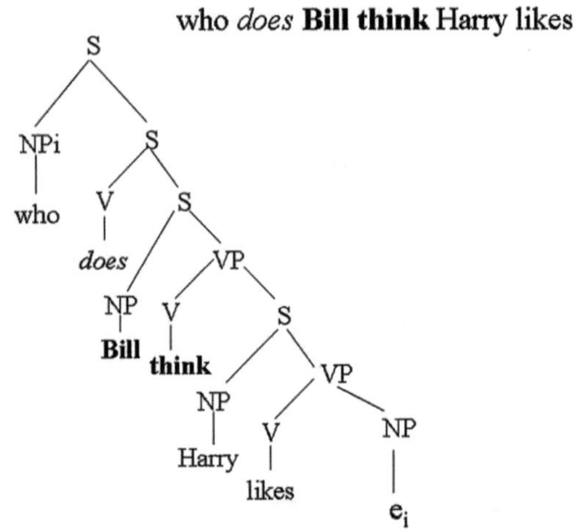

in principle the derivation could begin with any node with the directions of the arrows reversed appropriately. The order of derivation is flexible in a sense; see also Sect. 1 and (Chomsky 2007).

Multi-component TAG's (MC-TAG's)

MC-TAG's are like LTAG's except that in an MC-TAG the elementary trees can be a set of trees (usually a pair of trees in the linguistic context) and at least one of the trees in the multi-component set is required to have a lexical anchor. In this sense, MC-LTAGs are lexicalized. In a tree-local MC-TAG when a multi-component tree (say with two components) composes with another tree, say t, the two components compose with the same elementary tree, t, by targeting two nodes of that elementary tree. This is why this composition is called tree-local. The two components target two nodes in the local domain of an elementary tree, preserving the notion of locality that is inherent in LTAG. Tree-local MC-LTAGs are known to be weakly equivalent to LTAGs, i.e., they generate the same set of strings but not necessarily the same set of structural descriptions. Instead of formally describing the composition operation in MC-LTAG, we will illustrate it by means of a simple example. In Fig. A.5 we have shown a simple tree-local MC-LTAG derivation.

The derivation proceeds as follows; a3 is substituted in a2 as shown and then it is (reverse) adjoined to the b21[22] component of b2. The multi-component tree b2 is composed with a1 as follows. b21 is substituted at the NP node of a1 and the b22 is adjoined to a1 at the root node S of a1.

[22]By reverse adjoining we mean reversing the direction of composition. The weak generative capacity of Tree-Local MC-TAGs is not affected by such reverse compositions. They do provide more flexibility which is very crucial as we will see in Appendix B.

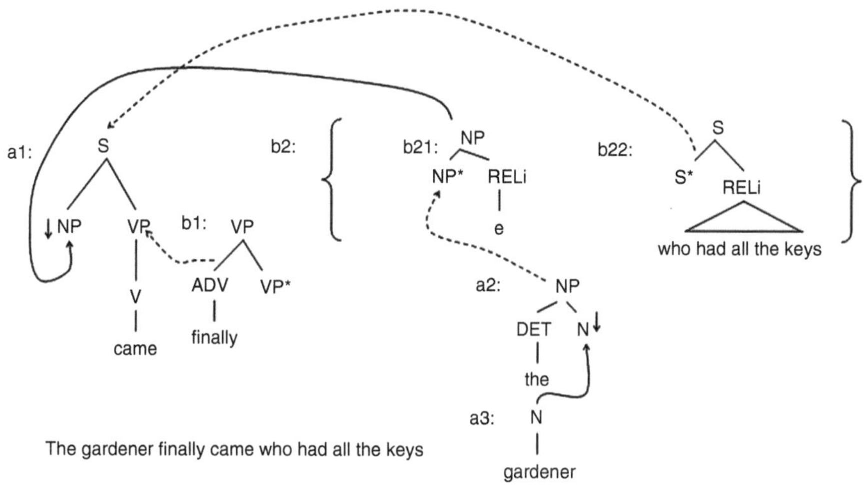

Fig. A.5 A derivation in a tree-local MC-TAG

In a multi-component composition in the linguistic context there is usually a constraint between the two components (besides any co-indexing across the components). For example, in our example above there is a constraint (not shown in the Fig. A.5) that the foot node of b22 immediately dominates the NP root node of b21. This constraint is obviously satisfied in the derivation in Fig. A.5.

Schabes and Shieber (1993) introduced an alternate definition of derivation for LTAG in which they allowed multiple adjoining of certain kinds of elementary trees at the same node during the derivation. They showed that multiple adjunctions of the kind described above do not affect the generative capacity of LTAGs, i.e., they are weakly equivalent to LTAGs. This notion of multiple adjunctions can be extended to the derivations in tree-local MC-LTAG where we can allow multiple adjunctions of the NP auxiliary trees that correspond to the scrambled NP's (see Appendix B). The generative power of tree-local MC-LTAG is not affected. We use such derivations for certain scrambling patterns. In Appendix B, we will show some derivations for sentences with scrambling using tree-local MCTAG.

Appendix B

Finally, we will describe some derivations for the scrambling (permutation) patterns using LTAGs or MC-LTAGs.[23] First consider

[23]See Joshi (2004) for some early discussion of these issues.

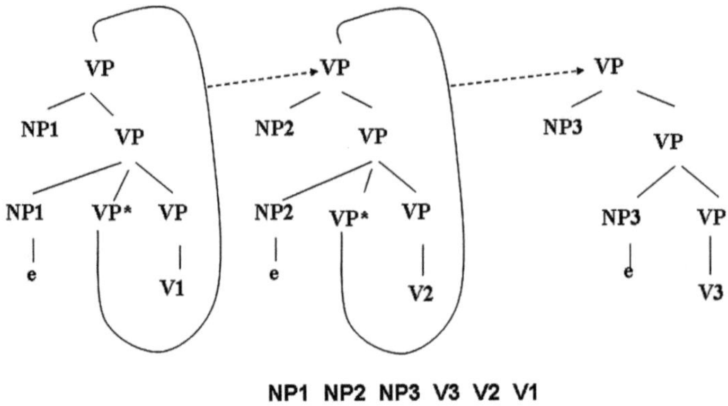

Fig. B.1 Derivation for NP1 NP2 NP3 V3 V2 V1

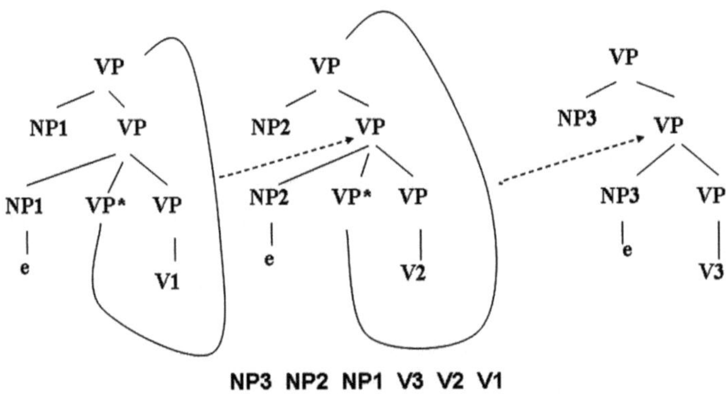

Fig. B.2 Derivation for NP3 NP2 NP1 V3 V2 V1

(1)	...dass	Hans	Peter	Marie	schwimmen	lassen	sah
		Hans	Peter	Marie	swim	let	saw
		NP1	NP2	NP3	V3	V2	V1

This canonical sequence and (2) below can be easily derived as shown in Figs. B.1 and B.2 above. Trees for V1, V2, and V3 are all LTAG trees (all nodes on the spine (except the lowest node) are labeled as VP nodes for convenience).

(2) NP3 NP2 NP1 V3 V2 V1

In Fig. B.1, the V1 tree is adjoined at the root node of the V2 tree and then the derived tree is adjoined to the root node of the V3 tree. In Fig. B.2, the V1 tree is adjoined to the interior VP node of the V2 tree and the resulting tree is then adjoined into the interior VP node of the V3 tree. Finally in Fig. B.3, the V1 tree is adjoined to the interior VP node of the V2 tree and then the resulting tree is adjoined to the root node of the V3 tree.

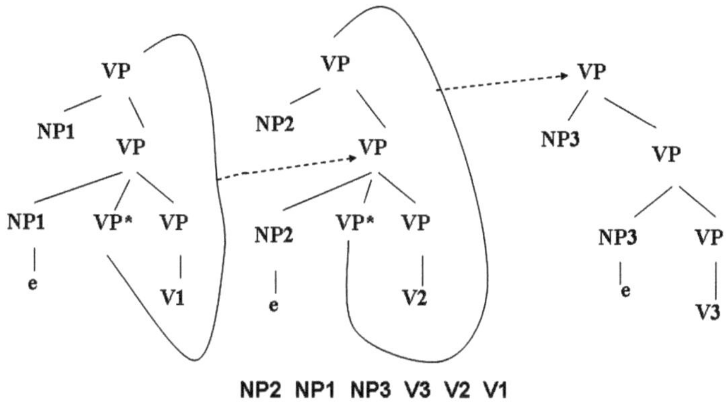

Fig. B.3 Derivation for NP2 NP1 NP3 V3 V2 V1

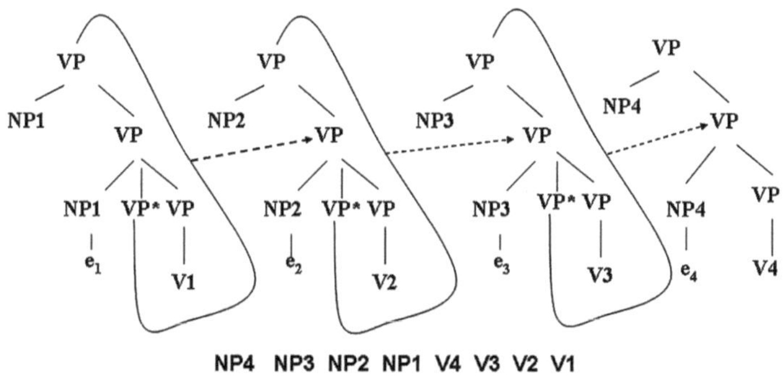

Fig. B.4 Derivation for NP4 NP3 NP2 NP1 V4 V3 V2 V1

Finally, in Figs. B.4 and B.5 we show some derivations for sequences with four NP's and four VP's. Figure B.4 shows a derivation where the V1 tree is adjoined to the interior VP node of the V2 tree and the resulting tree is adjoined to the interior VP node of the V3 tree. Finally, the resulting tree is adjoined to the interior VP node of the V4 tree.

In Fig. B.5 the V1 tree is adjoined to the root node of the V2 tree and the resulting tree is adjoined to the root node of the V3 tree. The V4 tree is a multi-component tree. It has two components. The lower component of the tree has V4 and its subject argument, NP4 which has been scrambled and hence shown with a trace. The upper component has the NP4 argument which is scrambled. There is also a constraint between the two components (not shown explicitly in Fig. B.5). The VP foot node of the upper component must dominate the VP root node of the lower component when the multicomponent is composed with an elementary host tree. The two components of the multi-component tree for V4 must combine with the same elementary tree, i.e., each component must target an elementary tree in order to compose with it, either by substitution or by adjoining. This assures the locality of the composition,

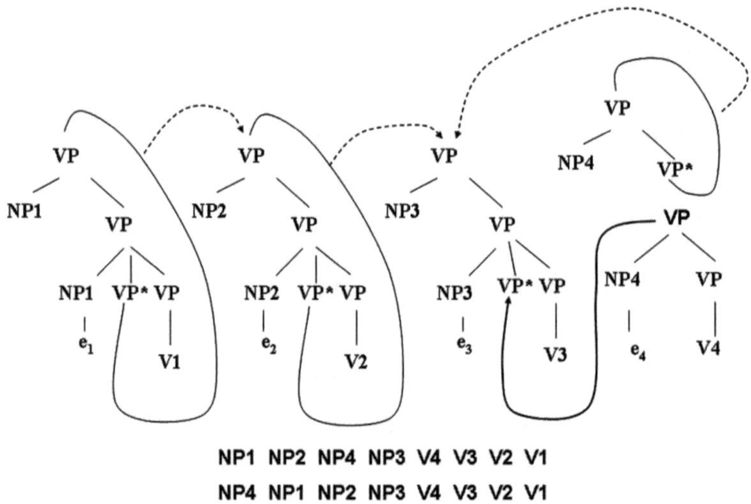

Fig. B.5 Derivations for two sequences NP1 NP2 NP4 NP3 V4 V3 V2 V1 and NP4 NP1 NP2 NP3 V4 V3 V2 V1

as required by the definition of a Tree-Local Multi-Component TAG (TL-MCTAG), which is weakly equivalent to the standard TAG. The lower component of the V4 tree is "attached" (reverse adjoined in a sense[24]) to the VP foot node of the V3 tree (as VP4 is the complement argument of V3) and the upper component of the V4 tree is adjoined to the root node of the V3 tree. Note that we have multiple adjunctions at the root node of the V3 tree.[25] Figure B.5 thus represents two possible sequences as indicated. It can be thought of as an underspecified representation of these two sequences.

Note that in Fig. B.5, the elementary 'tree' for V4 is in two parts (components), the lower and the upper part. The lower part has the trace for NP4 and the upper tree has the scrambled NP4 argument. There is an implied constraint (not shown here) that after these two components of the tree for V4 compose with the tree for V3 at the nodes VP (root node) and the frontier node labeled VP* respectively, these adjoining sites are in a dominating relation. Such constraints are a part of the Tree Local MCTAG composition and further these constraints are specified for each elementary object in a TL-MCTAG.

[24]In MC-TAG there is the possibility of changing the direction of the derivation under certain conditions, similar in a sense to the change of the direction of the derivation in a categorial grammar by the use of type raising. This sort of flexibility of composition has to maintain the semantic coherence of the composition, i.e., for example, a Vi tree can only compose with a tree of the Vi + 1 type. We omit these details here; see also Chiang and Scheffler (2008) for Delayed Composition.

[25]Multiple adjunctions in LTAG are discussed in Schabes and Shieber (1993). These notions can be extended to MC-TAG.

With 4 verbs and 4 NP's there are 24 possible sequences, permuting the NP's and keeping the VP's in a fixed order. It can be shown that 22 of these sequences can be generated by Tree-local MCTAG and two sequences cannot be generated. These two sequences are

(1)	NP3	NP1	NP4	NP2	V4	V3	V2	V1
(2)	NP2	NP4	NP1	NP3	V4	V3	V2	V1

Thus there is a gap of two sequences in the set of the non-canonical transforms of the canonical sentence forms with three levels of recursion. For every depth of recursion beyond three or more embeddings gaps will continue to exist, for example, with four levels of embedding there will be six gaps, etc.

In a TAG or MC-TAG the composition is always binary. However, the elementary trees are not required to be binary. In the figures above the lower components are non-binary. If one imposes the binary form restriction on the elementary trees of TAG and MC-TAG then it can be shown that the first time the gaps appear is when we have four levels of embedding.

References

Bach, E., Brown, C., & Marsen-Wilson, W. (1986). Crossed and nested dependencies in German and Dutch: A psycholinguistic study. *Language and Cognitive Processes, 1*(4), 249–262.

Chiang, D., & Scheffler, T. (2008). Flexible composition and delayed tree-locality. In *Proceedings of the TAG + 9 Workshop*, Tübingen.

Chen-Main, J., & Joshi, A. K. (2008). Flexible composition, multiple adjoining, and word order variation. In *Proceedings of the TAG + 9 Workshop*, Tübingen.

Chomsky, N. (1957). *Syntactic structures*. The Hague: Mouton.

Chomsky, N. (2007). Approaching UG from below. In U. Sauerland & H. M. Gärtner (Eds.), *Approaching UG from below* (Interfaces + Recursion = Language? Chomsky's minimalism and the view from syntax-semantics). Berlin: Mouton de Gruyter.

Frank, R. (2002). *Phrase structure composition and syntactic dependencies*. Cambridge: MIT Press.

Gärtner, H.-M., & Michaelis, J. (2007). Some remarks on locality conditions and minimalist grammars. In U. Sauerland & H. Gartner (Eds.), *Interfaces + Recursion = Language?* (pp. 161–196). Berlin: Mouton de Gruyter.

Harris, Z. S. (1957). Co-occurrence and transformation in linguistic structure. *Language, 33*(3), 283–340.

Hollebrandse, B., & Roeper, T. (2007, March). Recursion and propositional exclusivity. In *Proceedings of the workshop on recursion*, University of Illinois, Normal, IL.

Joshi, A. K. (1985). Tree adjoining grammars: How much context sensitivity is required to provide reasonable structural descriptions? In D. Dowty, L. Karttunen, & A. Zwicky (Eds.), *Natural language parsing* (pp. 206–250). Cambridge: Cambridge University Press.

Joshi, A. K. (1990). Processing crossed and nested dependencies: An automaton perspective on the psycholinguistic results. *Language and Cognitive Processes, 5*, 1–27.

Joshi, A. K. (2004). Starting with complex primitives pays off: Complicate locally, simplify globally. *Cognitive Science, 28*(2004), 637–668.

Joshi, A. K., & Schabes, Y. (1997). Tree adjoining grammars. In G. Rosenberg & A. Salomma (Eds.), *Handbook of formal languages* (pp. 99–123). Berlin: Springer.

Kallmeyer, L., & Joshi, A. K. (2003). Factoring predicate argument and scope semantics: Underspecified semantics with LTAG. *Research on Language and Computation, 2003*(1–2), 3–58.

Kroch, A. S., & Joshi, A. K. (1986). Analyzing extraposition in a tree adjoining grammar. In G. Huck & A. Ojeda (Eds.), *Discontinuous constituents, syntax and semantics*. New York: Academic.

Peters, S., & Ritchie, R. W. (1973). On the generative power of transformational grammars. *Information Sciences, 6*(49–83), 1973.

Schabes, Y., & Shieber, S. (1993). An alternative conception of tree adjoining derivation. *Computational Linguistics, 20–1*, 91–124.

Vijay-Shanker, K. (1988). *A study of tree-adjoining grammars*. Ph.D. dissertation, University of Pennsylvania, Philadelphia, PA, USA.

Recursion in Grammar and Performance

Edward P. Stabler

Abstract The deeper, more recursive structures posited by linguists reflect important insights into similarities among linguistic constituents and operations, which would seem to be lost in computational models that posit simpler, flatter structures. We show how this apparent conflict can be resolved with a substantive theory of how linguistic computations are implemented. We begin with a review of standard notions of recursive depth and some basic ideas about how computations can be implemented. We then articulate a consensus position about linguistic structure, which disentangles what is computed from how it is computed. This leads to a unifying perspective on what it is to represent and manipulate structure, within some large classes of parsing models that compute exactly the consensus structures in such a way that the depth of the linguistic analysis does not correspond to processing depth. From this unifying perspective, we argue that adequate performance models must, unsurprisingly, be more superficial than adequate linguistic models, and that the two perspectives are reconciled by a substantial theory of how linguistic computations are implemented. In a sense that will be made precise, the recursive depth of a structural analysis does not correspond in any simple way to depth of the calculation of that structure in linguistic performance.

Keywords Computational linguistics • Parsing • Recursive depth • Processing depth

In the last 50 years of cognitive science, linguistic theory has proposed more and more articulated structures, while computer science has shown that simpler, flatter structures are more easily processed. If we are interested in adequate models of human linguistic abilities, models that explain the very rapid and accurate human recognition and production of ordinary fluent speech, it seems we need to come

E.P. Stabler (✉)
Department of Linguistics, UCLA, Los Angeles, CA, USA

T. Roeper and M. Speas (eds.), *Recursion: Complexity in Cognition*, Studies in Theoretical Psycholinguistics 43, DOI 10.1007/978-3-319-05086-7_8,

to some appropriate understanding of the relationship between these apparently opposing pressures for more and less structure. Here we show how the apparent conflict disappears when it is considered more carefully. Even when we regard the linguists' project as a psychological one, there is no pressure for linguists to abandon their rather deep structures in order to account for our easy production and recognition of fluent speech. The deeper, more recursive structures reflect insights into similarities among linguistic constituents and operations, but a processor can compute exactly these structures without the extra effort that deeper analyses might seem to require. To show how this works, we review Sect. 1 standard notions of recursive depth and some basic ideas about how computations can be implemented, Sect. 2 a consensus position about linguistic structure, and Sect. 3 some large classes of parsing models that compute exactly the consensus structures in such a way that the depth of the linguistic analysis does not correspond to processing depth. The apparent tension is resolved in this last step, with a proper understanding of how computations can be implemented. The needed perspective on what it is to represent and manipulate structures, to implement a computation, is completely standard. From this unifying perspective, we will argue that adequate performance models must be more superficial than adequate linguistic models, that this is unsurprising, and that the two perspectives are reconciled by a substantial theory of how linguistic computations are implemented. In a sense that will be made precise, the recursive depth of a structural analysis does not correspond in any simple way to depth of the calculation of that structure in linguistic performance.

1 Recursion and Depth

Ordinary fluent speech is perceived as having parts, in temporal sequence. Considering the natural patterns of parts found in these sequences, we find first that they do not have a principled upper bound on their length; in this sense, languages are infinite. So to define the patterns, we define an infinite set rather than imposing some arbitrary length bound. The standard definitions of infinite sets of sequences are recursive, in the sense that the set being defined is used in its own definition. This idea is familiar from the definitions of other infinite sets. For example, we can define the set of natural numbers as the smallest set containing 0 such that, for every natural number, its successor is also a natural number. This definition is sometimes presented with a notation like the following, in which we see that the set $\mathbb{N}$ of numbers that we are defining appears on the right, in the definition:

$$\mathbb{N} := 0 \mid s(\mathbb{N}).$$

This simply says that the set of natural numbers contains 0 and also the successor of every natural number (and nothing more). The successor function is also said to be recursive, since it applies to its own output. In a similar way, the theorems of standard logics are defined as the axioms together with everything derived from the theorems by the rules of inference $\mathcal{I}$,

$$\text{T} := \text{Axioms} \mid \mathcal{I}(\text{T}).$$

Here again the definition of the set of theorems T is recursive, as are the inference rules if they can apply to their own output. Languages can be defined by rewrite rules, but it is often more convenient to use an inductive definition similar to the previous examples:

$$\text{L} := \text{Lex} \mid \mathcal{F}(\text{L}).$$

This says that a language includes a lexicon and everything that can be built from lexical items by some structure building operations $\mathcal{F}$. Many generative grammars can be stated in this form, even many of those that are not explicitly presented this way (Keenan and Stabler 2003).

Given a recursive definition of a set S and a particular thing a, how can we tell whether a is in the set S? Clearly, one way to do this is by proving from the definition that a is included. For example, we can show that s(s(0)) is a number using our definition of $\mathbb{N}$ by observing first that s(s(0)) is a number if s(0) is, and that s(0) is a number if 0 is, and 0 is a number. In a sense, this establishes that s(s(0)) is a number by using the definition three times. We can present this proof as a 'derivation tree', where $\bullet$ is the successor function:

This tree is often said to have depth 3, since that is the number of steps from the root to the leaf. Returning to our question, how can we tell whether something is in a set defined by a recursive definition, the answer is: by finding a 'derivation' of that element, in this sense.

Suppose we have a propositional logic with this set of four axioms,

$$\text{Axioms} := \{p, q, r, (p \to (q \to (r \to s)))\},$$

and with the inference rule *modus ponens*, $\mathcal{I} = \{\text{mp}\}$. Then, with the definition T := Axioms | $\mathcal{I}$(T), we can show that s is a theorem in T with the derivation standardly presented as shown on the left below, which is just another form of the derivation tree shown on the right, where $\bullet$ is mp:

$$\dfrac{\dfrac{\dfrac{p \to (q \to (r \to s)) \quad p}{q \to (r \to s) \quad q}}{r \to s \quad r}}{s}$$

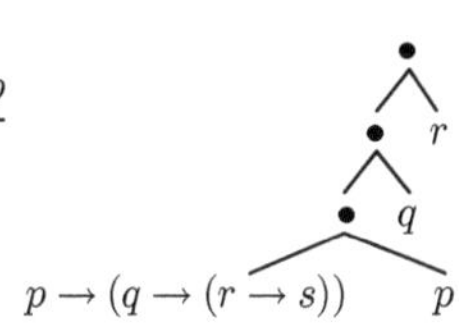

The depth of this derivation tree is 4, since that is the length of its longest path from root to leaf. In effect, on this path, we use the recursive definition four times to get this tree.

Now a first point of interest is that there are other definitions of the same set of theorems T, where many theorems have much shallower derivations. For example, suppose that in addition to mp, we have an inference rule which takes 4 premises instead of 2, and, in effect, applies mp 3 times at once:

$$\frac{\alpha \rightarrow (\beta \rightarrow (\gamma \rightarrow \psi)) \qquad \alpha \qquad \beta \qquad \gamma}{\psi}$$

With this rule, even though the set T is not changed in any way, we can derive s in 1 step instead of 3 (here using $\circ$ to represent our new rule),

$$\frac{p \rightarrow (q \rightarrow (r \rightarrow s)) \qquad p \qquad q \qquad r}{s}$$

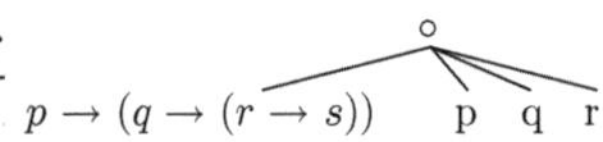

With this proof we have saved just a couple of steps, but obviously for longer proofs, the number of steps we save can increase without bound. That is, in the new system with two inference rules, many results can be established with much less recursive depth. There is a large literature in logic about how to 'speed up' proofs by adding new mechanisms like this (Gödel 1936; Statman 1974; Buss 1994).

Consider the sequence of actions: drink, drive, and go to jail. We might assume that there is no recursion in a sequence like that, but now we see that the question depends on how the sequence is defined. For example, if at each decision point, a similar decision process • is operative, then we could have a recursive depth of 3 as shown on the left below. But if the three actions just follow one after the other, then there is just one step:

(T)

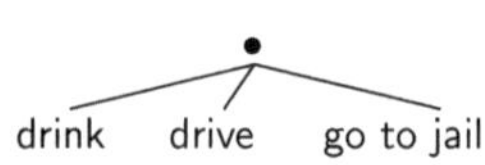

(T')

In the absence of a theory of action or any other application, there may be no reason to prefer one description over the other. But turning now to human language, we argue for the following points: a methodological point M1, and two rather secure empirical hypotheses H1 and H2.

(M1) There can be strong reasons to favor a highly recursive definition like (T) over a less recursive one like (T′), or vice versa.

(H1) There is a significant consensus about the kinds of recursive operations needed to define human languages.

(H2) Performance models need to flatten where linguists don't.

Before explaining and defending these points, two more quick preliminaries are needed: in Sect. 1.1 we define a couple of notions based on the previous discussion, and in Sect. 1.2 we briefly consider a computer science perspective on definitional complexity and implementation.

1.1 Complexity: Size, Depth, Depth$_{\circ}$

Let's say definition of L is recursive if the definition uses L, as in the examples above. And an operation (like the successor function, or *modus ponens*, or a rule of grammar) is recursive if it can apply to its own output. Taking a slightly more complex example, in a formal grammar inspired by recent Chomskian syntax, there is an operation called *merge* • and an operation called move ∘, so a language L is defined as follows (Stabler 1997; Michaelis 2001a; Stabler and Keenan 2003):

$$L := \text{Lex} \mid \mathcal{F}(L), \text{ where } \mathcal{F} = \{\bullet, \circ\}.$$

A grammar of this kind can establish that *who Juliet loves* is a "complementizer clause" (CP) with a derivation like the one shown below right, which is more conventionally shown with the corresponding derived tree on the left:

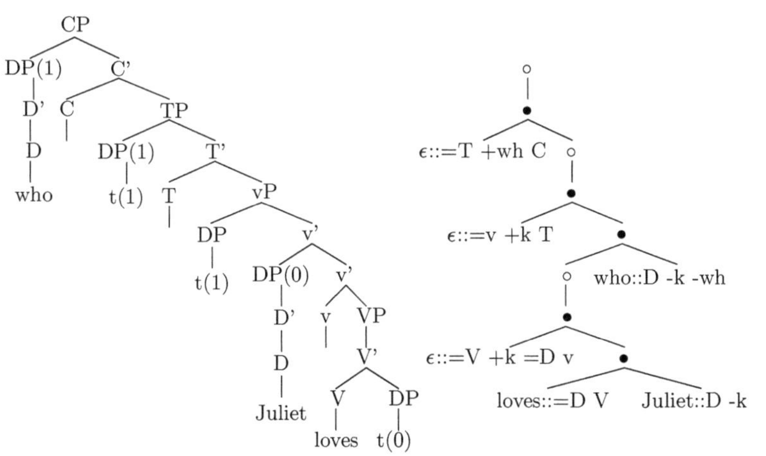

size 31, depth 11, depth$_{CP}$ 1, depth$_{v'}$ 2 size 14, depth 9, depth$_{\bullet}$ 5, depth$_{\circ}$ 3

The features associated with lexical items at the leaves of the derivation tree determine what derivational steps can apply,[1] but the relevant thing here is the recursion in the derivation. As indicated in the captions above, let the *size* of a tree be the number of nodes; let the *depth* of a tree be the length of its longest path, and for each operation f in the tree (or for each node label when the tree is not a derivation tree), let the $depth_f$ be the greatest number of occurrences of f on any path from root to leaf. Let's say that a tree is recursive if for some f, depth_f is greater than 1. Recursion in a derivation tree then indicates that some operation f is applying to its own output (perhaps mediated by other operations).

Linguists and psycholinguists usually give most attention to derived trees, like the one on the left, above, but for most processing problems, that is a mistake! For parsing, for example, it is the derivation that is more fundamental, since the existence of a derivation is what determines whether a structure is assigned at all, and the derivations immediately determine what the derived structures are (while the converse does not generally hold).[2] We will accordingly attend primarily to recursion in the operations, in the derivation tree, rather than to recursion in the associated derived structures.

1.2 Computational Preliminaries

In computer science, a problem is often approached by, first, giving a clear and perspicuous definition of what needs to be computed, and then carefully searching for efficient ways of computing it. These two objectives pull different directions! This happens because, roughly, (T) the simplest definitions often require many steps to derive a result, while (P) the most time-efficient computations are those that take rather few steps.

(T) To specify what is computed, we aim for simplicity.
(P) For efficiency, aim to keep calculations flat.

The first point is very important! Computer scientists have slogans emphasizing the importance of simplicity and that a loop or recursion should always be used whenever some step or sequence of steps needs to be repeated[3]:

> Inside every big problem is a little problem trying to get out.
> Two or more, use a `for`.

---

[1] See Stabler (1997), Harkema (2001b), Michaelis (2001a), and Stabler and Keenan (2003).

[2] A certain arbitrariness in the derived structures can be revealed in a precise way by showing that quite different derived structures can yield isomorphic sets of derivations, with identical leaves – a key idea behind the results established in Theorems 1 and 2, mentioned below.

[3] The first of these is sometimes attributed to Tony Hoare, the latter to Edsger Dijkstra.

If a specification is too complex, it becomes infeasible to assess whether it is right! Linguists emphasize the analogous point too: we would like to find as simple a perspective as possible on as much as possible in the language. For example, Chomsky (1995) notes that there are fundamental similarities between merge $\bullet$ and move $\circ$, and he accordingly frames his theory in terms of one merge operation $\bullet$. Note that if all the operations in the derivation tree above were instances of the same rule $\bullet$, then $\text{depth}_\bullet$ is increased.

However, the second point, (P), is also important! Obviously, computer scientists often want solutions as quickly as possible, and to model natural computations that are fast and easy, we would like a model which makes them fast and easy.

The conflict between (T) and (P) is tamed in computer science by the theory of how an algorithm can be implemented. When we can specify a function easily, in a 'high level language', we can sometimes convert that specification into an efficient implementation with a 'compiler'. Many of the steps in the conversions automated by optimizing compilers serve precisely to decrease recursive depth. Some steps of this kind include: recursion unfolding, loop unrolling, partial evaluation, inlining, macro expansion, deforestation, and many others.[4] There is a substantial body of theory here, but the basic idea is similar to the speed-up already mentioned above. For example, given a specification that requires a calculation of depth_f 2, as shown on the left below, a first conversion might specialize the preterminal steps of the calculation:

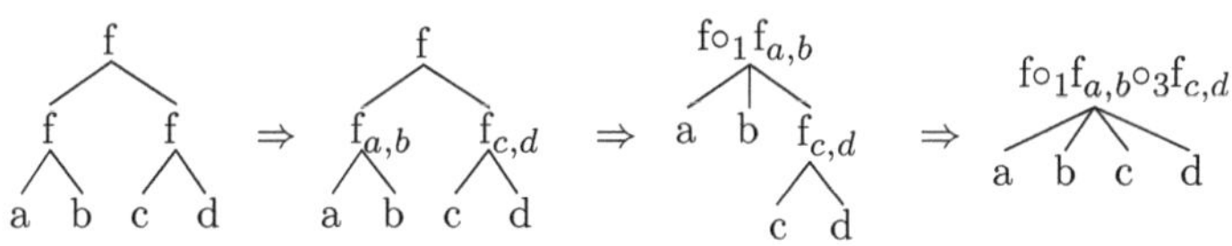

This first conversion of the problem decreases the depth_f from 2 to 1. The overall size and depth of the tree remains the same, but in many cases $f_{a,b}$, $f_{c,d}$ can be more efficient than two applications of f just because the specialized functions need not handle such a wide range of arguments. The second and third conversions shown above 'compose' the functions together, 'unfolding' the computation, as we did earlier when we composed mp with itself several times. These latter steps decrease the size and depth of the calculation overall. Typically,

[4]See Wadler (1990), Appel (1992), Jones et al. (1993), Marlow and Wadler (1992), Seidl and Sørensen (1997), Chitil (1999), Kühnemann (1999), Kennedy and Allen (2001), and Hamilton (2006). In compiling programming languages, loops need to be treated differently from recursion. These two mechanisms are expressively equivalent. Indeed this kind of equivalence was a significant motivation for the Church-Turing thesis. For present purposes, we will not introduce loops, and the difference between loops and recursion on standard machine architectures (or neurally plausible ones) will not be explored. The relevant point here is simply that both loops and recursion can define calculations of various recursive depths, and these depths are often decreased by optimizing compilers.

- Steps like these decrease recursive depth, but increase program size;
- Finding 'best' unfolding for a particular machine with size less than a fixed finite bound is typically intractable (Scheifler 1977); some recent 'profiling compilers' optimize those parts of the code that is most heavily used, a sensible "practice effect" strategy which could have an analog in natural systems;
- The 'best' unfolding also depends, of course, on architecture of the machine running it, on what steps are basic, as is often noted in linguistic contexts,[5] where we are often interested in the possible neural implementations.

In sum, when considering how a computation might be realized, for example a computation that decides whether a sequence has a derivation from some grammar, it is important to remember that this kind of problem can often be solved (i.e. solved exactly and correctly) with definitions that have been 'flattened' by methods like this.[6] This kind of flattening affects processing complexity without any change at all to the mapping computed.

2 Linguistic Structure

Now let's turn to the definition of linguistic structure. The previous discussion suggests that it may be useful to have this goal:

(M0) Disentangle what is computed from how it is computed, its realization.

Recursion matters in different ways for the specification of what is computed and for the performance model, since the recursion in the specification of what is computed often represents the recognition of simplifying regularities, but the optimal or psychologically appropriate computation may be flattened in the ways mentioned above. In fact, as (M0) suggests, we will see: (H2) performance models do in fact need to flatten where linguists don't. First, though, it is important to notice that there is significant agreement about some fundamental properties of the recursive mechanisms that define linguistic structures.

[5]See Smolensky and Legendre (2006), Parberry (1996), and Berwick and Weinberg (1984).

[6]What counts as an 'implementation' of an algorithm, exactly? Some theoretical proposals require a very fine grained correspondence between the steps of the algorithm and the steps of the implementation, while other require less (Moschovakis 2001; Blass et al. 2009). For some extremely simple neural mechanisms in sensory and motor systems, we can begin to see how relevant computations could be implemented, sometimes all the way down to the molecular level (Kempter et al. 2001; Franosch et al. 2005; Friedel 2008; van Hemmen and Schwartz 2008). But it is not yet clear what to expect from an ideally completed psychological/neurophysiological theory of human language abilities and their realization. As Berwick and Weinberg have pointed out in their excellent discussion of the situation, we of course aim to characterize the relation as simply and straightforwardly as possible (Berwick and Weinberg 1984, §2). But the implementation of programming languages is a rather large and complex story, and one expects the full story about the implementation of linguistic abilities to be even more so.

Although some of the early grammars and processing models were extremely powerful,[7] there were challenges to the idea that such powerful models were necessary (Pullum and Gazdar 1982), and in the late 1980s a surprising consensus began to emerge. It turned out that many independently proposed grammar formalisms were 'equivalent' in the sense that they could define exactly the same languages, including some non-context free languages, but remaining efficiently parsable. The proofs of these equivalences were generally rather easy, revealing some fundamental similarities among the recursive mechanisms. It later turned out that formal grammars inspired by Chomskian syntax, mentioned above, were also weakly equivalent to other independently proposed and well known formalisms. This yielded a surprising convergence, established in a number of papers, which can be summarized like this, writing 'CFG' for 'languages defined by context free grammars', 'TAG' for 'languages defined by tree adjoining grammars', 'CCG' for 'languages defined by combinatory categorial grammars', 'MG' for the 'languages defined by minimalist grammars', 'MCTAG' for 'languages defined by set-local multi-component tree adjoining grammars', 'MCFG' for 'languages defined by multiple context free grammars', 'CS' for 'languages defined by context sensitive grammars', 'RE' for 'recursively enumerable languages', 'Aspects' for 'languages defined by the transformational grammar of Peters and Ritchie (1973)', and 'HPSG' for 'languages defined by head-driven phrase structure grammar'[8]:

Theorem 1. $\text{CFG} \subset \boxed{\text{TAG}=\text{CCG} \subset \text{MG}=\text{MCTAG}=\text{MCFG}} \subset \text{CS} \subset \text{RE}=\text{Aspects}=\text{HPSG}$.

The boxed language classes are, to use the term of Joshi (1985), "mildly context sensitive," and Joshi proposes that human languages are in one of these. Following the results of Theorem 1, it was also discovered that many variants of minimalist grammars are equivalent: adding head movement (MGH), adding directional selection with 'head parameters' (DMG), adding asymmetrical feature checking (CMG), adding phases (PMG), adding sidewards movement (SMMG)[9]:

Theorem 2. MG=MGH=DMG=CMG=PMG=RMG=SMMG.

All these grammars define the same class of mildly context sensitive languages. This consensus is extended to some optimality systems by recent work based on earlier studies of OT phonology.[10] This is the consensus (H1) mentioned earlier:

---

[7] E.g. Chomsky's 'Aspects' theory (Peters and Ritchie 1973), and ATN processing models (Woods 1970).

[8] Theorem 1 is established in a series of rather many results, including Vijay-Shanker et al. (1987), Michaelis (1998, 2001b), Harkema (2001a), Seki et al. (1991), Peters and Ritchie (1973), Berwick (1981), Johnson (1988), Torenvliet and Trautwein (1995), Trautwein (1995), and Jaeger et al. (2005).

[9] Michaelis (2001b, 2002), Kobele (2002), and Stabler (2001, 2003, 2011b,a, 1999, 1997, 2006, 2004, 2006).

[10] Kepser and Mönnich (2006), Kanazawa and Salvati (2007), Kanazawa (2009), Graf (2010), Frank and Satta (1998), and Jäger (2002).

(H1) There is a significant consensus about the kinds of recursive operations needed to define human languages: they are 'mildly context sensitive' (Joshi 1985).

This claim still faces some challenges, but the challenges are rather minor in the sense that, even if they are successful, they require only a slight weakening of the claim, staying in the polynomially parsable class, properly included in the context sensitive languages.[11]

One respect in which minimalist grammars (MGs) are stronger than some similar, earlier proposals is that they allow a phrase to move after something has been extracted from it. This kind of 'remnant movement' has been proposed now for quite a wide range of constructions[12]:

[$_{VP}$ t_1 Gelesen]$_2$ hat [das Buch]$_1$ keiner t_2
read has the book noone
[$_{AP}$How likely [t_1 to win]]$_2$ is John$_1$ t_2?
[$_{VP}$Criticized by his boss t_1]$_2$ John$_1$ has never been t_2.
John [$_{VP}$reads t_1]$_2$ [no novels]$_1$ t_2.

Most earlier "slash-passing" and "store a filler, wait for a gap" parsing models cannot handle remnant movement, but standard parsing methods (CKY, Earley, ...) have been shown sound, complete, efficient for the whole range of MGs (Harkema 2001b), handling remnant movement and everything else that can be defined with this grammar formalism. Without remnant movement, the move and merge operations of MGs are strictly weaker, defining only context free languages (Kobele 2010).

3 Performance and Variation

With Theorems 1, 2 and ongoing work that is still enlarging the scope of those claims, the 'linguistic wars' may not be over, but we can now see that, behind the endless skirmishes, a significant consensus lurks in the background. In many cases,

[11]The remaining controversy concerns, for example, the status of copying-like operations in grammar, which threatens the "linear growth" property (Michaelis and Kracht 1997; Bhatt and Joshi 2004; Kobele 2006), and the appropriate treatment of dependencies in scrambling (Rambow 1994; Becker et al. 1992; Perekrestenko 2008).

[12]Cf. Webelhuth and den Besten (1987), Müller (1998), Kayne (1998), Koopman and Szabolcsi (2000), Lee (2000), Baltin (2002), and Abels (2007).

we have automated recipes for converting grammars of one type into equivalent grammars of other types, so significant proposals expressed in one framework can often be adapted fairly easily to others. With this mind, let's consider how minimalist grammars (MGs) could be incorporated into a performance model. It is sometimes held that this is impossible, that MGs are unsuitable for performance models.[13] But since MGs participate in the broadening consensus, this becomes implausible: the ways of accounting for influences on human online syntactic analysis in one of these formalisms can be adapted to the others.

When an utterance is structurally ambiguous, human subjects will often notice one structure before, or instead of, the others. Here we coarsely classify some of the proposals about "first pass" parsing preferences, in order to point out briefly how all of these performance models must draw finer distinctions among derivational steps than the linguistic theory does. Consequently, even when the model computes exactly the structures defined by the grammar, the depth of depth of recursion for the computation is lower than the depth of recursion for the grammatical derivation.

Categorial frequencies. Let's say that a proposal about parsing preferences is *categorial* if it says one or another (partial) analysis is preferred because of the categories of the expressions involved. It is rather difficult to experimentally distinguish purely categorial influences from influences that are based on the particular lexical identities of elements of the same category, as discussed for example in Gibson and Pearlmutter (1998) and Jurafsky (2003), but many models of human recognition complexity and ambiguity resolution encode category-based preferences. This includes, for example, models based on probabilistic context free grammars (Jurafsky 1996; Hale 2006; Levy 2008; Levy et al. 2009).

The merge steps of a minimalist derivation can be specialized for each category by marking each step with the features that trigger its application. This is achieved by mapping an MG with a particular lexicon into a strongly equivalent multiple context free grammar, as is done, in effect, by the conversion of an MG to a strongly equivalent MCFG (Michaelis 1998).

---

[13]For example, Ferreira says in 2005,

> . . . many psycholinguists are disenchanted with generative grammar. One reason is that the Minimalist Program is difficult to adapt to processing models. Another is that generative theories appear to rest on a weak empirical foundation. . . [I]t is now clear that no one interested in human performance can ignore the possible effects of things such as frequency and exposure on ease of processing. (Ferreira 2005, pp. 365, 370)

Old questions about the relation between grammar and performance (Bever 1970; Fodor et al. 1974) are still alive today, prompted in particular by instances where simple grammatical sequences are not recognized as acceptable, and instances were ungrammatical sequences seem to be acceptable (Ferreira 2005; Phillips and Wagers 2007; Phillips et al. 2011). Here we aim only to defend the very general methodological and structural points already indicated: (M0,M1,H1,H2).

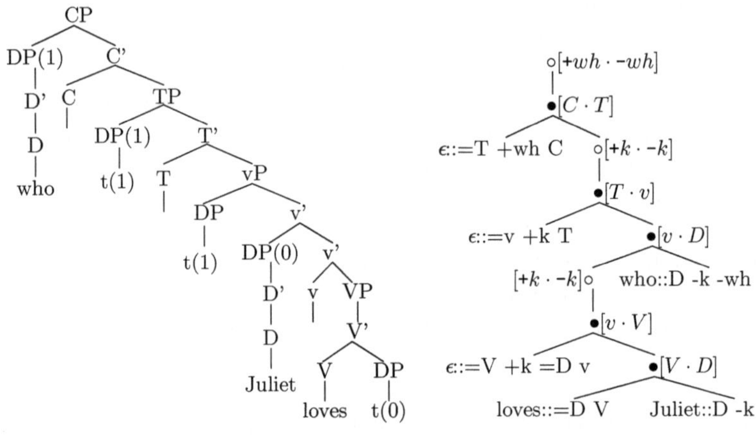

$\text{depth}_{\bullet[x,y]}$ 1, $\text{depth}_{\circ[x,y]}$ 1

The steps, now classified by the categories they operate on, can be assigned different probabilities in a manner exactly analogous to the assignment of probabilities to the rules of a context free grammar. This kind of step was taken by Hale (2006, 2003), for example, in a model of processing complexity.

This step, while it increases grammar size, can decrease processing time very significantly exactly because each step is more specialized. The following graph compares the processing times required to compute the same derived structures with a "naive" implementation compared to one where each rule is specified for particular categories:

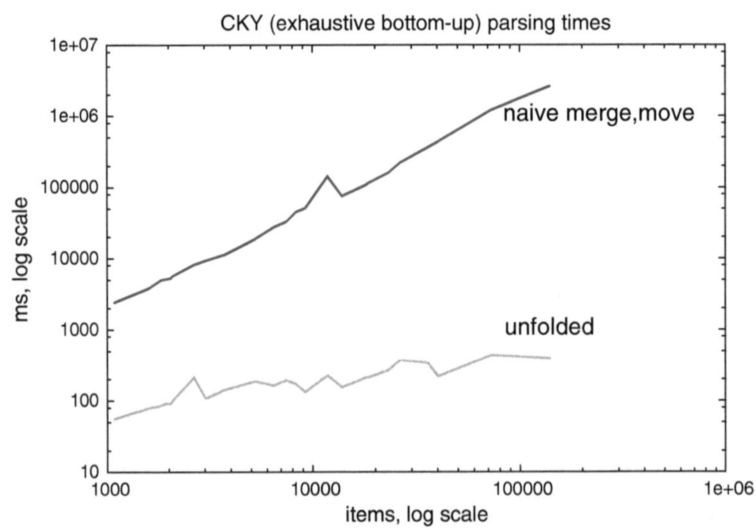

This graph plots processing time versus the number of "items" computed by a CKY parser, just to illustrate how significant the effect can be on a standard computing device. Notice that with the operations split into separate categories, the depth_f for these new operations f will tend to be significantly less in each analysis than the original $\text{depth}_\bullet$ and $\text{depth}_\circ$, but the output of the analysis is not changed in any way.

Lexical frequencies. More common in psychological models is a much finer tuning: not by category, but by particular lexical items.[14] It is familiar at least since Zipf (1935) that some lexical items combine much more often than others in the same category. One simple computational model of this kind of preference is a 'bilexical' grammar which, in effect, treats each category as including the particular lexical item which is its 'head' in some sense of that term (Magerman 1994; Nederhof and Satta 2000; Eisner 2000). When this refinement is applied to MGs, in most discourses it becomes rare to see depth_f greater than 1 for any (categorially and lexically specific) operation f.

Left context and discourse effects. Even more specific operations are obtained when we consider not just the pair of heads being combined, but the whole linguistic left context that has already been seen, together with the discourse situation. It is perhaps still controversial how much of this material really conditions first pass parsing preferences, but the evidence shows that it is applied quite quickly (Crain and Steedman 1985; Altmann et al. 1998; Chambers et al. 2004). In parsing models where the full left context is available, it has been found useful in prediction (Roark 2004, 2001) – an unsurprising result!

Modular perspectives. The previous models were presented as involving refinements of the rules of grammar, so that more specialized rules can apply in more specialized contexts (to achieve exactly the same analysis). But these models need not be described (or implemented) that way. The reason for replacing the original operations $\bullet$, $\circ$ of the grammar with a large set much more specific $\bullet_C$, $\circ_C$ operations that apply only in particular contexts C is to allow for different probabilistic assessments of derivations in those different contexts. But obviously, this can be done without specializing the operations: we can calculate the probability of the derivations using the original operations in a model where those probabilities are conditioned by context. For example, as we take each step (with one of the original operations), we can obtain the probability of the extended partial derivation by multiplying the previous probability times the probability of this step, *in context.* This provides a ranking of the partial derivations at each step, which could be used for example in "beam parsing" strategies that prune away the extremely improbable analyses incrementally (Jurafsky 1996; Roark 2004, 2001), or in one-parse-at-a-time strategies that pursue the most probable or most expedient option at each point (Hale 2011). In these models, the "finest" classification of derivations comes not from the derivation steps, which are very general, but from the ranking steps, rank_C which will assign each derivation a probability or weight in context.

[14]Ford et al. (1982), Trueswell (1996), and MacDonald et al. (1994).

Note that instead of having a large family of ranking functions $rank_C$ each specialized for some context C, we could treat the context C as an argument to a very general ranking function. In fact, this is the usual perspective: the ranking function is given some representation of context C and of the derivation D. So could depict a derivation tree with very general operations $\bullet$, $\circ$, *rank* but then the leaves of rank would include not just linguistic material but also contexts, changing the nature of the comparison between this evaluation tree and the grammatical derivation trees. In any case, the effect of the context here is clearly to classify derivations very finely, along dimensions that matter to performance but not to the syntactic structural options. It is uncontroversial that this must be done, somehow, to model context effects on human sentence processing. The review of options here emphasizes the fact that such effects can be obtained with standard grammars, either by refining the operations of those grammars or by adding separate ranking steps. And of course all these steps could all be collapsed by "optimizing" or "practice" effects in other ways, analogous to methods used in compiling programming languages, without any change in what is computed.

4 Summary: How Depth of Recursion Matters

The study of how linguistic structure is recognized takes place in the setting of significant agreement about what that structure is:

(H1) Human languages are mildly context sensitive.

When we consider any non-trivial computation, it is not hard to understand the importance of these basic methodological precepts:

(M0) Disentangle what is computed from how it is computed, its realization.
(M1) Even in the study of a single domain, there can be strong reasons to favor a highly recursive definition over a less recursive one, and vice versa.

Considering the study of human language recognition and production in particular,

(H2) Performance models need to flatten where linguists don't.

No reasonable linguist has argued that there are human languages with grammars that are non-recursive in the senses defined here. And there is very little controversy (mentioned in footnote 11) about whether the operations defining human languages are ones that define "mildly context sensitive" languages. The controversy raised by Everett (2005) and others concerns (interesting but) more superficial questions about derived structure: e.g. are there languages in which we never find clauses containing clauses?

(H2) is true simply because the recursion in the theory of language allows us to state regularities that are independent of many influences on the performance of linguistic tasks. It is hard to see how reasonable progress could be made without separating these issues, just as it is hard to see how to implement a complex high level program on a manufactured computer without carefully separating and detailing the implementation. Adding non-structural factors preserves structure but hides it, reducing the depth of recursion and overwhelming us with detail unless it is approached systematically.

References

Abels, K. (2007). Towards a restrictive theory of (remnant) movement: Improper movement, remnant movement, and a linear asymmetry. *Linguistic Variation Yearbook 2007, 7*, 53–120.

Altmann, G. T. M., van Nice, K. Y., Garnham, A., & Henstra, J.-A. (1998). Late closure in context. *Journal of Memory and Language, 38*(4), 459–484.

Appel, A. W. (1992). *Unrolling recursions saves space* (Tech. Rep. tr-363-92). Department of Computer Science, Princeton University.

Baltin, M. R. (2002). Movement to the higher V is remnant movement. *Linguistic Inquiry, 33*(4), 653–659.

Becker, T., Rambow, O., & Niv, M. (1992). *The derivational generative power of formal systems, or, scrambling is beyond LCFRS* (IRCS Tech. Rep. 92–38). University of Pennsylvania.

Berwick, R. C. (1981). Computational complexity of lexical functional grammar. In *Proceedings of the 19th annual meeting of the Association for Computational Linguistics, ACL'81*, Stanford (pp. 7–12).

Berwick, R. C., & Weinberg, A. S. (1984). *The grammatical basis of linguistic performance: Language use and acquisition*. Cambridge: MIT.

Bever, T. G. (1970). The cognitive basis for linguistic structures. In J. R. Hayes (Ed.), *Cognition and the development of language*. New York: Wiley.

Bhatt, R., & Joshi, A. (2004). Semilinearity is a syntactic invariant: A reply to Michaelis and Kracht. *Linguistic Inquiry, 35*, 683–692.

Blass, A., Dershowitz, N., & Gurevich, Y. (2009). When are two algorithms the same? *Bulletin of Symbolic Logic, 15*(2), 145–168.

Buss, S. R. (1994). On Gödel's theorems on lengths of proofs I: Number of lines and speedup for arithmetics. *Journal of Symbolic Logic, 59*(2), 737–756.

Chambers, C. G., Tanenhaus, M. K., Eberhard, K. M., Filip, H., & Carlson, G. N. (2004). Actions and affordances in syntactic ambiguity resolution. *Journal of Experimental Psychology: Learning, Memory and Cognition, 30*(3), 687–696.

Chitil, O. (1999). Type inference builds a short cut to deforestation. In *International conference on functional programming*, Paris (pp. 249–260).

Chomsky, N. (1995). *The minimalist program*. Cambridge: MIT.

Crain, S., & Steedman, M. (1985). On not being led up the garden path. In D. R. Dowty, L. Karttunen, & A. Zwicky (Eds.), *Natural language parsing*. New York: Cambridge University Press.

Eisner, J. (2000). Bilexical grammars and their cubic-time parsing algorithms. In H. Bunt & A. Nijholt (Eds.), *Advances in probabilistic and other parsing technologies* (pp. 29–62). Dordrecht/Boston: Kluwer.

Everett, D. L. (2005). Cultural constraints on grammar and cognition in Piraha: Another look at the design features of human language. *Current Anthropology, 46*, 621–646.

Ferreira, F. (2005). Psycholinguistics, formal grammars, and cognitive science. *Linguistic Review, 22*, 365–380.

Fodor, J. A., Bever, T. G., & Garrett, M. F. (1974). *The psychology of language: An introduction to psycholinguistics and generative grammar*. New York: McGraw-Hill.

Ford, M., Bresnan, J., & Kaplan, R. M. (1982). A competence-based theory of syntactic closure. In J. Bresnan (Ed.), *The mental representation of grammatical relations*. Cambridge: MIT.

Frank, R., & Satta, G. (1998). Optimality theory and the generative complexity of constraint violability. *Computational Linguistics, 24*, 307–315.

Franosch, J.-M. P., Lingenheil, M., & van Hemmen, J. L. (2005). How a frog can learn what is where in the dark. *Physical Review Letters, 95*(7), 078106.

Friedel, P. (2008). *Sensory information processing: Detection, feature extraction, and multimodal integration*. Ph.D. thesis, Technical University of Munich.

Gibson, E. & Pearlmutter, N. J. (1998). Constraints on sentence processing. *Trends in Cognitive Science, 2*, 262–268.

Gödel, K. (1936). Uber die Länge von Beweisen. *Ergebnisse eines mathematischen kolloquiums* (pp. 23–24) [On the length of proofs. In *Kurt Gödel: Collected works,* (Vol. 1, pp. 396–399), 1986. New York: Oxford University Press].

Graf, T. (2010). Reference-set constraints as linear tree transductions via controlled optimality systems. In P. de Groote & M.-J. Nederhof (Eds.), *Formal Grammar 2010/2011* (Lecture notes in computer science 7395). Heidelberg: Springer.

Hale, J. (2003). *Grammar, uncertainty, and sentence processing*. Ph.D. thesis, Johns Hopkins University.

Hale, J. T. (2006). Uncertainty about the rest of the sentence. *Cognitive Science, 30*(1), 609–642.

Hale, J. T. (2011). What a rational parser would do. *Cognitive Science, 35*(3), 399–443.

Hamilton, G. W. (2006). Higher order deforestation. *Fundamenta Informaticae, 69*(1–2), 39–61.

Harkema, H. (2001a). A characterization of minimalist languages. In *Logical aspects of computational linguistics* (Lecture notes in artificial intelligence, Vol. 2099, pp. 193–211). New York: Springer.

Harkema, H. (2001b). *Parsing minimalist languages*. Ph.D. thesis, University of California, Los Angeles.

Jaeger, E., Francez, N., & Wintner, S. (2005). Guaranteeing parsing termination of unification grammars. *Journal of Logic, Language and Information, 14*(2), 199–234.

Jäger, G. (2002). Gradient constraints in finite state OT: The unidirectional and the bidirectional case. In J. Greenberg (Ed.), *More than words: A Festschrift for Dieter Wunderlich* (pp. 299–325). Berlin: Akademie Verlag.

Johnson, M. (1988). *Attribute value logic and the theory of grammar* (CSLI lecture notes, Vol. 16). Chicago: CSLI Publications.

Jones, N. D., Gomard, C. K., & Sestoft, P. (1993). *Partial evaluation and automatic program generation*. Englewood Cliffs: Prentice-Hall.

Joshi, A. K. (1985). How much context-sensitivity is necessary for characterizing structural descriptions. In D. Dowty, L. Karttunen, & A. Zwicky (Eds.), *Natural language processing: Theoretical, computational and psychological perspectives* (pp. 206–250). New York: Cambridge University Press.

Jurafsky, D. (1996). A probabilistic model of lexical and syntactic access and disambiguation. *Cognitive Science, 20*, 137–194.

Jurafsky, D. (2003). Probabilistic modeling in psycholinguistics: Comprehension and production. In R. Bod, J. Hay, & S. Jannedy (Eds.), *Probabilistic linguistics* (pp. 39–96). Cambridge: MIT.

Kanazawa, M. (2009). A pumping lemma for well-nested multiple context free grammars. In *13th international conference on developments in language theory, DLT 2009*, Stuttgart.

Kanazawa, M., & Salvati, S. (2007). Generating control languages with abstract categorial grammars. In *Proceedings of the 12th conference on formal grammar (FG'07)*, Dublin. Stanford: CLSI Publications.

Kayne, R. S. (1998). Overt vs. covert movment. *Syntax, 1*(2), 128–191.

Keenan, E. L., & Stabler, E. P. (2003). *Bare grammar: Lectures on linguistic invariants*. Stanford: CSLI Publications.

Kempter, R., Leibold, C., Wagner, H., & van Hemmen, J. L. (2001). Formation of temporal-feature maps by axonal propagation of synaptic learning. *Proceedings of the National Academy of Sciences of the United States of America, 98*(7), 4166–4171.

Kennedy, K., & Allen, R. (2001). *Optimizing compilers for modern architectures: A dependence-based approach*. San Mateo: Morgan Kaufmann.
Kepser, S., & Mönnich, U. (2006). Properties of linear context free tree languages with an application to optimality theory. *Theoretical Computer Science, 354*, 82–97.
Kobele, G. M. (2002). Formalizing mirror theory. *Grammars, 5*, 177–221.
Kobele, G. M. (2006). *Generating copies: An investigation into structural identity in language and grammar*. Ph.D. thesis, UCLA.
Kobele, G. M. (2010). Without remnant movement, MGs are context-free. In *MOL 10/11* (Lecture notes in computer science, Vol. 6149, pp. 160–173). Berlin/Heidelberg: Springer.
Koopman, H., & Szabolcsi, A. (2000). *Verbal complexes*. Cambridge: MIT.
Kühnemann, A. (1999). Comparison of deforestation techniques for functional programs and for tree transducers. In *Fuji international symposium on functional and logic programming*, Tsukuba (pp. 114–130).
Lee, F. (2000). VP remnant movement and VSO in Quiaviní Zapotec. In A. Carnie & E. Guilfoyle (Eds.), *The syntax of verb initial languages*. Oxford: Oxford University Press.
Levy, R. (2008). Expectation-based syntactic comprehension. *Cognition, 106*, 1126–1177.
Levy, R., Reali, F., & Griffiths, T. (2009). Modeling the effects of memory on human online sentence processing with particle filters. In *Proceedings of the twenty-second annual conference on neural information processing systems*. Vancouver: Canada.
MacDonald, M., Pearlmutter, N. J., & Seidenberg, M. S. (1994). Syntactic ambiguity as lexical ambiguity resolution. In C. Clifton, L. Frazier, & K. Rayner (Eds.), *Perspectives on sentence processing* (pp. 155–180). Hillsdale: Lawrence Erlbaum.
Magerman, D. M. (1994). *Natural language parsing as statistical pattern recognition*. Ph.D. thesis, Stanford University.
Marlow, S., & Wadler, P. (1992). Deforestation for higher-order functions. *In Proceedings of the fifth annual Glasgow workshop on functional programming. Glaskow, UK* (pp. 154–165). London: Springer.
Michaelis, J. (1998). Derivational minimalism is mildly context-sensitive. In *Proceedings, logical aspects of computational linguistics, LACL'98*, Grenoble (pp. 179–198). New York: Springer.
Michaelis, J. (2001a). *On formal properties of minimalist grammars*. Ph.D. thesis, Universität Potsdam. *Linguistics in Potsdam 13*, Universitätsbibliothek, Potsdam.
Michaelis, J. (2001b). Transforming linear context free rewriting systems into minimalist grammars. In *Logical aspects of computational linguistics* (Lecture notes in artificial intelligence, Vol. 2099, pp. 228–244). New York: Springer.
Michaelis, J. (2002). Notes on the complexity of complex heads in a minimalist grammar. In *Proceedings of the 6th international workshop on tree adjoining grammars and related frameworks, TAG+6*, Venice (pp. 57–65).
Michaelis, J., & Kracht, M. (1997). Semilinearity as a syntactic invariant. In *Logical aspects of computational linguistics* (Lecture notes in computer science, Vol. 1328, pp. 37–40). New York: Springer.
Moschovakis, Y. N. (2001). What is an algorithm? In B. Engquist & W. Schmid (Eds.), *Mathematics unlimited – 2001 and beyond* (pp. 919–936). New York: Springer.
Müller, G. (1998). *Incomplete category fronting*. Boston: Kluwer.
Nederhof, M.-J., & Satta, G. (2000). Left-to-right parsing and bilexical context-free grammars. In *Proceedings of ANLP-NAACL 2000*, Seattle.
Parberry, I. (1996). Circuit complexity and feedforward neural networks. In P. Smolensky, M. C. Mozer, & D. Rumelhart (Eds.), *Mathematical perspectives on neural networks*. Mahwah: Erlbaum.
Perekrestenko, A. (2008). Minimalist grammars with unbounded scrambling and nondiscriminating barriers are NP-hard. In *Proceedings of the 2nd international conference on language and automata theory and applications, LATA2008* (Lecture notes in computer science, Vol. 5196, pp. 421–432). Berlin: Springer.
Peters, P. S., & Ritchie, R. W. (1973). On the generative power of transformational grammar. *Information Sciences, 6*, 49–83.

Phillips, C., & Wagers, M. (2007). Relating time and structure in linguistics and psycholinguistics. In G. Gaskell (Ed.), *Oxford handbook of psycholinguistics* (pp. 739–756). Oxford: Oxford University Press.

Phillips, C., Wagers, M., & Lau, E. (2011). Grammatical illusions and selective fallibility in real-time language comprehension. In J. Runner (Ed.), *Syntax and semantics, Volume 37: Experiments at the interfaces* (pp. 153–186). Bingley: Emerald.

Pullum, G. K., & Gazdar, G. (1982). Natural languages and context free languages. *Linguistics and Philosophy, 4*, 471–504.

Rambow, O. (1994). *Formal and computational aspects of natural language syntax*. Ph.D. thesis, University of Pennsylvania. Computer and Information Science Technical report MS-CIS-94-52 (LINC LAB 278).

Roark, B. (2001). Probabilistic top-down parsing and language modeling. *Computational Linguistics, 27*(2), 249–276.

Roark, B. (2004). Robust garden path parsing. *Natural Language Engineering, 10*(1), 1–24.

Scheifler, R. W. (1977). An analysis of inline substitution for a structured programming language. *Communications of the Association for Computing Machinery, 20*, 647–654.

Seidl, H., & Sørensen, M. H. (1997). Constraints to stop higher-order deforestation. In *ACM symposium on principles of programming languages* (pp. 400–413). New York: ACM.

Seki, H., Matsumura, T., Fujii, M., & Kasami, T. (1991). On multiple context-free grammars. *Theoretical Computer Science, 88*, 191–229.

Smolensky, P., & Legendre, G. (2006). *The harmonic mind: From neural computation to optimality-theoretic grammar, Volume I: Cognitive architecture*. Cambridge: MIT.

Stabler, E. P. (1997). Derivational minimalism. In C. Retoré (Ed.), *Logical aspects of computational linguistics* (Lecture notes in computer science, Vol. 1328, pp. 68–95). New York, Springer.

Stabler, E. P. (1999). Remnant movement and complexity. In G. Bouma, E. Hinrichs, G.-J. Kruijff, & D. Oehrle (Eds.), *Constraints and resources in natural language syntax and semantics* (pp. 299–326). Stanford: CSLI Publications.

Stabler, E. P. (2001). Recognizing head movement. In P. de Groote, G. Morrill, & C. Retoré (Eds.), *Logical aspects of computational linguistics* (Lecture notes in artificial intelligence, Vol. 2099, pp. 254–260). New York: Springer.

Stabler, E. P. (2003). Comparing 3 perspectives on head movement. In A. Mahajan (Ed.), *From head movement and syntactic theory, UCLA/Potsdam working papers in linguistics*, UCLA (pp. 178–198). Los Angeles: California.

Stabler, E. P. (2004). Varieties of crossing dependencies: Structure dependence and mild context sensitivity. *Cognitive Science, 93*(5), 699–720.

Stabler, E. P. (2006). Sidewards without copying. In *Formal grammar'06, proceedings of the conference* (pp. 133–146). Stanford: CSLI Publications.

Stabler, E. P. (2011a). After GB theory. In J. van Benthem & A. ter Meulen (Eds.), *Handbook of logic and language* (2nd ed., pp. 395–414). Amsterdam: Elsevier.

Stabler, E. P. (2011b). Computational perspectives on minimalism. In C. Boeckx (Ed.), *Oxford handbook of linguistic minimalism* (pp. 617–641). Oxford: Oxford University Press.

Stabler, E. P. & Keenan, E. L. (2003). Structural similarity. *Theoretical Computer Science, 293*, 345–363.

Statman, R. (1974). *Structural complexity of proofs*. Ph.D. thesis, Stanford University.

Torenvliet, L., & Trautwein, M. (1995). A note on the complexity of restricted attribute-value grammars. In *Proceedings of computational linguistics in the Netherlands, CLIN5* (pp. 145–164) Department of Computer Science, University of Twente, Twente.

Trautwein, M. (1995). The complexity of structure-sharing in unification-based grammars. In *Proceedings of computational linguistics in the Netherlands, CLIN5* (pp. 165–180) Department of Computer Science, University of Twente, Twente.

Trueswell, J. (1996). The role of lexical frequency in syntactic ambiguity resolution. *Journal of Memory and Language, 35*, 566–585.

van Hemmen, J. L., & Schwartz, A. B. (2008). Population vector code: A geometric universal as actuator. *Biological Cybernetics, 98*(6), 509–518.
Vijay-Shanker, K., Weir, D., & Joshi, A. K. (1987). Characterizing structural descriptions produced by various grammatical formalisms. In *Proceedings of the 25th annual meeting of the Association for Computational Linguistics*, Stanford (pp. 104–111).
Wadler, P. (1990). Deforestation: Transforming programs to eliminate trees. *Theoretical Computer Science, 73*(2), 231–248.
Webelhuth, G., & den Besten, H. (1987). Adjunction and remnant topicalization in the Germanic SOV-languages. Paper presented at the GLOW conference, Venice.
Woods, W. A. (1970). Transition network grammars for natural language analysis. *Communications of the Association for Computing Machinery, 13*(10), 591–606.
Zipf, G. K. (1935). *The psychobiology of language: An introduction to dynamic philology*. Boston: Houghton-Mifflin.

Empirical Results and Formal Approaches to Recursion in Acquisition

Bart Hollebrandse and Tom Roeper

Abstract We argue that the move from Direct recursion with conjunctive interpretation to Indirect recursion, where the Strong Minimalist Thesis requires that, at Phase boundaries, a semantic interpretation is necessary, provides the blueprint for the acquisition path. We provide an overview of experimental results on the early acquisition (3–4 years) of recursion for PP's ("on the shelf in the jar"). Adjectives ("big little tractor") simple compounds ("christmas tree cookie"), and later acquisition (5–7 years) for sentences ("I think you said they gonna be warm") and verbal compounds (tea-pourer-maker) where language particular factors play a role. Various other factors, branching direction, Relativized MInimality, and morphology must be integrated by the child into the grammatical mechanism. We argue that they cause PP, Adjective, and simple Compounds to be acquired early (3–4 years) and Sentential and Verbal compounds to be acquired late. The fact that subtle steps in acquisition can be captured by very abstract syntactic principles should be seen as a strong source of support for linguistic theory, and an important basis for applied work.

Keywords Direct recursion • Indirect recursion • Acquisition • Compounds

---

B. Hollebrandse (✉)
Dutch Department, University of Groningen, Groningen, The Netherlands
e-mail: b.hollebrandse@rug.nl

T. Roeper
Department of Linguistics, University of Massachusetts, Amherst, MA, USA
e-mail: roeper@linguist.umass.edu

T. Roeper and M. Speas (eds.), *Recursion: Complexity in Cognition*, Studies in Theoretical Psycholinguistics 43, DOI 10.1007/978-3-319-05086-7_9,

1 Introduction: Recursion in Acquisition

Recursion has been at the heart of generative grammar from the outset. Our goal is to engage acquisition evidence by showing how steps on the acquisition path connect to UG representations of recursion.

The joint development of theoretical and empirical understanding involves, appropriately, an evolving back-and-forth between formal representations and its empirical consequences.

The ultimate goals are to capture:

(1) a) How do Syntactic principles constrain acquisition?
b) The Acquisition path: what are the micro-steps acquisition follows?
c) Interface theory: how does the child link to semantics and pragmatics?
d) Parsing: how does acquisition reflect parsing?

We concentrate on how several intuitively simple grammatical concepts delineate the acquisition path.

A) Recursion types: conjoined recursion and embedded recursion,
Conjunction: the father and son and friend came.
Embedded recursion: the father's son's friend came.
B) Substitution of a complex phrase for a simple Head
N => NP
[I want cookie (N) => I want the cookie (NP)]
C) Syntactic constraints on movement: limits on moving
over identical material.
[a machine that makes something that makes coffee
=> coffee-maker-maker]

First we will discuss broader theoretical and acquisition challenges.

1.1 Non-linguistic Recursion

There exist several claims about the relevance of non-linguistic recursion to grammar. They are often not sufficiently developed to reveal empirical grammatical consequences. One claim by Jackendoff (2010) and Corballis (2012) as well as Everett (2012) – is that recursion is just a feature of general cognition, and a given language may or may not have recursion. Such claims have much less force when specific properties of linguistic recursion are examined, such as how Merged structures are Labelled, which has no immediate translation into, for instance, vision. We return to non-grammatical recursion when we address apparent "discourse recursion" where new propositions are inserted into conversation before old ones are finished.

At another extreme, for instance, some have claimed that recursion is simply a generalization from lexical evidence (Tomasello 2003; Lieven 2009), while others claim, for instance Tenenbaum, that the distribution of frequency-defined evidence is sufficient (see Perfors et al. 2010) to inform the child without specific grammatical assumptions, nor grammar-specific learning assumptions, although critical recursive structures are notoriously infrequent.

This last perspective does yield an important initial option for acquisition: the evidence suggests that very specific lexical subcategorizations might, indeed, capture first-order recursion (single embedding) in child grammar. The shift to truly recursive generative power (clear with 3-level embedding) becomes the critical step as in: I know that you said that Bill is here. Such a step must be supported with evidence for the child to distinguish recursive languages from those where a single embedding occurs.

1.2 The Cross-Linguistic Challenge

The first question is simply empirical: when do children recognize language specific forms of recursion, since there is substantial variation across languages? (see Koopman 2014, this volume for an example). A central case is the possessive: there are recursive prenominal possessives in English (John's friend's father's hat) while German allows only a single, non-recursive "Saxon" genitive (Marias Haus. *Marias Hunds Haus (=Maria's dog's house)). This contrast already entails an important lesson: a single example should not be sufficient to trigger recursion because were it sufficient, then children would generate recursive possessives in German, which they do not.

The same arguments hold for all of the constructions in (2):

(2)	a. Compounds:	Germanic languages => recursion	
		Romance languages => no recursion	
	b. Possessives:	English =>	recursive possessives (Saxon Genitive)
		German =>	no recursive possessives
	c. Adjectives:	English =>	recursive prenominal adjectives
			no recursive post-nominal adjectives[1]
		French =>	no recursive prenominal adjectives
			recursive post-nominal adjectives

[1]This is an oversimplification. French allows more than one prenominal adjective, but only from a very lexically restricted set [grande, pauvre, certaine etc.]. See Trueswell (2004) for a careful and relevant discussion of the syntax/semantics mapping, whose effect, from our perspective complicates the acquisition problem and reveals that the trigger for lexically-independent recursive productivity will be more complex than we present here.

d. Serial verbs: Bantu => recursion
English => no recursion
e. PP recursion: English => recursion
Not present if a language has incorporated prepositions.
f. Clausal recursion: Germanic, Romance
Piraha (disputed (Everett 2012; Sauerland 2012)) => no recursion

This variation is, of course, linked to deeper parameters in each language: Is it VO or OV, left-branching or right-branching? These parametric decisions will have important implications for the acquisition path for recursion in each language.

1.3 Recursion Recognition

In addition, we must ask: where UG allows recursion in every language, does the child immediately recognize it? For instance, even if all adult grammars allow sentence-initial repeated adjunction as in PP-recursion, such as according to John according to Bill the earth is flat, will children recognize it and do it?[2] To get a sense of relevant data here, consider here a few of the more than 100 recursive adjuncts in CHILDES produced at 2 years or less where "I think" may be just an adjunct:

(3) Naomi (Sachs) at 2.1 years:
nai46.cha:*CHI: I think maybe she's looking at her potty.
nai46.cha:*CHI: I think maybe she's a big girl.
bates@ 2;4. [I think I go to the store]
nelson @ 1;11.20 [I think I want to sleep on a boat]

Other children produce sentences with several sentence-initial adjuncts like:

alltp3.cha:*CHI: well I guess # maybe I have to use this

By contrast, cases with "you think" seem to involve real complements:

(4) kucjaz@ 2;10.6 [I dont know what do you think it was]
bloom @ 2;7.13 [see you what do you think of that]
manchester2;9.15 [you think it dont belongs to me]
manchester@ 2;10.28 [what do you think of it now, robber?]

[2]See Sowalsky et al. (2009) for preliminary experimentation on the topic.

In general, there exists quite infrequent naturalistic data with recursion, such as the possessive: “you’re my mom’s mom” (Danilo Azcarate (3.9 years), or a novel but natural extension “there’s another another box” (Danilo Azcarate (5.1 years)).

How might the input suggest recursion? English might seem to have evidence for recursive serial verbs (come help cook), but it is lexically limited and does not trigger what might look like natural extensions *try help cook/*help start sing which need an infinitive (help start to sing). In contrast, Adone (2012) shows that serial verbs in Bantu languages are quickly treated recursively by children. However English children simply do not do produce such recursive extensions.

Snyder and Roeper (2004)[3] examined naturalistic data and experimentation on recursive compounds, adjectives, and serial verbs. They were led to the conclusion that children must “experience” recursion because in each instance children began with single instances, but no recursion unless the adult language exhibits it. This conclusion raises a question: at what formal level do they experience recursion?[4] It remains an important question: how exactly does the child represent and recognize both the structure and meaning of rare recursive constructions?[5]

2 Representation at the Formal Level

We will ultimately delineate three kinds of recursion: Direct recursion, indirect recursion, and recursion via substitution or Generalized transformations Frank (2006) (see also Joshi 2014, this volume) and we will point toward other Minimalist options like the use of Pair-Merge (Chomsky 2012). We begin with Merge as minimal Recursion (see Chomsky (this volume)).

2.1 Merge as Recursion

Chomsky, Hauser, and Fitch (CHF) (2002) made the strong claim that recursion is the universal core of grammar (see Chomsky (this volume)). The claim applies to the core operation of Merge, which is evident as soon as a language compositionally

[3]These papers preceded Chomsky et al. (2002) but they argue as well, for the pivotal role of recursion in acquisition.

[4]Our philosophy is that all data is relevant to strong but abstract hypotheses. It is their combined power, not separately methodologically “proven” status, which should determine their value, though in each instance careful exploration is warranted.

[5]We will not discuss formal options for which empirical consequences are so far not evident. For a review see Lobina (2011) and papers in Biolinguistics. We will also not address philosophical and psychological implications beyond language where formal connections are obscure. See Corballis (2012) and Hinzen (2007) for discussion.

combines more than two elements. It is therefore present when a child moves from two-word to three-word sentences, which introduces the subsequent "explosion" to 3,4,5 word utterances, which is precisely what recursive Merge predicts.

2.2 *Acquisition Triggers*

CHF did not address the question of language-specific recursion.[6] If some languages lack a form of recursion, then in every language where it is present, it must be specifically triggered.[7]

A first hypothesis could be that a single example would suffice to create a Labelled Node which automatically has the capacity to be recursive, but this possibility is not supported in the acquisition data. For instance, a single instance (in fact a class of examples) of N-N compounds in French (homme-grenouille (frogman)) does not lead French children to an English-like N-N-N compound (frogman suit) which step English children easily take, as Snyder and Roeper (2004) show. To put the question in the context of English recursion, how and when does a single adjective node (A) change into an adjective phrase (AP) that can generate further A's?[8] Is just one enough?

---

[6]The term language-specific recursion is useful because it differentiates the concept since the literature often assumes that if recursion is universal, then every language should exhibit every form of recursion. The term is also misleading because one could say that there is one form of recursion, Merge, but it applies to different Labelled structures in different languages (thanks to N. Chomsky (pc) and see his contribution to this volume). We use the term language-specific recursion to emphasize the fact that it must be triggered by language-specific properties.

[7]It is also possible (Speas pc) that all forms of recursion are universal, and the child must learn restrictions on them. That is, the child assumes every new Labelled structure is recursive at first. This interesting hypothesis does not find support in naturalistic data. We do not find, for instance, that German children can say: "the car's roof's color", but then learn to restrict prenominal possessives to a single human possessor. Snyder has shown that "grammatical conservatism" is pervasive in acquisition production evidence and calls for a variety of explanations. See Roeper (2007) for discussion of very limited cases of overgeneralized comparatives "I'm not as tall as you as Mom" or "it is the same as that as that". Widespread overgeneralization of non-recursive structures to recursive uses has simply not been found in production.

[8]A sketch of how recursion has been represented historically provides some background perspective. The first generative grammars (Chomsky (1956)) captured recursion by Generalized Transformations (GT), where two independent forms (kernel sentences) were generated, and then an operation inserted one inside the other [e/g. John has a book. The book is green => John has a book which is green.] Then in Chomsky's Aspects-approach (1965), recursion was treated as a universal property of Phrase Structure Rules [e.g. VP => V (S)] When PS rules were eliminated, recursion became essentially an emergent property resulting from lexical subcategorization [e.g. believe [S]] and the generality of X-Bar principles [XP => X (XP)]. Eventually GT reappeared with the necessity of Late Merge for, in for instance relative clauses, to explain a number of acquisition and binding theory phenomena (Lebeaux 2000; Bhatt and Pancheva 2004). In modern theory, GT is any instance where two pre-existing internally complex phrases are allowed to Merge.

This question engages a fundamental issue: what is the exact content of the Labels on a node (e.g. is it a Noun, Nounphrase, or Determiner Phrase) and how does a child recognize them?[9] Hornstein (2009) makes an interesting claim:

(5) It is an innate Universal Grammar requirement that every Node must have a Label.[10]

This plausible assumption does not inform the child what the label is. Recent work on Labelling algorithms (Chomsky 2012) would make this innate connection less straightforwardly recognizable by a child. The skeptical reader might ask: where exactly does this formal question become empirical?

One approach is to examine the acquisition path itself for clues. The first stage could be entirely lexical (see Lieven 2009) whose data of initial stages supports this view. The child who hears:

(6) a. walk on
walk yesterday
walk home
walkdog

may not have decisive information to decide what the complement of walk can be. However it is clear when functional categories arise that a decision has been made:

(6) b. walk the dog

We can use the approach behind Generalized Transformations to capture this critical transition on the acquisition path: the substitution of a Maximal Projection for a Lexical category in a mental grammar. Thus we substitute a Maximal Projection like NP or DP for a simple N, when the child moves from utterances like:

(7) "Boy want cookie" = N V N
to: "the boy wants the cookie" DP V DP.

The idea that the child moves from simple to complex via operations of substitution is what ordinary intuition would suggest. The formal character of substitution, we argue, may be the same for recursion in forms like: John's sister's friend's father, to which we return.[11]

---

[9] In fact in recent work Chomsky (2012) claims that movement may be a means to determine a Node Label, as proposed by Moro's approach to anti-symmetry (See Bauke (2012) for a good exposition.)

[10] There are interesting variants here dealing with temporarily unlabelled nodes and how XP = YP merge is resolved. See Narita (2011) and Bauke (2011) for discussion.

[11] See Joshi (current volume) and Stabler (current volume) for relevant background discussion.

3 Direct Recursion

The ASPECTS Phrase-structure formalism allows a natural statement of a critical contrast between (Roeper 2011) direct recursion and indirect recursion:

(8) Direct Recursion
X => X (and X)
for example: S => S (and S)

This rule allows a category to generate itself. Roeper (2011) argues that this form of recursion is the first stage default of grammar because it captures the early use of "conjunction" to cover many relations that are later captured with so, then, but or causal or temporal connectives. The ease of "and" is evident in many early examples:

(9) 2;8.8 [unhun and its raining on him and I drawed the rain]
2;8.22 [unhun that's my fishing pole and its longest]
2;8.25 [unhunh I pushed it and it popped]

Nevertheless, conjunction is a restricted part of grammar. It limits both transformational and interpretive processes. The conjoined NP's in (10) do not allow syntactic extraction:

(10) *who did you see John, Bill and *t*

and no cyclic (Phase-based) interpretation is imposed, because the DP's (John, Bill, Fred) do not stand in an embedded relation, hence (11a) and 11b) mean the same thing (with some additional implicatures based on ordering)[12]:

(11) a. John, Bill, and Fred arrived.
b. Bill, Fred, and John arrived.

These contrast sharply with possessives where stagewise interpretation changes reference, which we will capture with Indirect recursion below:

(12) John's father's friend's hat =/= John's friend's father's hat

Direct recursion can be applied at any level of grammar, from sentences to nouns, verbs, prepositions, and prefixes: pre- and post-operative care or John knows why and how Bill won.[13]

[12]Interesting variations include an "and then" and/or causative interpretation of and which are among the implicatures, see Noveck and Chevaux (2002) study of contrasts like: Mary got pregnant and got married/Mary got married and got pregnant.

[13]Not only does it show no syntactic operations, like extraction, but it lifts binding constraints (John told Bill to invite Mary and himself [himself = John or Bill]). In addition, to the extent that

Our acquisition claim has led to Arsenijevic and Hinzen's stronger claim that Direct recursion is outside of grammar altogether (Arsenijevic and Hinzen 2012). From our perspective the "outside" status of Direct recursion supports the notion that the Narrow Faculty of Language does not exhaust language, and the Broad Faculty of Language, including Direct Recursion, is different. We make the following hypothesis:

(12) Direct Recursion is a default representation in children's grammars.

It appears as a child's first representation of complex structures in:

(13)

a) Possessives [Jane's father's bike = Jane's and father's bike] and
b) Adjectives [second, green ball = second and green ball]
c) Verbal compounds [tea-pourer-maker = tea-pourer and maker]
d) Prepositional Phrases[14]
e) Sentential complements

The child's capacity to move beyond the default conjunctive reading for each construction depends upon distinct semantic and pragmatic conditions.

3.1 *Indirect Recursion*

Indirect recursion can be simply conceptualized in terms of PS rules (14) but its representation within minimalism is more revealing and articulates the core grammatical interface with semantic interpretation (see below). The S-node must go through a VP node before it reappears:

(14) S => NP VP
VP => V (S)

Sentence-level recursion does not directly generate itself but must pass through another node. A cursory glance at recursive structures shows that this is

---

arguments for multi-dominance depend upon Across-the-board operations, they are often mediated by *and:*

(i) to go is what Bill wanted and Fred hated suggesting again that quite different aspects of non-core grammar may be involved where Direct Recursion and hence conjunction applies.

[14]In addition, other domains beyond the scope of this article provide support for this concept:

(i) as a first parsing form: relative clauses are seen as conjoined before being interpreted
(ii) a default representation in L2

commonly the case (15). Of critical importance is that, unlike conjunction, each element must be interpreted in order. *mother's cook's car* does not mean *cook's mother's car.*

(15) a. Sentence: John thinks that Bill thinks that Fred . . .
V CP V CP . . . .
b. Nominal: John's knowledge of Bill's knowledge of
DP PP DP PP
c. Adjective: "a big little truck"
Adj [NP [ADj N]]
d. Possessive: John's friend's father's car
NP DP NP DP NP DP NP

The relationship arises for recursive possessives in a less visible way through recursive Determiner Phrases (DP) which generate a POSS node which then generates another DP:

(16) DP => Spec D
D => 's NP
Spec => DP

```
                     DP4
                    /    \
                   Spec    D
                   /     /     \
                  DP3   's     NP4
                 / \             |
            Spec  D              |
              |   / \            |
            DP2 's  NP3         car
           /   \      |
        Spec    D   father
          |    / \
         DP1  's  NP2
        /  \       |
    Spec    D    friend
            |
           NP1
            |
           John
```

John's friend 's father 's car

Interpretation must occur at each NP, creating a semantic difference before the final Head is reached: John's friend's father's =/= father's friend's. As we shall see, these consequences for semantic interpretation can be explained in terms of the Strong Minimalist Thesis (Chomsky 2008).

3.2 Generalized Transformations

Generalized Transformations allow the substitution of one structure for another. The potential of such a substitution is really implicit in many accounts of acquisition.[15] Thus early expressions like *eat raisin* are naturally captured if we assume, as mentioned, that the child has only the structure V N, but these forms must give way – be substituted for – by Maximal Projection (MP): VP NP and then V DP. This follows if there are, as Boskovic (2008) argues, languages without DP's, but only NP's.[16] The prediction would be that an NP language would not have recursive possessives. It also makes sense to expect a child to pass through an NP stage before acquiring Roeper et al. (2012) provides evidence that English children pass through a "Japanese" stage without DP. While it is possible to assume that the MP's are present from the outset, we argue that GT substitutions, which are really forms of expansion, are a natural means to capture the acquisition path and differentiate languages where smaller projections may occur.[17] Thus we imagine that an NP structure VP => V NP has a DP built on top of it which then substitutes for a subcategorized Argument of the verb, as in (10):

```
(17)    CP   VP
             /\
          V   NP =>  DP
              |      / \
              N    Spec  D
                   /     /\
                 DP        NP
```

This acquisition act will introduce recursion automatically because the SPEC of DP can generate another DP.[18] The strength of this proposal emerges if it captures subtle data. As a case study let us examine the emergence of the possessive. We begin one step earlier.

---

[15]See Frank (2006) who explains the absence of object-relatives in early relative clause acquisition as a failure to substitute: NP (S) for a simple NP. This can explain why children give a subject interpretation to the final relative in: the rat hit the cat that pushed the dog.=> rat pushes the dog. Tavakolian (1981) suggests that the child interprets the that as and which is another example of reversion to a conjoined structure and Direct recursion.

[16]Suggested by Speas (pc).

[17]See Stickney (2009) where she examines how children acquire partitive structures like: a lumpy bowl of the soup in contrast to a lumpy bowl of soup (where either the bowl or the soup are lumpy). Under the notion that a child can reduce a DP to a default NP, the results are predictable. Nakato (2011) likewise shows that children may erroneously fail to represent DP in plural constructions at a certain stage.

[18]We assume that UG provides this information, but our formulation abstracts away from many questions about the Feature structure of DP's which may involve Case, Number, and Gender information.

3.3 Possessive Acquisition Sequence

Eisenbeiss et al. (2009) shows that there is a stage where children have either a Determiner, as in *the hat* or an Adjective, as in *big hat* before they get a combination *the big hat*. We could capture this stage by imagining a more abstract notion of Modifier at the outset (11), evidence that Labelling could be quite abstract:

(18)
```
         NP
        /  \
    MOD      N
     |
    the
    Adj
```

There is a hint that children might first analyze lexical possessives as adjectives (Brown 1973).

(19) "a my pencil"

This suggests that the Adjectival Modifier includes lexical possessives. We predict that there will be no cases like: *"a John's pencil" because the Proper Noun John's is not a possible adjective, and none have been attested.[19] This misanalysis is not surprising if the first phase of possessives is exclusively lexical, as Galasso (1999) has shown.

(20) Galasso:

I want me bottle. Where me Q-car? That me car. (2;6–2;8)
No you train. It's you pen. It's you kite. (3;2)
I want to go in him house. Him bike is broken.
Mine banana. Mine bott1le. Mine car. Mine apple. (2;4)
Mine pasta

This fits the idea that the child has Modifier N as the first form, where Mod => {ADj, Det} which notably does not generate a structure that can carry recursion. Now the next stage may still fall within a non-recursive D-projection that enables ellipsis:

(21) **Single Poss:** Jensen and Thornton (2007):
[whose hat is that]
"Mrs. Wood's" (2.7)

Note that one cannot do ellipsis on the lexical possessives: whose hat is that? => * my__ without a nominal place-holder/n/as in mine.

[19]Even more subtle phenomena are involved since we have a men's room where the compound men's room is treated as a Head.

At this stage the Possessive might be essentially German, where only a single Human possessive is allowed[20](Maria's Hut). It is noteworthy that all of these examples involve human possessors, just as if the German restriction held.[21]

In fact, the NP-possessive may stay in the adult grammar. Munn (1999) and van Hout et al. (2012) argue that there must be two possessive positions, an NP-possessive and a DP-possessive, to account for cases like 0).

(22) Bloomingdale's men's clothing

In (22) the inner form (men's) is limited to Noun heads, carries generic meanings, and does not allow phrases, being an NP projection, while the outer possessive, under a DP allows phrases and further recursion [John's Bloomingdale's men's clothing].

(23) [the man]'s hat as well as the [man's hat]
[the man on the corner]'s hat

It is the phrasal possessive which contains a DP and therefore automatically generates indirect recursive forms as well. Thus we can argue that the inner possessive is not eliminated, but rather extended by being embedded in a DP. Frank (2006) provides a TAG-grammar system of constraints that leads him to a proposal along these lines for acquisition.[22] A DP substitutes for the NP by building a DP on top of it, where we see the DP as a projection of NP following Frank (2006) (call it DNP), producing forms, as in:

---

[20]It seems to be limited to Humans for children at the first stage as well, but there is no careful study to our knowledge. We shall not provide a full account of the German form beyond the suggestion that it can be seen as a Case-marker (Sauerland pc) that applies in a complex way to Determiners and Nouns (des Mannes [the man's]). See Kremers (2000) for an analysis of genitives as raising from N to D.

[21]See Eisenbeiss et al. (2009) for a comprehensive discussion of the emergence of possessive in English and German. They discuss there the intricate interaction with case-assignment, which allows forms like: des alten Mannes where the possessive is expressed on the Determiner and the noun. While German does not allow recursion of the pronominal Saxon genitive, it allows post-nominal recursion through the articles and no Prepositions, as in Der Hund der Schwester der Frau [the dog (of) the + gen sister (of) the (gen) the woman]. Their acquisition remains unexplored.

[22]Here is his formulation: "I will adopt the two operations of the Tree Adjoining Grammar (TAG) formalism: *substitution* and *adjoining* [...]. Given two independently derived pieces of structure, the substitution operation inserts one along the periphery (or frontier) of the other at a node, called the substitution site. One can think of substitution as an operation which rewrites a node along the frontier of one structure as another piece of structure (called an *elementary tree*). (Frank 2006: 149)"and "This approach obeys a principle of Label Identity: I assume the Condition on Elementary Tree Minimality (CETM), according to which the heads in an elementary tree must form part of the extended projection of a single lexical head, following the notion of extended projection of Grimshaw (2000). As a result, the lexical array underlying an elementary tree must include a lexical head and may include any number of functional heads that are associated with that lexical head. "(Frank 2006: 151).

(24)

```
              DP-NP
               / \
              Spec  D
               /    / \
             Poss 's  NP
              /        /
          DP2         N
          /\          |
      Spec D         hat
       /   / \
      DP 's NP
       |      \
       D       \
      / \     friend
    the   NP
          / \
        man  PP
              |
        [In the middle]
```

=> <u>the man in the middle'</u>s friend's hat

This is the crucial acquisition step. If the NP possessive position remains in the grammar, then it is clear that the child will need unambiguous evidence to justify the recursion in the DP. This can now be either examples of recursive possessives – which is likely – or phrasal possessives. Green (2011) provides evidence that the morphology is not required in AAE for either recursive or phrasal possessives. Thus one can form:

(25) "John friend hat" [=John's friend's hat]
"the boy in the back bible" [=in the back's bible] (Green's attested examples)

If the recursive structure is present with POSS nodes, then the morphological absence of the/'s/is not surprising.

It now becomes natural to argue that the child's move to recursion involves an important and independent step. In fact there is now strong independent evidence that children resist recursion with possessives:

This dialogue is one of many (Roeper 2007):

Stage 2

MOTHER: What's Daddy's Daddy's name?
SARAH: uh.
MOTHER: What's Daddy's Daddy's name?
SARAH: uh.
MOTHER: What is it? What'd I tell you? Arthur!
SARAH: Arthur! Dat my cousin.

Fig. 1 Recursive and conjunctive options for possessives

MOTHER: Oh no, not your cousin Arthur. Grampy's name is Arthur. Daddy's Daddy's name is Arthur.
SARAH: (very deliberately) No, dat my cousin.
MOTHER: Oh. What's your cousin's Mumma's name?
What's Arthur's Mumma's name?
And what's Bam-Bam's daddy's name?
SARAH: Uh, Bam-Bam!
MOTHER: No, what's Bam-Bam's daddy's name?
SARAH: Fred!
MOTHER: No, Barney.

Such children find no path to comprehension – not even an extra-grammatical pragmatic guess.

3.4 *Experimental Evidence of Direct Recursion Conjunction*

Experimental evidence indicates that recursive possessives are difficult. First Sarah Gentile (2003) gave children three pictures based on familiar Sesame Street characters, but no story was presented (adults were tested in the next study (18).

(26) A. Picture of Cookie Monster
B. Picture of Cookie Monster and his sister
C. Picture of his sister
=> Experimenter: "Can you show me Cookie Monster's sister's picture?"

These first results showed that about one third of the 3–4-year-olds took the conjunctive reading (Cookie Monster's *and* sister's picture) and chose Picture B. Nevertheless many can do it particularly by 5 years.

Maxi Limbach gave children and L2 German speakers learning English recursive examples, although again German allows only a single possessive. After a presentation of all bikes and actors (Fig. 1), the bikes were then shown in separate pictures and participants chose which fit "Jane's father's bike".

Fig. 2 4-level recursive and conjunctive options for possessives

(27) Context story example for screen setting:
Jane has a nice blue bike and Jane's father Gordon has a racing bike. When they do a tour together, they have another bike which they can ride together. Sam has a red bike and his father Paul has a silver bike.

Bikes Racing tandem blue = both father's and Jane's
blue =Jane's, green = father's

Subjects who were either native (NS) or non-native speakers (NNS) were involved: 25 American English-speakers and 23 German university L2 students. 26 children were divided into three age groups – nine 3-year-olds (average age: 3;7), eight 4-year-olds (average age: 4;5), and nine 5-year-olds (average age: 5;7). NNS adults gave a conjoined reading or dropped one of the possessives (38 %, compared with 37 % for the 5-year-olds). It is noteworthy that the 5-year-olds gave 22 % conjoined readings, while the NNS adults gave only 8 %, preferring 30 % of the time to drop the first or second possessive. Here are overall results (see Limbach and Adone (2010) for further analysis): Children generally acquire possessive recursion by the age of 5–6. The "conjunction" option is much stronger among children than adults.

This suggests that some types of recursion may remain a permanent, generally undetected, impossibility for L2 speakers. In fact, informal questioning of L2 speakers of English persistently leads to declarations that that recursive possessives are difficult, and triple recursion almost impossible to understand and significantly more difficult to produce. This suggests that where understanding succeeds, it depends upon heavy contextual inference beyond grammar.

Recent results (Roeper et al. 2012) from Japanese children extend the experimentation to 4-level recursion with the following kind of pictures and subjects in (28) (Fig. 2):

(28) Story (told in Japanese): This (pointing at the leftmost girl) is Shiro. This (pointing at the man in the middle) is Shiro's father, Ieroo. This (pointing at the rightmost man) is Ieroo's older brother, Guriin. Shiro and Ieroo (pointing at the two people) have the same flower on their shirts as a sign of the parent-child relationship. This (pointing at the leftmost rabbit) is Shiro's rabbit. This (pointing at the rabbit in the middle) is Ieroo's rabbit. This (pointing at the rightmost rabbit) is Guriin's rabbit. Ieroo's rabbit (pointing at the rabbit in the middle) is the father of Shiro's rabbit (pointing at the leftmost rabbit). They have the same flower on their shirts as a sign of the parent-child relationship.

(29) Shiro-chan-no otoosan-no oniisan-no usagi-no kasa-wa nani-iro kana?
Shiro-Gen father-Gen older brother-Gen rabbit-Gen umbrella-Top what-color Q
'What color is Shiro's father's older brother's rabbit's umbrella?'(4-POSS)

(30) Ieroo-san-no oniisan-no usage-no kutsu-wa nani-iro kana?
Ieroo-Gen older brother-Gen rabbit-Gen shoe-Top what-color Q
'What color are Ieroo's older brother's rabbit's shoes?' (3-POSS)

The names were mnemonic (e.g. Orenji = orange) to help the children keep track of them. The table in (31) gives the general pattern of errors (see Roeper 2011) for details.

(31)

	Correct	Conjunction	Dropping	Dropping or misanalysis	Other
1-POSS	73/78 (93.6%)	0/78 (0%)	0/78 (0%)	0/78 (0%)	5/78 (6.4%)
2-POSS	91/130 (70%)	5/130 (3.8%)	22/130 (16.9%)	0/130 (0%)	12/130 (9.2%)
3-POSS	43/104 (41.3%)	0/104 (0%)	34/104 (32.7%)	12/104 (11.5%)	15/104 (14.4%)
4-POSS	43/104 (41.3%)	3/104 (2.9%)	10/104 (9.6%)	25/104 (24%)	23/104 (22.1%)

They tested 26 children (3;4.15–6;1.7) (and 13 adults) and found that younger children had great difficulty, but 11 5 year olds were able to get 50–75 % of 4-level recursive structures correct. Obtaining 50 % correct is far better than chance, since there were six potential choices.[23,24]

[23]The errors were diverse, and interesting, involving unanticipated compound readings, but we will not explore them here.

[24]For similar work see Fujimara (2010).

A remarkable feature of these results is the fact that 3 and 4 level recursion were equally difficult. This provides an important hint about how to look at the Generalized Transformation account. While we have not explored the precise syntax of Japanese, if GT applies as we have suggested, then its impact is upon the accessible grammar and not upon the incrementally increasing parsing challenge. It leads to this speculation in (32):

(32) GT possessive substitution:
is an operation on the grammar[25]

Thus recursion seems to provide a significant challenge on the acquisition path, but once overcome, the access to the recursive grammatical representation is automatic. Modeling these facts carefully may lead to important insights about the relation between grammar and parsing. It remains true that failure on complex recursion by young children may have a parsing component as well, but the solution to the parsing problem does not by itself advance the grammar to include recursive structures.

The results raise a further question: why are two level possessives easier than 3–4 level recursive expressions. Our proposal offers a hint here as well. The highest instances of possessives seem to involve a lexical element (my): "you're my mom's mom" which extends the first possessives which are uniformly lexical (my, your).[26] This fits our argument that there are two possessive positions, which would then entail the possibility that children and adults might understand the two possessive cases as two kinds of possessive: NP-possessive, and DP-possessive. Then it is only the 3-level possessives which absolutely demand a recursive structure. This is exactly what our proposal predicts and we find supported by the data.

4 Adjective Recursion

We will now review a variety of other constructions that reveal the same pattern, but not in the same depth as our discussion of possessives. Adjective recursion provides a special acquisition challenge because the adult grammar regularly allows both conjunctive and indirect recursive forms. A sentence like (25a) is easily paraphrased as (25b):

(33) a. the big, round, funny balls
b. the big and round and funny balls

[25] A further speculation is that recursion is related to Operator-linked identity in this respect. Similarly NPI concord does not seem to add difficulty in parsing when a series of negatives are linked via polarity items (any): John doesn't want any food at any time for any reason. We leave this possible connection for future research.

[26] Eisenbeiss et al. (2009) report that Sarah (Brown corpus) uses exclusively lexical possessives until age 3.1.

Required non-conjunctive recursion is less frequent. Matthei (1982) showed 3–4 year old children the array of balls in 0) with the instruction below (following C. Chomsky suggestion).

(34)

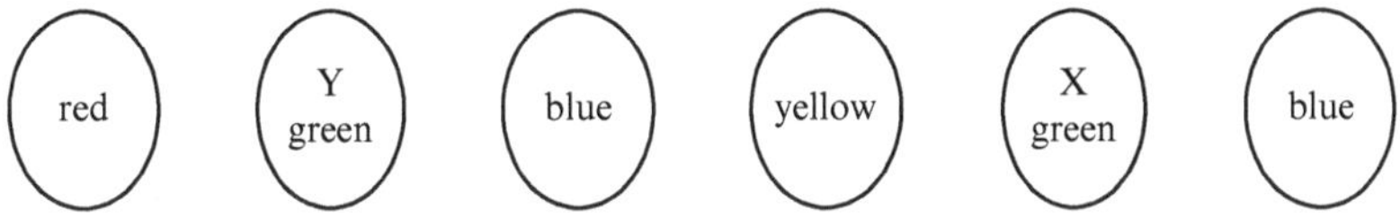

"show me the second green ball"

More than 50 % of 3–4 year olds chose (Y) instead of (X), giving a conjoined. reading "second and green ball". The structure they needed was essentially indirect, second [green ball], not directly modifying another adjective as in crystal-clear water, which is notably non-recursive (unless there is a meaning for? crystal-water).[27]

(35)

```
      NP
     /  \
  Adj   NP
  /     / \
2nd   Adj  N
       |   |
     green ball
```

Adults in fact will give priority to the recursive forms where children do not. In a small experiment, Bryant (2006) found that when 4 year old children were asked to choose they choose the big balls and the black balls, and not the ones that would entail a recursive structure of the form in (37). We conclude that the conjunction-option is the default here.[28]

(36) two big balls (white), two black balls (little), and two black ball (big)
"the big black balls"

(37) big [N [A black] [balls]

When do children produce recursive adjectives? Chloe Gu (2008) found early conjunctive forms [Adam 2. 3] "I sleepy tired" "you big tired" "I funny little boy" Only at a later age are there intersective adjectives: [Adam 3.4] "he got a new little big trailer" (small version of a big trailer). It does not have the contradictory meaning "little and big trailer" and therefore must have a representation of the form: little [N big trailer]. It is notable that in the naturalistic data and the experimental

---

[27] Hamburger and Crain (1982) and recent discussion (Marciles et al. in press) about the extra cognitive demands of recursive forms do not change the argument that the conjunctive form is syntactically simpler and the first level of analysis by the child.

[28] See Hubert (2009) for sentences where there are two adjectives, but they are not carried over under ellipsis.

data, while the conjunctive reading is a default, children acquire the recursive adjective fairly early.[29]

Although the surface shows just the addition or, merge of another adjective, the structure involves a further NP node: [AP big [NP [AP little [NP truck]] because the word big modifies the whole NP [little truck] and not the N alone [truck] nor the adjective alone [A big [little]]. So an NP projection above little truck is generated before a further higher AP projection is merged. (See Trueswell (2004) for an analysis along these lines.)

4.1 PP-Recursion

Like adjective recursion, much of PP recursion can be represented as conjunctive. Therefore, again, the environments where PP recursion is required may be less evident to a child than one might imagine. Just like resistance to recursive possessives, Gu (2008) found these examples of resistance to recursive PP's:

(38) Father: Up in the shelf in the closet
Child: yeah
Father: can you say that
Child: up in the shelf in the closet
Father: very good, up in the shelf in the closet in the kitchen, can you say that
Child: yeah, up in the # up in the # what
Father: up in the shelf in the closet in the kitchen
Child: up in the shelf in the # what
Father: closet
Child: in the closet in the kitchen
Father: in the jar up in the shelf? can you say that?
Child: I can't
Father: you can
Child: in the jar # say in the jar
Child: up in the shelf in the jar in the closet in the kitchen

Note that the PP's are now conjoined in interpretation: *in the shelf* and *in the jar*, rather than recursively embedded with the impossible meaning the shelf is in the jar.

Perez et al. (2011) explored production of recursion with PP's as well as possessives. in an experiment with 46 children, 25 younger (up to 3.6) and 21 older

[29]Snyder (2001) and Snyder and Roeper (2004) show that recursive bare noun compounds arise fairly early as well.

(up to 5 years), she found no difficulties with coordination when children were asked the question in (39):

(39) what is the boy pulling? "a dog, a cat, and a bike"

She found that young children would often comprehend (with picture help) recursive sentences, but avoided producing them (40). However, children below 3 would often answer (40b):

(40) a. This is Elmo and this is his sister.
whose ball is flat? => Elmo's sister's ball (correct)
b. "the sister" or "Elmo's big sister"

Similar examples were found for PP's (41) and (42):

(41) which girl? => the girl with the dog with the hat (correct)

(42) "the girl with the hat dog",
"the girl with, the dog girl, but not the same as the other one"

We take this to indicate that they understood, but seek complex paraphrases to avoid recursion in production. The production/comprehension difference in recursion may be one key to understanding the common intuition that real grammatical distinctions underlie obvious discrepancies[30] between comprehension and production.[31]

In sum, across many domains, each of which deserves further focused research, we repeatedly find that (a) children resist Indirect recursion, and (b) that Direct recursion, with a conjunctive interpretation, operates as a first representation and therefore a Default representation.

The conjoined representation has other potential dimensions in grammar. Stabler (2014, this volume) argues, conjunction is a first pass parsing analysis for adults which is restructured as indirect recursion at the semantic level. We turn to the SMT below which in part explicates this perspective. Our explanation for the non-automatic nature of recursion in acquisition, thus far, is that it involves an independent act of substitution, following TAG grammar, along the acquisition path.[32]

---

[30]See Amaral and Roeper (2014) for discussion of how the theory of Multiple Grammars can accommodate such differences.

[31]Optimality theory has offered one technical solution to the question.

[32]Work on recursive serial verbs exists, see Adone (2012), and Snyder and Roeper. While English children are exposed to lexically constrained V-V combinations (come help/go sing/let go) they do not generalize nor produce recursion with novel verbs. In addition, Wilbur (pc) shows that deaf sign language allows recursive possessives as well.

4.2 Sentential Recursion

We turn now to sentential recursion which has been the nexus of controversy. Important questions about the interface between grammar, discourse, and cognition become relevant here (see De Villiers et al. this volume).

Since interface theory remains in its infancy, our questions about the nature of discourse and cognition remain broad. Could there be other cognitive capacities – utilized in general discourse and conversation – that achieve the same interpretation by other means that are captured in grammar by actual recursive embedding? Everett (2012) and others (Jackendoff and Pinker 2005) argue explicitly that general cognitive abilities are involved and allow for "discourse recursion". It is certainly true that we can embed one topic inside of another in conversation or prose, as we shall see. Nonetheless, we argue that "discourse recursion" is in no way comparable to recursion in sentence grammar.

5 So-called "Discourse" Recursion and "Cognitive" Recursion

The boundaries of grammar are not self-evident. Jackendoff and Pinker (2005) have explicitly compared linguistic recursion to non-linguistic recursion, like visual[33] self-embedding (for instance our grasp of Russian dolls inside each other). In contrast, we argue that the formal basis for recursion in various mental domains can and should be different because they must meet different interface demands. Recursion in grammar involves a translation between a hierarchical into a linear structure.

Nonetheless the connection between linguistic and non-linguistic cognitive representations are an important interface. Where recursion appears, could there be a non-linguistic representation, which influences stages in acquisition? An interface claim could be: the child must extract from a pragmatic situation a semantic form of recursion which is then mapped onto an isomorphic syntax. Before successfully mapping the meaning onto sentential syntax, it could exist via inference in discourse that receives a distinct semantic representation.[34]

What would "discourse recursion" look like? It arises through a potential similarity between grammar-guided interpretations and our broad sense of how we organize meaning in our minds. We argue that the connection is essentially a form of metaphorical recursion which refers broadly to the fact that we can embed one

[33]Speas (pc) makes the interesting observation that a representation of visual recursion might treat Russian dolls as "Labeled" since we could also have recursive containment of different objects (can inside a box inside a bag) which could have different "labels". In general, we think that recursion may be separately represented in different mental modules, much as stereoscopy is a common concept, but separately represented for eyes (sight) and ears (sound).

[34]Thanks to Jeremy Hartman (pc) who suggested this view of the acquisition path.

phenomenon in another. Our construals – our inferences – are much looser than strictly compositional, and therefore unlike sentence recursion, as we shall illustrate. Consider a digression in a conversation, characterized as a form of recursion where we have two halves of a conversation, but embed one inside another as in (36):

(43) half conversation 1 [conversation 2] half conversation 1

But such embeddings may not obey any strict constraints however, and we easily find examples as in (37), which is exactly a kind of crossing structure that language virtually disallows.[35]

(44) half conversation 1 [half conv2] half conv 1
"John is – I really like him – coming over today. He always brings treats."

It is natural to infer a connection between "I like him" and "he brings treats", but no strict embedding is required to achieve this meaning.[36] However when grammatical structures of reference-tracking cross discourse boundaries, we find grammatical constraints in force. Joshi (2014, this volume) observed that do so constructions which require reconstruction across discourse, do not allow recursion (38):

(45) The earth is flat
John doesn't think so
and Mary doesn't think so

Here both VP-ellipsis markers refer back to the initial the earth is flat. It is impossible to obtain the reading (39):

(46) Mary doesn't think John doesn't think the earth is flat.

The constraint is slightly less clear when we use the deictic term that which usually refers to the last-mentioned item (40):

(47) a. John is a villain
b. Bill believes that
c. Mary doesn't accept THAT.

---

[35]Except for highly marked and limited forms like respectively:

(i) John, Bill and Fred went home, to church, and to school respectively

The construction itself does not allow recursion:

(ii) *John and Bill went to home and school respectively and Mary and Susan went to the market and lake respectively.

[36]The study of co-relatives suggests that real syntactic embedding, via intonation, may be involved in such cases. See Rodriquez (in prep) for this perspective.

The discourse in 0) could mean: Mary doesn't accept that Bill believes John is a villain, but it certainly can also mean: Mary doesn't accept that John is a villain. If we add a further (41) even with strong emphatic intonation, it becomes very obscure and virtually impossible to reconstruct the meaning which is readily available through sentence recursion (42):

(48) Susan doesn't condone THAT

(49) Susan doesn't condone that Mary doesn't accept that Bill believes John is a villain.

Discourse, like thought in general, allows us to build up complex meanings with cross-connections that are not easy to model and that do not obey a compositional deductive structure. It is the compositional structure enforced by systematic syntax which allows us to generate nonsense and be counter-intuitively imaginative. Discourse recursion seems like recursion – hence we call it "metaphorical recursion". To sense how diverse discourse reference-tracking can be, consider the discourse in (50):

(50) John was very humorous in saying that a black democrat could win and Susan said she didn't believe it. Mary liked that.

that = a. John was very humorous.
b. John said a black democrat could win.
c. a black democrat could win
d. Susan said she didn't believe it.

The interpretation of that can not only take a cumulative reading but pick out many sub contexts depending upon how we construe the situation. It follows naturally that embedded, apparently recursive propositions, is one of the meanings we can arrive it, but just one.

Levinson (2013) has recently argued specifically that discourse allows recursion and that it does not obey constraints on center-embeddedness. We take his evidence to provide striking support for our view that Discourse Recursion via inference is specifically lacking in the kind of syntax/semantics interface which delivers the scopally focussed and precise syntax-semantics interface produced by sentential grammar. His evidence provides support for the view articulated by Chomsky (this volume) that grammatically disallowed structures are still easily thinkable ("fine thoughts") even when syntactically excluded.

5.1 Experimental Evidence on Discourse

Systematic reference-tracking, via that, appears to be very difficult and therefore we predict that it will be difficult for children, even when the context makes this meaning quite natural. Hollebrandse et al. (2008) gave children the following context and sought a meaning (51).

(51) Jimmy and his sister live next to a bridge. The bridge is broken.
Jimmy knows that.
His sister doesn't think that.[37]

We then asked either question (A) or (B) with sample answers following (52):

(52) A. Will his sister warn Jimmy? Why?
B. Will his sister cross the bridge? Why?

We anticipated answers with the following entailments (53):

(53) Type A: Will his sister cross the bridge?
yes -> His sister doesn't think that the bridge is broken
no -> His sister doesn't think that the Jimmy knows that ....
Type B: Will his sister warn Jimmy?
yes -> His sister doesn't think that the Jimmy knows that ....
no -> His sister doesn't think that the bridge is broken

We even expected actual recursive responses as in (48):

(54) "She will warn Jimmy because she thinks he doesn't know the bridge is broken".

The results show that 50 % of the answers are single embedding answers "No, she knows the bridge is broken" and 50 % gave irrelevant or obscure answers "yes, because she is nice". None gave recursive answers, nor clear yes/no answers that corresponded to the recursive meanings. Likewise 12/13 adults did not give recursive construals. We conclude that discourse recursion is available as a metaphorical construal of meaning, but often unavailable when systematic rules of reference-tracking are applied across discourse.

5.2 *Sentence Subcategorization and the Conditions for Recursion*

It is customary in the literature to consider a single instance of sentence subordination to be an instance of recursion (49), where one sentence is inside another sentence, and one verb requires another:

(55) John thinks that the earth is flat

[37] Some find "think that" in this context odd, but the same structure is present for "believe that".

Recursion seems to be unstoppable if one assumes that there is a phrase structure rule which allows a second VP to arise (50), producing examples as in (51):

(56) S => NP VP
VP => V (S)

(57) John thinks that Bill thinks that the earth is flat.

Nevertheless, it is reasonable to propose that the first clause is not an instance of recursion but a complex form of subcategorization. We will explore this claim in some depth because this perspective is not widely appreciated and its implications for acquisition are important (see Hollebrandse and Roeper (to appear)) and Hollebrandse (2010) for further discussion). In essence, the claim is that like single adjectives and single possessives, the single complement clause does not necessarily entail the presence of a recursive rule. We know that verbs subcategorize their complements and allow considerable lexical specificity and narrow choices of complements (58):

(58) John craned his neck/*his elbow

The conservative learning child (see Snyder 2007) should be careful not to overgeneralize (58). The same conservatism should affect full sentential complements. Verbs vary in what complements they allow. In fact, some verbs take infinitives and others take that-complements. We also know that more than one element can be subcategorized for when verbs allow indirect objects (59).

(59) VP => V-Latinate NP (to NP) (donate a dollar to Bill)
V-Anglo-Saxon NP NP (give Bill a dollar)

Idioms have long been known to extend into complementation (53a) and resist recursion and extraction (53b) while non-idiomatic forms are allright (54c):

(60) a. John knows what is what.
b. *John knows what Bill knows __is what
c. Non-idiom: John knows what Bill knows—is good for you.

It is now a short step to the hypothesis that what appears to be a single complement for children may have a more detailed subcategorization and start like a slightly abstracted idiom. What could a "slightly abstracted idiom" be? Consider examples in which the idiom is an idiom but has open variables (the ADJ + er the ADJ + er) (61):

(61) d. the more the merrier => the bigger the better-> the fatter the nicer

In fact, Perfors et al. (2010) have argued against the instant realization of recursion with precisely such representations (62):

(62) S => NP VP (NP VP (NP))

The structure in (62) could be a child's initial representation where the whole string is represented as particular subcategories. For instance, early Small Clauses might have this structure for a verb like see (63):

(63) I see Bill eat cake = NP [see N V N]]

with virtually no substructure beyond the sequence of Heads or phrases and a partly non-linguistic way to fix meaning (as with headlines: President shot Now dead]). Such a highly restricted syntax could, for instance, rule out passives in complement clauses even when they are allowed in the matrix clause because they did not fit the template. In fact there are instances where only passive is allowed even in matrix sentences, calling for very specific lexical restrictions:

(64) How are you supposed to do that?

A rare child overgeneralized and inferred that the active was possible (58):

(65) "How do you suppose me to do that?"

To capture such exceptions as (65) the child should allow each verb initially to have a very specific subcategorization before broader evidence forces a revision which introduces recursion. Thus only when the child hears both examples in (66) does the child seek a generalization that links operations on matrix sentences, to those on complements, which justifies the introduction of a rule in (67) with recursion:

(66) I saw that John pushed Bill
I saw that Bill was pushed by John

(67) see => that (S)

A micro-analysis of early complementation has not been done to see if it is true that initial representations are limited to a highly specific sequence of categories, which might exist even if short-lived. Early analysis of Subject Auxiliary Inversion, for instance, shows that children start with just one or two ["can I", "do I"] before the full range of auxiliaries becomes productive.

5.3 Substitution Algorithm

If children do begin with a rather non-adult structure, then the next step calls for a substitution on the model of GT substitution. We can state it as in (68).[38]

[38]This may require a more complex form of substitution than we outlined above where a category is expanded, but still within the larger range of Tag-grammar options.

(68) Acquisition Substitution Algorithm
Substitute a UG Category for strings of phrases that can be transformationally related

When a child hears a series of subcategorizations for a verb, as in (69), the acquisition mechanism seeks a simplification in the form of VP => V (S) at which point recursion is guaranteed.

(69) VERB: NP aux Verb NP
[see John can help Bill]
VERB: ADV NP M Passive-be Verb NP by NP
[see suddenly Bill was helped by John]
substitution: VP => V (S)

This perspective has really been implicit in most acquisition theories. The step is critical to understanding the acquisition path and exactly how UG can be mapped onto a piecemeal acquisition process.

5.4 *Acquisition Data*

What does the naturalistic data reveal about sentence recursion? Parents are familiar with endless conjoined relative clauses like: "this is the toy that I like that I got for Xmas that has six wheels...". This phenomenon, called tail recursion in computer science, is a common feature of nursery rhymes whose hidden goal may be promoting recursion: "This is the cow that tossed the dog that worried the cat that chased the rat...", but when do children exhibit real sentence complement sentence recursion?

Infinitives

For infinitives, we find fairly early examples and the absence of Tense, as we discuss below, may be a critical factor. Here is a sample of early infinitival-recursion from a cursory search that reaches as young as 2 years:

(70) Naomi 2.4 years
@ 2;9.2
@ 2;6.5
@ 3;11.11 [No it isn't going to start to bleep.]
@ 3;3. [now I'm going to start to touch your knee]

Adam 3.5
Paul want to go to sleep and stand up

5.5 *Tensed Clauses*

Diessel (2004) argues plausibly that the very earliest forms of "I think" mean just "maybe", as in:

adam 3.0 years 20.cha:*CHI: I going make a trailer # I think

The earliest clear case of recursion with finite clauses we are aware of is given in (64).

(71) Danilo 3.4 "I think Daddy said he wears underpants"

We find as well that at 3–4 years Adam appears to use recursive clauses (65):

(72) adam 3.8 years 37.cha:*CHI: I thought you said the animals gonna have space
adam 4.3 years 45.cha:*CHI: he thought those guns were coming from outside him
adam 4.3 years 45.cha:*CHI: he thought I said something (a)bout window
adam 4.7 years 50.cha:*CHI: she thought that was tiger # that what she told me # before she went upstairs.
adam 5 years 52.cha:*CHI: I thought you said they gonna be warm[39]

These could either be a recursive adjunct which is conjoined (I think and Daddy said), or it could be a recursive adjunct with a single subordinated clause, as if it were like:

(73) according to John, according to Bill, the earth is flat

or, quite possibly, it is real sentential recursion. One signal fact is that there are no complementizer <u>that</u> phrases (they would be like: "I think Mom said <u>that</u> she . . . ". These are omitted even in French and German where they are obligatory in the adult language (Weissenborn et al. 1995). Second some cases "you said don't do anything . . . " suggest direct quotation, while others suggest subordination of

[39]Other children at this age have suggestive examples (1).

(1)	nai63.cha:*CHI:	you must've thinked [= thought] Betsy drew that I'm Betsy.
	acelw5.cha:*CHI:	I think Mom said that she [//] um she's coming home.
	anl.cha:*CHI:	I thought you said you's gonna let me have it.
	anl.cha:*CHI:	she said [//] know what Vanessa said?
	sandra.cha:*CHI:	you said I thought you weren't friends.
	anc.cha:*CHI:	you know [//] you heard what that guy said?
	anl.cha:*CHI:	I thought she said she go with her own

indirect questions or headless relatives "you heard what that guy said". These examples are just hints of what should be carefully studied to determine the micro-steps of complement acquisition.

5.6 Comprehension of Recursion

A crucial acquisition test of recursion lies in comprehension. Do children grasp the subordinating interpretation of recursion across two clauses? We developed an experiment to test this question and the results again suggest that the introduction of recursive productivity is delayed (see deVilliers et al. this volume).

Hobbs et al. (2008) with 18 children (6;3–6;11, avg: 6;9) have shown that they have no difficulty giving single complement answers for situations with sentences like: "Dad is talking to Billy about moving his tools. Dad tells Billy that Jane said that hammers are too heavy. What did Jane say?" Children easily respond "hammers are too heavy".

However when the higher verb is needed to make sense of the question, implying recursive subordination, correct answers fall off sharply. This experiment is designed to make the single clause answer improbable: Billy's sister saying that comic books are stupid when she is fighting with Billy over what he said. The example in (74) is from an experiment.

(74) Jane talks to mom. She is having a fight with Billy on the phone. Jane tells mom that Billy said that comic books are stupid.
"What did Jane tell mom?"

Nevertheless, 1/3 gave exactly the "Single" answer 0a). Only 1/3 of the 6 year olds were able to produce the recursive complement in repeating the story (68):

(75) a. Single complement: [said] "that comic books are stupid"
b. Recursive: [tell Mom]"that Billy said that comic books are stupid."
Results: 23 % irrelevant 34 % single, 33 % recursive

These results suggest that, just like with possessives, adjectives, and PP's the step to allow recursive complements is a significant one.

An important question now arises: why should Tensed clause recursion occur at a different point from infinitive clauses? deVilliers et al. (2011) have argued that Tensed clauses are special because Chomsky's Strong Minimalist Thesis (SMT) applies to them and requires that they be interpreted at each Phase (where the definition of Phase itself is a part of this argumentation). Here is Chomsky's formulation of the SMT:

(76) Strong Minimalist Thesis (SMT):
"Transfer.....hands Syntactic Object to the semantic component, which maps it to the Conceptual-Intentional interface. Call these SOs *phases*. Thus the Strong Minimalist Thesis entails that computation of expressions must be restricted to a single cyclic/compositional process with phases."
Chomsky (2005)[40]

This Transfer is another form of Indirect Recursion if we follow the Phase Alternation Constraint (see Boeckx 2008) : a PhaseHead and a Phase-Complement must occur in an alternating sequence. The Phase-Head does not Transfer until the next higher Phase. This formulation allows the familiar "escape hatch" for wh- when it occurs in the Phase Head CP leaving behind the Phase-Complement IP which undergoes Transfer (77).

(77)	PhaseHead-	Phase-comp	PhaseH	Phase-comp	Phase H	Phase-Comp
	CP	IP	CP	IP	CP	IP
	that	[John said	that	[Bill said	that	[John sings

Thus it is another form of the Indirect Recursion we have for Possessives and DP, and Adjectives and NP. Under this view, each complement must be immediately interpreted, which produces the meaning differences again between: father's friend's hat/friend's father's hat.

Applying the SMT to Tensed clauses may give us a basis for predicting that Tensed clauses will emerge later. In effect, for Tensed clauses the child must not only provide a syntactic representation but also interpret that representation: the link between Tense and factivity (truth) gives it a semantic complexity not found elsewhere. This in turn suggests, as others have, that infinitives might not form separate Phases, hence they are acquired earlier.

The demand of computing *opacity*–where the matrix verb changes the truth value of the complement present a special challenge. Thus our few naturalistic examples with apparent complementation may be closer, as we suggested, to adjuncts.

6 Compound Recursion and Syntactic Constraints on Cross-Over

We turn now to a case where recursion involves movement, not Tense. Nominal compounds are found in Germanic languages, Navajo, and other language families, but they are notably absent in Romance.

---

[40]Chomsky (2005, 2008), see also Richards (2011).

Where do English compounds start? Clark (1993) shows that children acquire verbal compounds at the age of 5 years, producing "wagon-puller". Hiraga (2010) shows that there are other early examples of verbal compounds as well (78):

(78) MOT: what do you think we make in our factory?
CHI: we don't make nothing.
CHI: I a cowboy maker.
CHI: A cowboy who shoots makers — Adam 3;8.26

Children after the age of 4 years and 4 months produce the correct form of verbal compounds. Some examples are shown below (79):

(79) CHI: something like a water pumper? — Adam 4;4.01
CHI: I'm gonna make it this is my dot maker. — Adam 4;4.01
CHI: one # two # three # four dinosaur killer! — Adam 4;6.24
CHI: that's a gun shooter. — Adam 4;7.16
CHI: that's a grain thrower. — Adam 4;10.23

While "dot-maker" and "water-pumper" are pretty clearly object-incorporated compounds, "cowboy maker" is explained by the child as an Agent-object relation, and "gun shooter" might also be an Instrument verb ("gun that shoots"). These seem strange to adults, but English compounds also allow denominal adjectives: Dodger pitcher (pitcher who is a Dodger, not someone who pitches Dodgers). At the early stage, then, the incorporated element might be "free" in its relational interpretation. It is within the recursive compounds that the object relation is uniquely fixed: window-wiper-fixer.

Such recursive verbal compounds are not mastered so early. Hiraga (2009) showed that not until the 6–7-year-old range did children show a clear ability to comprehend them. Her scenarios were designed to allow both a conjunctive and a recursive option (73) although only the latter is really productive for adults, as their results indicated.

(80) Kitty makes a great machine. The machine pours tea into many cups at once. Bunny bought the machine from Kitty, so Bunny only makes tea and doesn't have to pour it. The machine pours tea into five cups at once, so Bunny's sisters and brothers can drink it. Doggy doesn't have the machine, so he makes and pours tea himself. One of them said "I am a tea pourer maker" (Fig. 3).

Fig. 3 Compound recursion experiment: "Which one is the tea-pourer-maker? Why?"

The conjunctive reading is given by 6-year-olds (74)

(81) tea pourer maker
   a. "because he makes and pours tea"
   b. "because he doesn't have the machine"
  tea lover taker
   a. "because s/he likes tea and takes tea to school."
  bottle opener breaker
   a. "because he opens them and sometimes breaks them"
   b. "because he is too strong, so he breaks them"
  picture taker liker
   a. " because he takes pictures and likes them"
  pencil sharpener breaker
   a. "because he sharpens pencils and dropped them"
   b. "because he knocked it over".

Examples of recursive interpretation are also contested (75)

(82) tea pour maker (N.I. at 6;11.2):
    "because she pours, actually, she made the machine that pours tea"
  tea pourer maker (N.I. at 6;11.2):
    "because he makes the machine that pours tea"
  tea pourer maker (I.R. at 6;5.4):
    "because she made the machine that can pour tea"
  tea pourer maker (P.H. at 5;11.20):
    "because he made the machine that could pour it for you"

Other examples with stories included the following (83):

(83) pencil sharpener spiller, picture taker liker, bottle opener breaker, tea lover taker

These examples feel intuitively more difficult for adults as well, but 9 out of 10 adults gave 89 % recursive answers; only 2 out of 45 gave a conjunctive reading.[41] 5–7 year olds however showed a strong bias toward conjunctive readings: 74 % conjoined readings and 23 % recursive readings. and 22/37 children had 0/1 recursive readings from five examples, showing more than half of the children for whom the option was almost completely excluded.

A question arises: Why should this form of recursion be so much later than adjective recursion (*second green ball*) which also is left-branching? Why does it feel more difficult to adults? Snyder showed that recursive noun compounds [*school lunch box*] are evident in the 3–4 year range, while these verbal compounds are significantly delayed.

We need to look carefully at the representations to fashion an answer. Snyder (2001) established a connection to the presence of particles in Germanic languages and showed that the emergence of productive nominal compounds arises in children in nursery school after the appearance of particles. Snyder and Roeper (2004) argue that a syntactic operation of incorporation moves the verbal object to a position inside the verb.[42]In both instances, a leftward movement operation is involved[43]:

(84)
```
      N
     /\
    N clitic
   /    \
 cup   coffee
 ←============

        -er drink coffee => -er coffee-drink => coffee-drinker
```

We follow many accounts in arguing that recursive compounds are formed in two steps from a structure like this for the critical case: tea-pourer-maker

[41] A kind of conjunctive reading (i.e. *and* or *or*) is possible if one takes a 'slash' interpretation, as in *This is a printer/copier*. When the experiment was designed, we were unaware of this option, and it is interesting that it was not taken more often by adults.

[42] Snyder and Roeper utilize the Abstract Clitic position (Keyser and Roeper 1992), where particles are generated [eat-up] as the position from which incorporation occurs, which makes it possible to express the parametric connection between compounding and the presence of particles. Thus incorporation is a two-step process: move the object into the clitic position, and then it into a position inside the verb. The same mechanism is used for apparently simple compounds like: restaurant coffee cup See Bauke (2012) for a follow-up to this account.

[43] We use the Snyder/Roeper analysis for concreteness, though it is not critical to the claim we will make.

```
(85)   - er [pour tea] => -er [tea-pour] => tea-pour-er
        <====   <====
```

Now this process is repeated with a second incorporation:

(86) -er2 make [tea-pourer1] => -er2 [tea-pourer1-make] = tea-pourer-maker

In this process the first compound -er1 moves over the second -er2 for the second incorporation (87):

```
(87)
                                   N
                                  / \
                               - er  VP
                           /      \   |
                           [+obj] -er  V
                                  /\
                               make N
                                      \
      [tea-pourer-make      <=======   [+obj]
                                         / \
                                        V    -er
                                    /      \
                                 [+obj]    pour tea
                                        <=== [+obj]
                                tea-pour+er
```

What part of this complex derivation provides a challenge to the child?

What factor could be causing the delay to introduce recursion into verbal compounds?

6.1 *The Relativized Minimality Approach*

We advance the hypothesis that because the derivation of forms like tea-pourer-maker involves one compound moving inside another compound, it is subject to a general constraint on an element "crossing over" an identical element that has been as has been proposed by Rizzi in the tradition of Relativized Minimality and applied to experimental results in Friedmann et al. (2009).

They show that aphasics and children show a stronger sensitivity to Cross-over than adults do. In particular they (and others) show that object clefts, but not subject clefts, are difficult (88):

(88) it was the boy that the girl saw

The difficulty with object relatives has been found in Hebrew and in many experiments in the COST project.[44] Relativized Minimality (RM) (Rizzi 1990) states (89).

(89) in a configuration a local relation between X and Y cannot hold when Z intervenes between X and Y, and Z is a potential candidate for a local relation.

They suggest that animacy could be one critical feature that,, if it is present on one NP, will not allow another animate NP to move over it:

(90) X. . . . . . .Y. . . . . . .Z
animY animZ
←=========animZ

Thus the object cannot move over an NP with the same semantic relation. While this cross-over is allowed for adults, it is not allowed in the child grammar because the child, by hypothesis, requires strict disjointness; the target and the intervener should not share any identical features. Thus while gender is different in (80), animacy is the same, causing a block. Recursive verbal compounds show the same cross-over property. The recursive form involves two instances of incorporation from a verb into a higher –er NP as Hiraga (2009) notes. It could be simply a Noun-feature or a Verb crossing over another N feature which causes the problem.

The presence of cross-over is the essence of our explanation.. But exactly how does one element "see" another and what element is seen? This is an important theoretical question to which one can imagine several answers. (We outline one technical account in the footnote.)[45]

---

[44]The COST project funded by the EU studied the acquisition of relative clauses in numerous European dialects.

[45]In the original RM scheme for wh-movement an intermediate form of Agreement between Spec and Head blocked Further movement for wh-extraction of adjuncts.
Here is another possibility (suggested by Klaus Abels (pc)): in order to make the system "see" the crossing element, we can utilize a Probe-Goal (Chomsky 2005) system where we argue that the relative clause marker seeks an element with a certain feature content (i)

(i) it was the boy who the girl liked t
| [+ANIM] |
Probe+NPanim========→ NPanim

The wh-word in (91) sees girl which satisfies the criteria and blocks a further search. In our representation, cross-over appears to apply to the two –er cases, when one crosses the other. But how does one –er "see" the other? It could, moreover, be a more abstract feature on nominalizers [both -er and -ing carry a Nominal feature] which would predict that other compounds, without two -er's, would also be difficult, like:

pencil-sharpener-making

If true, this additional abstract cross-over constraint – a property of syntactic derivations – could be what makes the recursive possessive extra difficult, more difficult than similarly left-branching adjectives (the biggest second green ball). It is what we predict on general grounds as well. Once the Generalized Transformations extend the structure beyond any lexical representation, then the complexities of syntactic derivation come into play.

7 Conclusion

We surveyed the appearance of language-specific recursion across the major types available in English: Possessives, Adjectives, PP's, Verbal Compounds, Embedded Clauses. Our analysis moves toward incorporating specific properties of linguistic theory into an acquisition mechanism. We conclude with an outline of our approach in (91).

(91) General Perspectives:

1. There is a linguistically universal form of recursion found simply in the combination of Merge and Labelling,
2. It is distinct from other language-specific forms of recursion that are linked to parametric decisions.
3. Recursion in grammar is distinct from other forms of mental recursion.

Critical Technical Assumptions:

4. Direct Recursion which produces conjunctive interpretations
5. Indirect Recursion: it obeys the Strong Minimalist Thesis and utilizes Alternating Phase Heads to impose stepwise interpretations
6. Generalized Transformations: to model the move from simple to complex structures.

Another possibility is that it is the verb crossing another verb which is the culprit:

(i) [+verb] -er tea-pourer-maker

Now the tea-pourer-make must move to the left. If the system sees pour first and then attempt to satisfy the [+verb] requirement, producing the impossible form: tea-pour ± ± er -er make or if the whole form is pied-piped: tea-pourer-er make both of which are uninterpretable. Thus we can formulize in current theory what is quite evident to intuition and plausibly connected to the several year delay in the acquisition of recursive verbal compounds.

When Cross-over is violated, children again revert to the conjoined reading when they cannot overcome the constraint.

Based on this approach we made several observations about the Language Acquisition Path.

1. Direct recursion serves as a Default representation.
2. the experience of "recursion" is necessary to acquire it because single instances are insufficient.
3. The acquisition advances by the substitution of more complex categories for simpler ones.
4. 3-level recursion is critical because the first representation of recursion could be lexically restricted forms of subcategorization.
5. the acquisition path is determined in part by parametric decisions involving Head direction, branching direction, and other properties,
6. the acquisition path is influenced by syntactic constraints, like Cross-over.

In each instance we found the presence of Default Conjunctive interpretations precede the Indirect Recursion in the adult language. We also argued that the acquisition path is determined both by Parametric factors and derivational principles like a constraint against Cross-Over. Many questions remain:

1. Does one form of recursion trigger another?
2. Is there a semantic representation for recursion that precedes a mapping onto syntax?
3. Is the acquisition path across different languages the same?

And a last important question should be addressed when a topic so close to the lives of children is discussed: are there vital pedagogical implications that proceed from our understanding of how recursion works in particular languages?

References

Adone, D. (2012). *The acquisition of Creole languages: How children surpass their input.* Cambridge: Cambridge University Press.

Amaral, L., & Roeper, T. (2014). Multiple grammars and the acquisition of L2. *Second Language Research, 30(1)*, 1–36.

Arsenijevic, B., & Hinzen, W. (2012). On the absence of X-within-X recursion in human grammar. *Linguistic Inquiry Summer, 43*(3), 423–440.

Bauke, L. (2012). *Lexical and non-lexical symmetry breaking: An analysis of nominal root compounds, nominal gerunds and small clauses.* Dissertation, Frankfurt University.

Bhatt, R., & Pancheva, R. (2004, Winter). Late merger of degree clauses. *Linguistic Inquiry, 35*(1), 1–45.

Boeckx, C. (2008). *On the locus of asymmetry in UG.* Ms., Harvard University.

Bošković, Ž. (2008). What will you have DP or NP? In E. Elfner & M. Walkow (Eds.), *NELS 37: Proceedings of the 37th annual meeting of the North East Linguistic Society.* Amherst: GLSA.

Brown, R. (1973). *A first language.* Cambridge, MA: Harvard University Press.

Bryant, D. (2006). *Koordinationsellipsen in Spracherwerb: die Verarbeitung Potentieller Gapping Strukturn.* Dissertation, Humboldt University.

Chomsky, N. (1956). *Syntactic structures*. The Hague: Mouton.
Chomsky, N. (1965). *Aspects of a theory of syntax*. Cambridge, MA: MIT Press.
Chomsky, N. (2005). Three factors in language design. *Linguistic Inquiry, 36*, 1–22.
Chomsky, N. (2008). On phases. In R. Freidin, C. P. Otero, & M. L. Zubizarreta (Eds.), *Foundational issues in linguistic theory: Essays in honor of Jean-Roger Vergnaud* (pp. 133–166). Cambridge, MA: MIT Press.
Chomsky, N. (2012, to appear). Problems of projection. *Lingua, 130*, 33–49.
Chomsky, N., Hauser, M. D., & Fitch, W. T. (2002). The faculty of language: What is it, who has it, and how did it evolve? *Science, 298*, 1569–1579.
Clark, E. (1993). *The lexicon in acquisition*. Cambridge: Cambridge University Press.
Corballis, M. C. (2012). *The recursive mind: The origins of human language, thought, and civilization*. Princeton: Princeton University Press.
de Villiers, J. G., de Villiers, P. A., & Roeper, T. (2011). Wh-questions: Moving beyond the first phase. *Lingua, 121*, 352–366.
Diessel, H. (2004). *The acquisition of complex sentences*. Cambridge: Cambridge University Press.
Eisenbeiss, S., Matsuo, A., & Sonnenstuhl, I. (2009). Learning to encode possession. In W. McGregor (Ed.), *The expression of possession* (pp. 143–211). Berlin: deGruyter.
Everett, D. (2012). *Language: The cultural tool*. LLC, New York: Random House.
Frank, R. (2006). Phase theory and tree-adjoining grammar. *Lingua, 116*, 145–202.
Friedmann, N., Belletti, A., & Rizzi, L. (2009). Relativized relatives: Types of intervention in the acquisition of A-bar dependencies. *Lingua, 119*, 67–88.
Fujimara, C. (2010). *Acquisition of recursive possessives in Japanese*. Ms., University of Massachusetts.
Galasso, J. (1999). *The acquisition of functional categories: A case study*. University of Essex dissertation, Colchester.
Gentile, S. (2003). *On the acquisition of left-branching recursive possessives*. University of Massachusetts honors thesis, Amherst.
Green, L. (2011). *Language and the African American child*. Cambridge: Cambridge University Press.
Gu, C. (2008). *Structural ambiguity and AP/PP recursion in language acquisition*. Ms., University of Massachusetts.
Hamburger, H., & Crain, S. (1982). Relative acquisition. In S. Kuczaj (Ed.), *Language development* (Syntax and semantics, Vol. 1). Hillsdale: Lawrence Erlbaum.
Hinzen, W. (2007). *An essay on naming and truth*. Oxford: Oxford University Press.
Hiraga, M. (2010). *Acquisition of recursive verbal compound nouns*. Paper presented at the 32nd annual conference Deutsche Gesellschaft für Sprachwissenschaft (DGfS), Berlin.
Hollebrandse, B., Hobbs, K., de Villiers, J., & Roeper, T. (2008). Second-order embedding and second order false belief. In A. Gavarro & M. J. Freitas (Eds.), *Language acquisition and development, proceedings of GALA 2007* (pp. 270–280). Newcastle: Cambridge Scholar Press.
Hornstein, N. (2009). *A theory of syntax: Minimal operations and universal grammar*. Cambridge: Cambridge University Press.
Hubert, A. (2009). The acquisition of ellipsis and the structure of the noun phrase Diplomarbeit Potsdam. MA-thesis.
Jackendoff, R. (2010). *What's in the human language faculty: Two views*. Ms., Tufts University.
Jackendoff, R., & Pinker, S. (2005). The faculty of language: What's special about it? *Cognition, 95*, 201–236.
Jensen, B., & Thornton, R. (2007). *Fragments of child grammar Macquarie*. Ms., Macquarie University in Australia.
Joshi, A. K. (2014). Recursion in language: Is it indirectly constrained. In M. Speas & T. Roeper (Eds.), *Recursion: Complexity in cognition*. Dordrecht: Springer.
Keyser, S. J., & Roeper, T. (1992). Re: The abstract clitic position. *Linguistic Inquiry, 23*, 89–125.
Koopmans, H. (2014). Recursion restrictions: Where grammars count. In T. Roeper & M. Speas (Eds.), *Recursion: Complexity in cognition*. Dordrecht: Springer.
Kremers, J. (2000). *Genitives: A recursive linear Approach University of Nijmegen*.

Lebeaux, D. (2000). *Language acquisition and the form of grammar*. Amsterdam: Benjamins.

Levinson, S. (2013). Recursion in pragmatics. *Language, 89*, 148–163.

Lieven, E. (2009). Developing constructions. *Cognitive Linguistics, 20*(1), 191–199.

Limbach, M., & Adone, D. (2010). Language acquisition of recursive possessives in English. In K. Franich, K. M. Iserman, & L. L. Keil (Eds.), *Proceedings of the 34th annual Boston University Conference on Language Development (BUCLD)* (pp. 281–290). Somerville: Cascadilla Press.

Lobina, D. (2011). *What recursion could not be*. Ms., Universitat Rovira i Virgili.

Marcilese, M., Corrêa, L. M. S., & Augusto, M. R. A. (in press). Recursive pre-nominal odifiers interpretation in language acquisition. In S. Stavrakaki, P. Konstantinopoulou, & M. Lalioti, M. (Eds.), *Advances in language acquisition*. Cambridge Scholar Publishing.

Matthei, E. H. (1982). The acquisition of prenominal modifier sequences. *Cognition, 11*, 301–332.

Munn, A. (1999). First conjunct agreement: Against a clausal analysis. *Linguistic Inquiry, 30*(4), 643–668.

Noveck, I., & Chevaux, F. (2002). The pragmatic development of and. In B. Skarabela, S. Fish, & A. H.-J. Do (Eds.), *Proceeding of the 26th annual Boston University Conference on Language Development (BUCLD)* (pp. 453–463). Somerville: Cascadilla Press.

Perez, A., Castilla, A., Bejar, D., & S. Massam. (2011). *Recursion as a learning problem: Developing a paradigm for recursion in language development*. Ms., University of Toronto.

Perfors, A., Tenenbaum, J. B., Gibson, E., & Regier, T. (2010). How recursive is language? A Bayesian exploration. In H. van der Hulst (Ed.), *Recursion and human language* (pp. 159–175). Berlin: Mouton DeGruyter.

Richards, M. (2011). *Deriving the edge: What's in a Phase*. Ms. Frankfurt.

Rizzi, L. (1990). *Relativized minimality*. Cambridge, MA: MIT Press.

Rodriquez, C. (in preparation). *Recursion and co-relatives*. Ms., University College London.

Roeper, T. (2007). *The prism of grammar: How child language illuminates humanism*. Cambridge, MA: MIT Press.

Roeper, T, (2011). The acquisition of recursion: How formalism articulates the child's path. *Biolinguistics*, 5.1–2: 057–086.

Roeper, T., & Snyder, W. (2004). Recursion as an analytic device in acquisition. In J. van Kampen & S. Baauw (Eds.), *Proceedings of generative approaches to language acquisition (GALA) 2003* (pp. 401–408). Utrecht: LOT.

Roeper, T., Terunuma, A., & Nakato-Mishiyata, T. (2012). *Acquisition of recursive possessives in Japanese*. Ms. University of Massachusetts, Amherst.

Sauerland, U. (2012). Paper presented at Complexity Conference, Konstanz.

Snyder, W. (2001). On the nature of syntactic variation: Evidence from complex predicates and complex word-formation. *Language, 77*, 324–342.

Snyder, W. (2007). *Child language: The parametric approach*. Oxford: Oxford University Press.

Snyder, W., & Roeper, T. (2004). Learnability and recursion across categories. In A. Brugos, L. Micciulla, & C. E. Smith (Eds.), *Proceedings of the 28th annual Boston University Conference on Language Development (BUCLD)* (pp. 543–552). Somerville: Cascadilla Press.

Snyder, W., Hiramatsu, K., & Roeper, T. (2001). Learnability in a minimalist framework: Root compounds, merger, and the syntax-morphology interface. In I. Lasser (Ed.), *The process of language acquisition*. Frankfurt: Peter Lang Verlag.

Sowalsky, E., Hacquard, V., & Roeper, T. (2009). Is PP opacity on the path to false belief. Cascadilla Proceedings Project. Somerville, Cambridge: Cascadilla Press.

Stabler, E. (2014). Recursion in grammar and performance. In M. Speas & T. Roeper (Eds.), *Recursion: Complexity in cognition*. Dordrecht: Springer.

Stickney, H. (2009). *The emergence of the DP in the partitive structure*. Dissertation, University of Massachusetts.

Tavakolian, S. (1981). The conjoined-clause analysis of relative clauses. In S. L. Tavakolian (Ed.), *Language acquisition and linguistic theory*. Cambridge, MA: The MIT Press.

Tomasello, M. (2003). *Constructing a language: A usage-based theory of language acquisition*. Harvard: Harvard University Press.

Trueswell, R. (2004). *Attributive adjectives and the nominals they modify.* Master thesis. Oxford University.

van Hout, A., Kamiya, M., & Roeper, T. (2012). Passivization, reconstruction and edge phenomena: Connecting English and Japanese nominalizations. *Natural Language and Linguistic Theory, 31*, 137–159.

Weissenborn, J., deVilliers, J., & Roeper, T. (1995). Wh-features in French and German: Connections between case, Wh-features, and unique triggers. *Recherches Linguistiques, 24*, 125–157.

Recursive Complements and Propositional Attitudes

Jill de Villiers, Kathryn Hobbs, and Bart Hollebrandse

Abstract The focus of this chapter is in what role syntactic recursion might play in the representation of propositional attitudes. Syntactic complements under mental and communication verbs are recursive, and so also are the propositional attitudes. There is strong evidence that children take some time to master the first order syntactic complementation typical of verbs of communication. When they do acquire these structures, the evidence suggests that this helps children reason about propositional attitudes such as false beliefs. In this chapter we seek to deepen our understanding of the crucial property of sentential and attitude embedding. Is the crucial aspect that the truth value of the complement differs from that of the embedded clause, or is it that both the sentence forms and the propositional attitudes are recursive? We show that new insight can be gained by examining higher order levels of both sentence embedding and propositional attitudes. Based on some empirical data on second order complements and propositional attitude reasoning in children, we propose that truth contrasts between clauses may provide a crucial trigger for recursive complements.

Keywords Language acquisition • False beliefs • Propositional attitude • Complement • Speech act verb • Attitude embedding

---

J. de Villiers (✉)
Department of Psychology, Smith College, Northampton, MA, USA
e-mail: jdevilli@smith.edu

K. Hobbs
Department of Psychology, Harvard University, Cambridge, MA, USA
e-mail: khobbs@fas.harvard.edu

B. Hollebrandse
Dutch Department, University of Groningen, Groningen, The Netherlands
e-mail: b.hollebrandse@rug.nl

T. Roeper and M. Speas (eds.), *Recursion: Complexity in Cognition*, Studies in Theoretical Psycholinguistics 43, DOI 10.1007/978-3-319-05086-7_10,

1 What Use Is Recursion?

The focus of our interest is in what role syntactic recursion might play in the representation of propositional attitudes. Syntactic complements under mental and communication verbs are recursive, and so also are the propositional attitudes:

1) Peter thought that Kate believed that Bart said that Jill would write it

There is strong evidence that children take some time to master first order complementation, as in:

2) The girl said there was a butterfly in her hair (but it is a leaf)
What did the girl say there was in her hair?
(correct: a butterfly, wrong: a leaf))

When they do, the evidence suggests that this helps children reason about propositional attitudes like *false beliefs,* namely that another person has a belief that is false relative to what you believe.

In this chapter we ask what is it that matters about this *first order* relationship: is the crucial aspect the embedding of a different truth-value in the complement than the matrix clause, or is it that both the sentence forms and the propositional attitudes are recursive? In this chapter we review the reasoning about the relationship and its interpretation. The obvious question emerges: what about higher level orders of both sentence embedding and of propositional attitudes? We review some empirical data on *second order* complements and propositional attitude reasoning in children, and propose that truth contrasts between clauses may provide a crucial trigger for recursive complements.

2 The Representational Medium

To represent propositional attitudes in the mind requires a medium that can capture all the subtle properties of intentionality, opacity and recursion (Fodor 1975, 2008; Segal 1998; Collins 2000). Intentionality refers to the fact that mental states have contents, and these contents can be not just singular things but whole propositions. One can imagine *an apple*, but one can also imagine, or believe, or forget, that *the apple fell behind the sideboard.* So the medium in question for representation must be capable of representing propositions, that is, structured strings of symbols with a truth-value. Opacity refers to the fact that the contents of mental states may in fact be nonexistent in reality. That means one can have beliefs about nonexistent objects, such as unicorns, or angels. Crucially, the contents of mental states look inwards to mental referents, not outwards to external referents. For instance, the statement in 3):

3) Peter believes there is a ball behind the sideboard

is not false, even though the object in question is an apple. To Peter, it is a ball. Finally, propositional attitudes are recursive, in that it is possible to have beliefs about beliefs:

4) Jill believes Peter believes there is a ball behind the sideboard

What kind of medium can these thoughts exist in? Images won't do. Although images can be recursive (e.g. fractals), and they can represent non-existent things (e.g. a unicorn), they do not have a truth value, and hence cannot be used to represent false or negative things (Fodor 1975).

Words won't do. Lexical items are not rich enough to capture propositions. If we want to say that John thought that his hat was his wife (Sacks 1985), we could say he was deluded, or mistaken, but that would not capture the content of his mistake. The contents of a propositional attitude are essential to its role in causal relations to other mental states and to behavior (Collins 2000).

A sentence, however, is just right. Several have argued that natural language sentences have the properties needed, although there is theoretical disagreement about whether the representational medium is a Language of Thought (Fodor 1975, 2008), Logical Form in natural language (Carruthers 2002) or Total Forms of natural language itself (Segal 1998; Collins 2000). But ordinary sentences have only a single proposition, and to capture the contrast between reality and the false belief of an individual, a complement clause is required if the meaning is to be conveyed in a single sentence (see Hollebrandse and Roeper 2014, this volume, for the question of whether discourse about false beliefs can be equivalently recursive).

The complements under verbs of mental state and communication have special semantic and syntactic status. For example,

5) Billy thought *that it was raining*

Even when the complement clause "it was raining" is false, the whole sentence is not false. This is unlike e.g. an adjunct clause, such as one headed by *when* under an ordinary, non-mental verb:

6) Billy left *when it was raining*

If the adjunct clause is false, the whole is also false. The truth shift may be an essential property that makes complement structures ideally suited for representation of false beliefs. Chomsky (2010) argues that Internal language [I-language] evolved primarily for the advantages it conferred in reasoning. Here we have a case in point. Whether or not this is true, the structures provided by I-language are ideally suited for reasoning about propositional attitudes, though of course this reasoning can go on unconsciously. Carruthers (2002) makes a related claim that reasoning could be formulated as LF, and that only when it is cloaked in phonological form do we experience it as conscious thinking (see also Jackendoff 2011). Critically for false belief reasoning, the reasoner must be able to manipulate the content of a false proposition not *just as* false but also as causally related to the intentions and goals of the holder of the propositional attitude. For example,

7) If
Billy thinks that it is raining
Then
Billy will put up his umbrella/cancel the barbecue/put on galoshes

If the person had no way to hold that representation, but instead dismissed the proposition "it was raining" as false, that person would be unable to either predict or explain Billy's behavior.

3 Developmental Data

Recursion is said to be a fundamental feature of human language (Hauser et al. 2002), and complements are also claimed to be universal by some linguists (Wierzbicka 1996), though this is now in contention (Everett 2008; Sauerland and Trotzke 2011). In principle, Universal Grammar? should provide recursion as a fundamental. In actuality, the forms in E-language develop over the preschool years, both in production and comprehension. This could be because not every language has recursion of every constituent type (Hollebrandse and Roeper 2014, this volume), so a child must proceed with caution. Or it may be that embedded structures exceed the performance capacities of young children.

When do children get complements on the special class of communication and mental verbs? Production data suggest that children start producing the forms between age 2 and 3 years (Bartsch and Wellman 1995). However, the sentences are often not false complements, but are more like "opinion markers", or evaluatives (see also Diessel and Tomasello 2001), and most often self-referent rather than referring to others' states of mind:

8)
*CHI: what he xxx name # I [/] I think ?
*CHI: I going make a trailer # I think
*CHI: I think I will use dis [: this] color
*CHI: I think you have sugar
*CHI: I think I will

In work on wh-questions, de Villiers (de Villiers 1995) had explored how children answer two-clause questions that require long distance movement:

9) What did the girl say was in her cereal t?

These questions were initially just "control" questions for our work on barrier effects on long distance movement (de Villiers et al. 1990). However, we realized that the children showed significant developmental growth in how they responded to these questions, too. Critically, we introduced stories in which the person said something false, because we needed to see if the trace of the wh-question was under the scope of both verbs:

10) The girl thought there was a bug in her cereal
But it was really a raisin
What did the girl say [there was t in her cereal]?

To get the right answer to the wh-question entails knowing that the clause is embedded, that is, that the wh-trace is embedded under the scope of BOTH verbs. In fact, children younger than about 4 years do not do this, but answer as if the trace was only connected to the second clause:

11) The girl thought there was a bug in her cereal
But it was really a raisin
What did the girl say [there was t in her cereal]?
Adults, 4-year-olds say "a spider"
Children below about 3.10 answer: raisin
[What] was in her cereal? = raisin

At first we thought the children's errors might be due to a recency effect, so we turned the stories around, but the same errors occurred (de Villiers 1999). Then we entertained the obvious possibility that the young children did not understand Theory of Mind yet. Perhaps they needed to pass false belief tasks before they could answer the questions correctly.

In fact, the reverse seems to be true. Children pass these tests of complementation *before* they can do standard Theory of Mind tasks. The **contingency** between the tasks has been shown for:

A) Preschool-aged typically developing children
(de Villiers and Pyers 2002; de Villiers 2005a, b; Milligan et al. 2007)
B) Oral and Signing Deaf children, whether language delayed or not
(de Villiers and de Villiers 2003; de Villiers 2005a, b; Schick et al. 2007)
C) Children with language delays (LI)
(de Villiers et al. 2003)
D) High functioning children with autism (Tager-Flusberg and Joseph 2005)
E) Bilingual Spanish-English children (Hobbs et al. in preparation)
F) Training studies (Hale and Tager-Flusberg 2003; Lohmann and Tomasello 2003)

It is necessary to emphasize the results with deaf children in particular. We found no delay in the development of false belief understanding in children who were native learners of American Sign Language, but serious delays in children who were learning English as their first language and thus quite delayed in their grammars. These children are bright, sociable and adept at learning, so one might expect that if there were an alternative way to represent others' beliefs (say by imagery, or acute attention to eyegaze cues) they might spontaneously invent it, but they did not. These findings are even more dramatic in the work of Jennie Pyers on the first cohort of Nicaraguan signers, whose pidgin Sign was apparently insufficient even as young adults for representing mental state language (Pyers and Senghas 2009).

A natural conclusion is that mastery of false beliefs rests on this aspect of language: children need the structures to represent false beliefs. This conclusion has been hotly disputed by other researchers who posit a role for language other than the representational one that we defend here (Woolfe et al. 2002; Nelson 2005; Harris 2005; Ruffman et al. 2003), for example, arguing that language is the medium by which we learn about the cultural theory of mind. It is also dismissed by infancy researchers who find evidence of some sensitivity to others' belief states using eye-gaze measures long before infants have any capacity for complex sentence representation (Onishi and Baillargeon 2005; Doherty 2006; Southgate et al. 2007). For analyses of these and other objections see de Villiers (2005a, b, 2009; Perner and Ruffman 2005; Hutto 2008; Low 2010; Apperly 2011). We proceed under the assumption that for false belief tasks that entail a *decision*, which may or may not be the only tasks that require representation, the evidence is strong for linguistic complementation being a powerful predictor.

We have argued that sentential complements provide the structure needed to represent the false beliefs of others, because they can embed false propositions. But questions remain about what children know when they answer a question such as 9) correctly. What kind of failure is it when children answer the question with respect only to the lower clause? In recent work (de Villiers et al. 2011, 2012; de Villiers 2009) de Villiers and Roeper have argued that children may obey an idealization suggested by Chomsky (2005) as an outcome of the Strong Minimalist thesis, that Phases should be shipped off to interpretation one Phase at a time. As a consequence, children may commit a premature semantic closure of the lower clause, a closure that is suspended in adult grammar to allow long distance wh-movement in circumscribed cases like complements. Importantly, it is tensed clauses that provide the occasion for this premature interpretation, not infinitives. The presence of tense is associated with an assertion, a proposition, therefore having a truth value:

Hypothesis 1: Mastery of a sentence with a false tensed complement clause is requisite for representing false beliefs.

4 When Is Embedding Recursion?

Complements under mental and communication verbs are also recursive, so can be used to report second, third, nth-level propositional attitude reports. The further question arises, do children in fact recognize at the first stage that the complement is recursive? Is embedding one complement enough to signal to the child that it is 'recursive'? Roeper (2007) and Hollebrandse and Roeper (2007) argue that

a) True recursion is infinite
b) Languages vary in which elements they allow to be recursive
c) Languages that disallow true recursion for some element may nevertheless allow one embedding. For example:
   Compounds in French: only ones are two nouns, and not free but idiomatic
   Possessives with 's in German: only one allowed
   Serial verbs in English: "come get" but not "come get eat"

The implication of their work is that being able to embed one complement does not necessarily mean that the language has recursion of complements, if the rule extends to all such cases.

If Roeper is right, then children may be conservative, that is, perhaps only evidence from *two* embedding complements may provide the trigger that they are recursive in English. These data are likely to be quite rare in the input. Alternatively, perhaps sentential recursion is universal, in which case the child may not be conservative and may generalize immediately. A third possibility is that recursion of complements may be universal but lexically restricted, either by type of verb or by type of complementizer, and children are conservative until the lexical properties are established. Therefore we need to ask, is recursive complementation available to English speaking children as soon as they get complements? Roeper and Hollebrandse would argue that recursion is not automatically triggered – for complements in particular – by mastery of one level of embedding.

Empirical data are thin when it comes to recursive complementation, and vanishingly rare in spontaneous speech of children and their caregivers. Several studies were designed to see when children could handle such structures in comprehension. Notice that mastery of the false complement structure can occur some months before mastery of the false belief task (de Villiers and Pyers 2002). Perhaps a more proximate cause of the change is establishing that complements are recursive. It might be that *recursion* is the key to really understanding and representing false beliefs at all. In other words, only when a child gets *recursive* complements does she have the structures for reasoning about others' false beliefs:

Hypothesis 2: Recursive complements are needed for level 1 false belief understanding

However, there is another alternative. It is of theoretical interest to ask whether there is any systematic relationship between mastery of recursive complements and mastery of recursive propositional attitudes. Roeper (2007) has proposed that the child may not in fact have recursion of either sentences OR propositional attitudes when they pass first order complements and first order false beliefs at age 4. Perhaps true recursion is later for both. One complement may be less clear in structure, but two complements may prove to the child that the form is recursive.

In addition, several theorists of Theory of Mind have criticized the standard false belief tasks as too low-level to really capture the knowledge of propositional attitude reasoning that adults have (Dennett 1987; Hutto 2008). That is, there are arguments on the other side about first order intentionality or propositional attitudes

(Dennett 1987) being different in kind from second order and up. Essentially, first order intentionality may not require propositional-level contents, but may be reflected in the reading of a behavioral disposition (the "killjoy" hypothesis in Dennett 1987). Imagine a 17-month-old human, watching a person fail to notice a change in the hiding place of an object. When the infant then expects the character to go to the place founded on their false belief/ignorance, this has been interpreted as false belief understanding, as if the infant had the representation "Jack thinks it's in the red box" and that fed their eyegaze direction. But it could also be an expectation based on intention understanding, such as expecting Jack to continue to have the behavioral direction he had before. Developmental psychologists are hard at work to distinguish the two alternatives empirically (Dienes and Perner 2002; Baillargeon et al. 2010; Perner 2010; Low 2010). It may be that the child can understand a situation by developing some mnemonic, or behavioral expectation, based on a rule simpler than a propositional attitude ascription. Infants seem to be doing just this as long as they do not have to make a decision: if they have to act on the information, then it appears that language competence becomes entailed. In this latter case, it is very difficult to tease apart what role language plays. Language could be serving as a tool, providing a more effective mnemonic on a mental notepad. For example, language might assist in the control processes necessary to select a response (Jacques and Zelazo 2005), or it could serve as a way of keeping track of abstract relations (Loewenstein and Gentner 2005) Or, it could be the critical representational medium (de Villiers 2005a, b).

However, consider second order intentionality, as expressed in:

"Jack thinks Jill thinks it's in the red box".

At this point, simple heuristics such as following eyegaze or keeping track of orientations seem very hard to track. Propositional attitude reasoning seems the most likely alternative: one has to have a way to represent the fact that the content of Jill's thought is not in itself the content of Jack's thought, and yet Jill as the holder of the belief in question, and the content of her belief, are under the scope of Jack's thinking. There is no clear alternative to the complex hierarchical structure: we have to know which agent has which belief and not confuse them. Philosophers such as Dennett (1987) consider the breakthrough point in intentionality to be from first to second order. It is clear that human language has the rich kind of structure necessary for generating such a representation, for holding it in memory, comparing and contrasting the contents of beliefs, and so forth.

First then, it might then be argued that *true* understanding of both complements and false beliefs is established when each one can be shown to be recursive. In addition, recursive complements are a most suitable medium for representing recursive propositional attitudes. For that reason, Hypothesis 3 is a strong contender:

Hypothesis 3: *Recursive* complements are necessary for *recursive* (level 2+) false belief reasoning.

One last consideration is necessary before we consider the evidence. Perhaps recursion is an incidental rather than significant feature of sentences such as:

"Jack thinks Jill thinks it's in the red box".

That is, what matters to its complexity is that the final clause is false, and therefore we need to monitor who believes that. In sentences with single complements, the falsity of the embedded proposition is a significant indicator that the clause is not an adjunct but an embedded form. Adjunct clauses (e.g. because, so, when) must have "true" as their truth value:

Jill thought that *because the lid was open.*

Extending this idea, the real trigger of *recursion* of complementation is the recognition that the complement is false.

Hypothesis 4: In order to recognize genuine complement recursion, children need truth-value contrasts between clauses.

5 Empirical Studies on Second-Order Recursion

In our first study (Hollebrandse et al. 2006) we tested whether second-order complements (i.e. recursion in the syntax) was a predictor of false belief reasoning. We tried to divorce recursion from falsity of the complements to see what was most important, testing the various Hypotheses 1–4.

Hypothesis 1: Mastery of a sentence with a false tensed complement clause is requisite for representing false beliefs.
Hypothesis 2: Recursive complements are needed for level 1 false belief understanding
Hypothesis 3: *Recursive* complements are necessary for *recursive* (level 2+) false belief reasoning.
Hypothesis 4: In order to recognize genuine complement recursion, children need truth-value contrasts between clauses.

We devised short stories in which different people spoke about what others said, and rather than varying truth value, we varied the likely *opinion* of different characters, to provide some pragmatic rationale for using recursive complements. For example, Fig. 1 provides the picture for a short story:

12) Billy is talking to Jane. Mom is cleaning up
Billy tells Jane that Mom said that comic books are stupid

Fig. 1 A sample picture from the second order complements about attitudes task (Hollebrandse et al. 2008) (Figure: author's own)

Just as in the first order complement task, we asked a wh-question to elicit an answer that would enable us to see if the child understood the embedding. We devised two different wh-questions because some adults told us that the first type of question was ambiguous:

13) **Type 1: What Question**
What did Billy tell Jane?
(Mom says that comic books are stupid/*comic books are stupid)

14) **Type 2: Who Question**
Who said that comic books are stupid?
(Mom/*Billy)

The results show that only by about 8 years of age were children successful at answering the two kinds of questions (Hollebrandse et al. 2008).

A mistake that children commonly made was to answer with *first* order instead of second order:

Fig. 2 The bake sale second order belief task (Hollebrandse et al. 2008) (Figure: author's own)

15) Billy tells Jane that Mom said that comic books are stupid
What did Billy tell Jane?
2nd order:
"Mom says comic books are stupid"
First order:
* "Comic books are stupid"

It is very clear from this first finding that *Hypothesis 2 is not supported* by these facts. The children aged 5–6 years were well above the age at which first order explicit false belief tasks are passed. Therefore passing first order false belief tasks does not depend on mastery of second order, recursive complementation. Furthermore, second order complementation is difficult for children of this age even when falsehood is not a factor.

But is second order complementation necessary for second order false beliefs (Hypothesis 3)? Our second order false belief task followed the design of other such tasks in the literature (Perner and Wimmer 1985; Sullivan et al. 1994) but we took care to distinguish the possible answers to zero order, first order and second order beliefs (Fig. 2).

16) **Second Order False Belief scenario**
Bake Sale
Picture 1: Sam and Maria are playing together and look out the window and see that the church is having a bake sale
Picture 2: Sam watches Maria grab her money and run out of the house and over to the church. "Ah ha," Sam says. "She's going to buy chocolate chip cookies, her favorite." And then Sam goes back to playing with his toys

<u>**Picture 3:**</u> On her way to the bake sale, Maria runs into the mailman and tells him, "I'm going to get a nice pumpkin pie for Grandma."
Probe: Does Sam know that Maria is going to get a pumpkin pie?
<u>**Picture 4:**</u> When Maria gets to the bake sale, she finds out that all they have are brownies. So she couldn't buy a pumpkin pie for her grandma, but buys a bunch of brownies to bring back to her family instead
Probe: Does Sam know that she bought brownies?
1st order FB: What did Maria think (she) was (going to buy) at the bake sale?
<u>**Picture 5:**</u> Back at the house, Mom comes in and says to Sam "I noticed that the church is having a bake sale." "Oh yes," Sam says. "Maria went there." Then Mom asks, "oh, what does Maria think they're selling at the bake sale?"
2nd order FB: What does Sam tell his mom?
Why does he tell her that?

What is required to answer correctly? The stories do not use second-order complements, so the task does not entail processing second order language. The child must instead make the connection across the discourse in order to answer. The steps are something like this:

17) "Ah ha," Sam says. "She's going to buy chocolate chip cookies, her favorite."
➔ Sam thinks Maria wants to buy chocolate chip cookies
➔ Sam thinks that Maria thinks they are selling chocolate chip cookies
➔ What does Sam tell his Mom?
In other words, they must *represent* the second order belief:
Sam thinks [that Maria thinks [they are selling x]]

Hollebrandse et al. (2008) tested eighteen 6-year-old children on the following tasks:

i. **Second order False Belief** (4 stories)
e.g. What does Sam tell his Mom? (chocolate chip cookies)
ii. **Second order Complements** (6 items)
e.g. What did Billy tell Jane? (His Mom said that comic books are stupid)

Notice in Figs. 3 and 4 that the children passed the first order false belief question within the second order task at 91 %. Since these same children mostly failed the second order "attitudinal" complements, understanding second order complements does not seem to be a prerequisite for first order FB understanding. In addition, we did not find that single complementation leads immediately to recursive complementation. There is at least a year or two between mastery of single complements with false propositions, and our second order "attitudinal" complements task. Second, and highly surprisingly, we found that the 6-year-old children were significantly **better** at 2nd order False Belief than at our 2nd order attitudinal complementation task. *This is contradictory to Hypothesis 3 that recursive complementation is needed to represent second order false beliefs.*

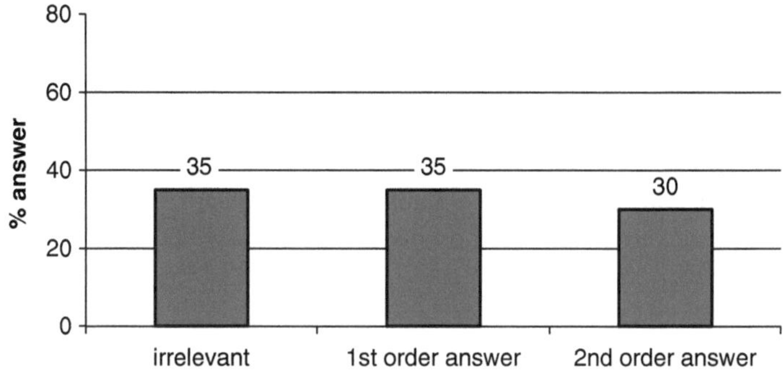

Fig. 3 Data from Hollebrandse et al. (2008) on understanding second order complements with attitudes (Figure: author's own)

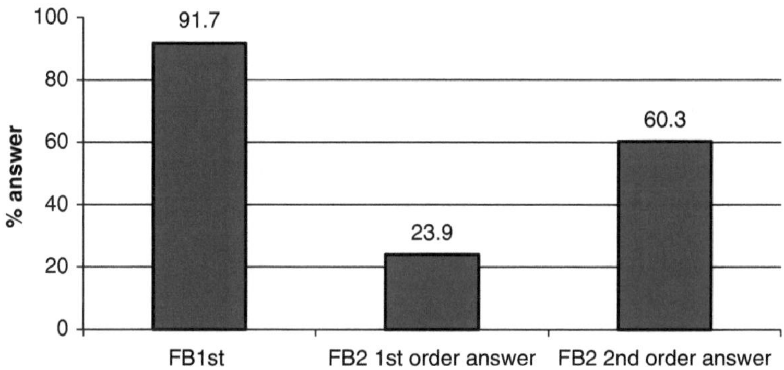

Fig. 4 Data from Hollebrandse et al. (2008) on understanding first and second order false beliefs (Figure: author's own)

Why is the second order attitudinal complementation task so hard? Some adults question our second order complementation task. Billy did "utter the words" that "comic books are stupid". Is this ambiguity the cause of the children's errors? A related explanation considers factivity: though *say* is not generally factive, a verb like *know* reveals the impact of factivity in the second verb:

18) Billy told Jane that Mom **said** that comic books are stupid
19) Billy told Jane that Mom **knows** that comic books are stupid

In 18), the factivity of the lower clause is inherited by the embedding clause, in that Billy could only say that Mom knows *p* if Billy also knew *p*. i.e. that Billy also knows that comic books are stupid. Suppose children (and some adults) consider "say" to have such inherited factive properties? In that case they might answer that Billy said that comic books are stupid, and our attempt to manipulate attitude differences was ineffective. This raises a further possibility that ties in with Hypothesis 4 and brings recursion back into the picture:

In order to recognize genuine complement recursion, children need truth-value contrasts between clauses.

Suppose it is the case that the sentences we constructed for second order complementation were not sufficiently marked as recursive because the *truth value* was not contrastive.

20) Billy told Jane [that Mom said [that comic books are stupid]].

The second complement could be argued not to have a truth value: it is Mom's evaluation, and we could merely disagree. Just as in the first order case, where false complements permit false belief reasoning, perhaps genuine second-order complements that *mark mistakes/changes in truth value* are the real prerequisite for second-order belief reasoning.

21) Billy told Jane [that Mom said [that it was raining]]. (It's not.)

There is some evidence that evaluative verbs are (sometimes) not recursive. The English verb "to consider" is not recursive:

22) Billy considered Jane to consider comic books stupid

and the Dutch verb *vinden* (another evaluative) is non-recursive (Hollebrandse and Roeper 2014, this volume). Perhaps only truth contrasts guarantee recursivity of particular verbs. We need to test the relationship between second order complements that mark truth-value changes, and second order false belief tasks.

6 Some Preliminary Evidence

The hypothesis is being subject to test on a large scale, serendipitously. As part of an intervention study exploring the effectiveness of an integrated preschool curriculum with a large sample of low-income children in Texas and Florida (de Villiers et al. 2009), we designed language and theory of mind tests to be repeated at intervals throughout the age range 2.5–7 years. As a consequence of the need to test the older children, we developed extended false belief measures, using second order belief stories, and we also expanded on our complement comprehension measures so that they too were second order. This will allow us to test whether second order complements that entail truth-value changes predict second order false belief reasoning. However, at the moment we only have preliminary data as the longitudinal study is not completely analyzed.

Fig. 5 A sample series of pictures from the new second order complement comprehension task, contrasting truth value (Figure: author's own)

At the time of writing we have data on 191 five- to six-year-olds in low-income preschools and daycares in Florida and Texas, being tested as part of a large NIH-funded study on curriculum intervention for school readiness. The children were assessed three times during the critical year of intervention, and then once a year later at follow-up. Ultimately we will have another year of follow-up data. These data are from the first follow-up testing, 1 year out of the program.

Here we only provide a brief report of the results of testing the children as the full results are still being compiled. In a large battery of cognitive, emotional and literacy/language assessments, the children all received the following tests relevant to the present questions:

a) 1st order complements (de Villiers and Pyers 2002)
b) 1st order false belief (a nonverbal analog of the standard task, see Schick et al. 2007)
c) 2nd order false belief (Hollebrandse et al. 2008)
d) 2nd order complements (contrasting truth (Hollebrandse and Van Hout 2009)

The latter task was devised to extend the single false complement task to second order, and goes as follows. Three pictures (Fig. 5) are shown to illustrate a story (23), and then a wh-question (24) is asked to diagnose the complements as before:

23) The woman called the policeman and said that there was *a rat* in her backpack.
The policeman did not hear her very well and said,
"What? The woman said there's *a cat* in her backpack! No, it must be *a scarf*."
But look. It's just *a stuffed toy*.

24) What did the policeman say the woman said t̲ was in her backpack?

First, consider the result in Fig. 6, demonstrating that the children have mastered first order complementation as indexed by the standard wh-question.

We checked whether we obtained the usual contingency between passing first order complementation and the nonverbal first order false belief, and the results

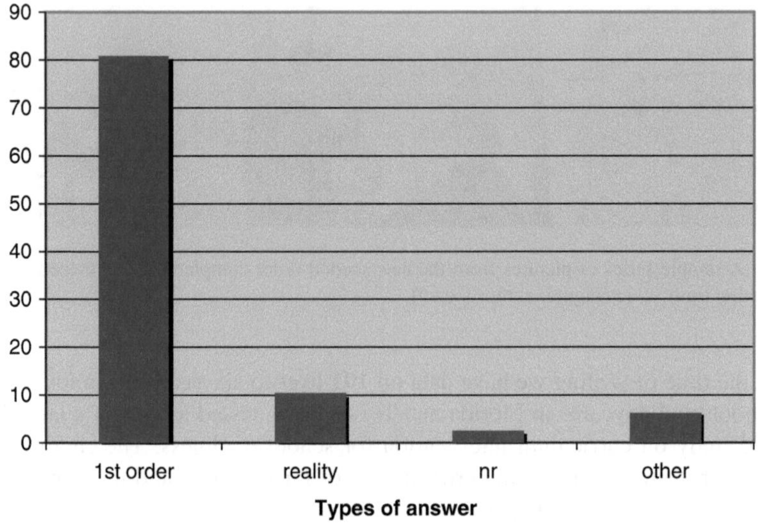

Fig. 6 Evidence of passing first order belief in the longitudinal study (Figure: author's own)

Table 1 First order complementation and nonverbal first order false belief

Contingency		Number of complements correct					
		0	1	2	3	4	Total
Number of nonverbal false beliefs correct	0	5	8	7	26	36	82
	1	1	4	9	10	38	62
	2	1	1	1	6	24	33
	3	0	0	1	1	12	14
	Total	7	13	18	43	110	191

are shown in Table 1. Passing the nonverbal false belief tasks is contingent on passing the complementation task, confirming previous results. As usual, we find an asymmetry: some children can pass first order complements but still not pass the first order false belief tasks.

What do the children do with the scenarios of second order complementation? The question was for example:

24) What did the policeman say the woman said was in her backpack?

The results so far indicate:

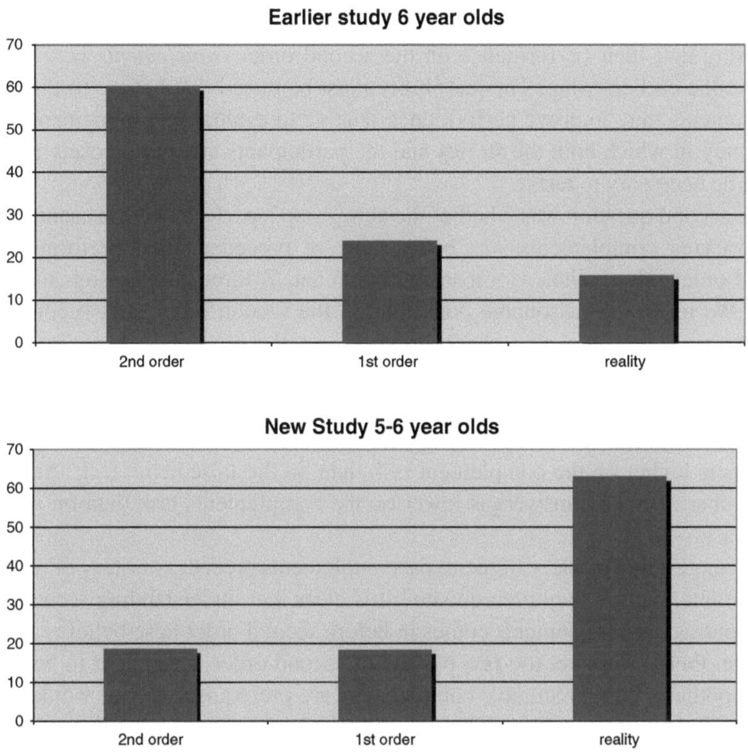

Fig. 7 Results on second order false beliefs comparing two different populations (Figure: author's own)

29 % correct 2nd order = cat
26 % 1st order policeman = scarf
16 % 1st order woman = rat
15 % reality = stuffed toy

The rate of second order answers, 29 %, is still outweighed by a preponderance of **first order answers (42 %)**. Nevertheless, there is a hint that the subjects are doing better on these second order complements than our earlier subjects did with the attitudinal complements. This larger group can be claimed to be at a less advanced stage of development than the sample of children tested previously; the large-scale study is composed of low-income children from much more impoverished educational backgrounds than children in our previous work (e.g. in Hollebrandse et al. 2008). The evidence comes from comparing the two participant groups on their performance on second order false beliefs in Fig. 7: clearly this second, larger, group is not as advanced as the previous groups of children in their theory of mind reasoning.

Given that the present subjects are less advanced in development, then it is surprising that their performance on the second order *complements* is at least as good as the earlier group. The possibility must be raised that having truth-varying complements may improve performance relative to evaluative complements, but a new study in which both the stories and the participants are more closely matched would be necessary to test it.

The second question was whether the new group's performance on second order, truth varying complements was better than, or preceded, their performance on second order false beliefs. Comparing Figs. 6 and 7, three comparisons make this point. We make the reasonable posit that on the second order task, second order answers are better than first order answers, and also that first order answers are more advanced than reality answers, even though they are wrong.

First, the percentage of success, that is, second order answers, is higher on the complements task than on the false belief task. Second, the number of first order answers is higher on the complement task than on the false belief task. And third, the number of reality answers is lower on the complements task than on the false belief task.

We suggest that mastery of recursive complements precedes mastery of recursive false beliefs, but are they prerequisite? It is clear that understanding second order truth-contrastive complements comes in before second order false belief reasoning. However there are as yet too few passers of second order false belief to be able to test Hypothesis 2 that recursive complements are prerequisite, as we would like to claim.

Why are second order complements needed for second order false belief reasoning? Remember that the stories do not use second-order complements: they are covert. The child must make the connection across the discourse in order to answer:

25) What does Maria think they are selling at the bake sale?
26) What does Sam tell his Mom?

We contend that the child must mentally represent the second order belief:

27) Sam thinks [that Maria thinks [they are selling chocolate chip cookies]]

Children at 5 or 6 cannot yet do this. Second-order complements are not yet securely represented (even with communication verbs) even when they are *overt.* Until they can handle them, indexed by our comprehension test, children will fail second-order false belief tests.

In another year, we will be able to trace whether there is a genuine contingency between passing second-order truth-contrastive complements and succeeding on second-order belief reasoning. All we can say at present is that success on complements is "leading the way".

7 Conclusion

It is not recursion of complements per se that allows children to use them for representing first order false beliefs. Hypothesis 4 finds support:

Hypothesis 4: In order to recognize genuine complement recursion, children need truth-value contrasts between clauses.

With that acknowledged, we may find evidence for Hypothesis 3:

Hypothesis 3: *Recursive* complements (i.e. truth value varying) are necessary for *recursive* (level 2+) false belief reasoning.

In both complementation and false belief reasoning, children first treat 2-level embedding as 1-level of structure. It is as if one piece of the hierarchy is flattened, or skipped over in parsing. Single level embedding, either of complements or of propositional attitudes, does not lead instantly to second level, or recursive, embedding.

We suspect that truth contrasts provide the key at two stages. Truth contrasts provide the key that certain clauses are complements rather than adjuncts, allowing the right structure to be constructed for first order complements under certain classes of verbs. We have speculated that truth contrasts may also enable children to recognize recursion of complementation more easily compared to propositions that vary in evaluative attitude. Are truth contrasts between clauses then critical triggers of sentence recursion? If this is right, then:

Truth contrasts -> sentence recursion -> false belief recursion.

This would occur at two levels, perhaps one at age 4 years and a second at age 6 years. Miller (2009), in a review of empirical work on second order theory of mind in children, calls for more research in this domain. In particular, he argues that the theories that explain the onset of first order theory of mind are often not easily adapted for explaining second order theory of mind. The story presented here is also incomplete and there remains much more to be explored.

References

Apperly, I. (2011). *Mindreaders: The cognitive basis of "Theory of Mind"*. Hove: Psychology Press.

Baillargeon, R., Scott, R., & He, Z. (2010). False belief understanding in infants. *Trends in Cognitive Science, 14*, 110–118.

Bartsch, K., & Wellman, H. M. (1995). *Children talk about the mind*. New York: Oxford University Press.

Carruthers, P. (2002). The cognitive functions of language. *Behavioral and Brain Sciences, 25*, 657–719.

Chomsky, N. (2005). *On phases*. Unpublished paper, MIT. Pdf available online at http://dmtr.ru/blog/omniling/

Chomsky, N. (2010). Some simple evo-devo theses: How true might they be for language? In R. K. Larson, V. Deprez, & H. Yamakido (Eds.), *The evolution of human language*. Cambridge: Cambridge University Press.
Collins, J. (2000). Theory of Mind, logical form and eliminativism. *Philosophical Psychology, 13*, 465–490.
de Villiers, J. G. (1995, March). *Steps in the mastery of sentence complements*. Paper presented at the meeting of the Society for Research in Child Development, Indianapolis, Indiana.
de Villiers, J. G. (1999). On acquiring the structural representations for false complements. In B. Hollebrandse (Ed.), *New perspectives on language acquisition*. Amherst: University of Massachusetts Press.
de Villiers, J. G. (2005a). Can language acquisition give children a Point of View? In J. W. Astington & J. A. Baird (Eds.), *Why language matters for Theory of Mind*. Oxford: Oxford University Press.
de Villiers, P. A. (2005b). The role of language in Theory-of-Mind development: What deaf children tell us. In J. W. Astington & J. A. Baird (Eds.), *Why language matters for Theory of Mind*. Oxford: Oxford University Press.
de Villiers, J. G. (2009, September). *On building up a sufficient representation for belief: Tense, point of view and Wh-movement*. Plenary address presented at the meeting of Generative Approaches to Language Acquisition, Lisbon, Portugal.
de Villiers, J. G., & de Villiers, P. A. (2003). Language for thought: Coming to understand false beliefs. In D. Gentner & S. Goldin-Meadow (Eds.), *Language in mind: Advances in the study of language and thought*. Cambridge, MA: MIT Press.
de Villiers, J. G., & Pyers, J. (2002). Complements to cognition: A longitudinal study of the relationship between complex syntax and false-belief understanding. *Cognitive Development, 17*, 1037–1060.
de Villiers, J. G., Roeper, T., & Vainikka, A. (1990). The acquisition of long distance rules. In L. Frazier & J. G. de Villiers (Eds.), *Language processing and acquisition*. Dordrecht: Kluwer.
de Villiers, P. A., Burns, F., & Pearson, B. Z. (2003). The role of language in the Theory of Mind development of language-impaired children: Complementing theories. In B. Beachley, A. Brown, & F. Conlin (Eds.), *Proceedings of the 27th annual Boston University conference on language development*. Somerville: Cascadilla Press.
de Villiers, J. G., de Villiers, P. A., Hobbs, K., Taylor, H., & Swank, P. (2009, April). *Language and Theory of Mind in a large low-income sample of preschoolers studied longitudinally*. Paper presented at the meeting of the Society for Research in Child Development, Denver, CO.
de Villiers, J. G., de Villiers, P. A., & Roeper, T. (2011). Wh-questions: Moving beyond the first phase. *Lingua, 121*, 352–366.
de Villiers, J., Roeper, T., Harrington, E., & Gadilauskas, E. (2012). Tense and truth in children's question answering. In A. K. Biller, E. Y. Chung, & A. E. Kimball (Eds.), *Proceedings of the 36th annual Boston University conference on language development*. Somerville: Cascadilla Press.
Dennett, D. C. (1987). *The intentional stance*. Cambridge, MA: MIT Press.
Dienes, Z., & Perner, J. (2002). What sort of representation is conscious? *Behavioral and Brain Sciences, 25*, 336–338.
Diessel, H., & Tomasello, M. (2001). The acquisition of finite complement clauses in English: A corpus-based analysis. *Cognitive Linguistics, 12–2*, 97–141.
Doherty, M. J. (2006). The development of mentalistic gaze understanding. *Infant and Child Development, 15*, 179–186.
Everett, D. (2008). *Don't sleep, there are snakes: Life and language in the Amazonian jungle*. New York: Pantheon Books.
Fodor, J. (1975). *The language of thought*. New York: Harvester Press.
Fodor, J. (2008). *LoT2: Language of thought revisited*. Oxford: Oxford University Press.
Hale, C. M., & Tager-Flusberg, H. (2003). The influence of language on Theory of Mind: A training study. *Developmental Science, 6*, 346–359.

Harris, P. L. (2005). Conversation, pretence, and Theory of Mind. In J. W. Astington & J. Baird (Eds.), *Why language matters for Theory of Mind*. New York: Oxford University Press.
Hauser, M., Chomsky, N., & Fitch, W. T. (2002). The language faculty: What is it, who has it, and how did it evolve? *Science, 298*, 1569–1579.
Hobbs, K., de Villiers, P., de Villiers, J., & Peña, E. (in preparation). *Evidence against a bilingual advantage in false belief reasoning*.
Hollebrandse, B., & Roeper, T. (2007, April). *Recursion and propositional exclusivity*. Paper presented at the Recursion in Human Languages conference, Normal, IL.
Hollebrandse, B., & Roeper, T. (2014). *Empirical results and formal approaches to recursion in acquisition*. In T. Roeper & M. Speas (Eds.), Recursion: Complexity in cognition. Dordrecht: Springer.
Hollebrandse, B., & Van Hout, A. (2009, February 7). *What did who say about who? The acquisition of double embedding*. Taalkunde in Nederland dag.
Hollebrandse, B., Hobbs, K., de Villiers, J. G. & Roeper, T. (2006, December). *Second order things*. Paper presented at UMass-UConn-Smith Language Acquisition Workshop, Amherst, MA.
Hollebrandse, B., Hobbs, K., de Villiers, J. G., & Roeper, T. (2008). Second order embedding and second order false belief. In A. Gavarró & M. J. Freitas (Eds.), *Language acquisition and development: Proceedings of GALA 2007* (pp. 270–280). Cambridge: Cambridge Scholar Press.
Hutto, D. (2008). *Folk psychological narratives*. Cambridge, MA: MIT press (Bradford books).
Jackendoff, R. (2011). What is the human language faculty? Two views. *Language, 87*, 586–624.
Jacques, S., & Zelazo, P. D. (2005). Language and the development of cognitive flexibility: Implications for Theory of Mind. In J. W. Astington & J. A. Baird (Eds.), *Why language matters for Theory of Mind* (pp. 144–162). New York: Oxford University Press.
Loewenstein, J., & Gentner, D. (2005). Relational language and the development of relational mapping. *Cognitive Psychology, 50*, 315–353.
Lohmann, H., & Tomasello, M. (2003). The role of language in the development of false belief understanding: A training study. *Child Development, 74*, 1130–1144.
Low, J. (2010). Preschoolers' implicit and explicit false-belief understanding: Relations with complex syntactical mastery. *Child Development, 81*, 597–615.
Miller, S. A. (2009). Children's understanding of second-order mental states. *Psychological Bulletin, 135*(5), 749–773.
Milligan, K., Astington, J., & Dack, L. A. (2007). Language and Theory of Mind: Meta-analysis of the relation between language ability and false-belief understanding. *Child Development, 78*, 622–646.
Nelson, K. (2005). Language pathways into the community of minds. In J. W. Astington & J. A. Baird (Eds.), *Why language matters for Theory of Mind*. Oxford: Oxford University Press.
Onishi, K. H., & Baillargeon, R. (2005). Do 15-month-old infants understand false beliefs? *Science, 308*, 255–258.
Perner, J. (2010). Who took the cog out of cognitive science? Mentalism in an era of anti-cognitivism. In P. A. Frensch & R. Schwarzer (Eds.), *Cognition and neuropsychology: International perspectives on psychological science* (Vol. 1, pp. 241–261). London: Psychology Press.
Perner, J., & Ruffman, T. (2005). Infants' insight into the mind: How deep? *Science, 308*, 214–216.
Perner, J., & Wimmer, H. (1985). John thinks that Mary thinks that: Attribution of second-order beliefs by 5- to 10-year-old children. *Journal of Experimental Child Psychology, 39*, 437–471.
Pyers, J., & Senghas, A. (2009). Language promotes false-belief understanding: Evidence from learners of a new sign language. *Psychological Science, 20*, 805–812.
Roeper, T. (2007). *The prism of grammar*. Cambridge: MIT Press.
Ruffman, T., Slade, L., Rowlandson, K., Rumsey, K., & Garnham, A. (2003). How language relates to belief, desire, and emotion understanding. *Cognitive Development, 18*, 139–158.
Sacks, O. (1985). *The man who mistook his wife for a hat, and other clinical tales*. New York: Summit Books.

Sauerland, U., & Trotzke, A. (2011). Biolinguistic perspectives on recursion: Introduction [special issue]. In U. Sauerland & A. Trotzke (Eds.), *Biolinguistics* *5*(1–2), 1–9. http://www.biolinguistics.eu

Schick, B., de Villiers, P. A., de Villiers, J. G., & Hoffmeister, R. (2007). Language and Theory of Mind: A study of deaf children. *Child Development, 78*, 376–396.

Segal, G. (1998). Representing representations. In P. Carruthers & J. Boucher (Eds.), *Language and thought*. Cambridge: Cambridge University Press.

Southgate, V., Senju, A., & Csibra, G. (2007). Action anticipation though attribution of false belief by 2-year-olds. *Psychological Science, 18*, 587–592.

Sullivan, K., Zaitchik, D., & Tager-Flusberg, H. (1994). Preschoolers can attribute second-order beliefs. *Developmental Psychology, 30*, 395–402.

Tager-Flusberg, H., & Joseph, R. (2005). How language facilitates the acquisition of false belief in children with autism. In J. W. Astington & J. A. Baird (Eds.), *Why language matters for Theory of Mind*. Oxford: Oxford University Press.

Wierzbicka, A. (1996). *Semantics: Primes and universals*. Oxford: Oxford University Press.

Woolfe, T., Want, T., & Siegal, T. (2002). Signposts to development: Theory of Mind in deaf children. *Child Development, 73*, 768–778.

Recursive Merge and Human Language Evolution

Koji Fujita

Abstract The recursive syntactic computation by means of Merge stands as one uniquely human capacity which in tandem with its interfaces with the meaning and sound systems (the C-I and S-M interfaces) defines a highly complex biological trait, the human language faculty. To explain language evolution and construct a theory of language which may attain evolutionary adequacy, it is mandatory that we first understand the origins and evolution of recursive Merge and the two interfaces. Largely drawing ideas from comparative psychology and cognitive archaeology, this paper outlines one possible evolutionary scenario, according to which (i) Merge evolved from a preexisting capacity for hierarchical and sequential object combination as typically observed in tool using and tool making (Action Merge), and (ii) the Merge-based syntax gave rise to (at least) the C-I interface and also the human lexicon.

Keywords Action Merge • Anti-lexicalism • C-I interface • Derivational recursiveness • External/Internal label • Evolutionary adequacy • Logical problem of language evolution • Sub-Merge • Tool making

1 Introduction

Biological evolution is largely a process of what Darwin called "descent with modification," that is, "change in the form, physiology, and behavior of organisms over many generations of time" (Ridley 2004: 19). So is language evolution, to the extent that we try to understand it in light of the hominid evolution. The creation of an evolutionary novelty has two major components: (A) diversification from a

K. Fujita (✉)
Graduate School of Human and Environmental Studies, Kyoto University,
Sakyo-ku, Yoshida, Kyoto 606-8501, Japan
e-mail: fujita.koji.3x@kyoto-u.ac.jp

T. Roeper and M. Speas (eds.), *Recursion: Complexity in Cognition*, Studies in Theoretical Psycholinguistics 43, DOI 10.1007/978-3-319-05086-7_11,

common origin (including functional shifts by exaptation/preadaptation; see Gould and Vrba 1982 for the distinction between these two notions) and (B) recombination of preexisting unrelated capacities into a new, interlocking system. Take the human language faculty to be a system of sound-meaning mapping mediated by the syntactic computational device. The two components of language evolution can be illustrated by asking questions such as (a) "Where did syntax come from? What was its precursor?" and (b) "How did syntax and semantics, or syntax and phonology, come to be connected with each other?" Let us call these questions the syntax puzzle and the interface puzzle of language evolution, respectively.

Admittedly, syntax and the interfaces are the two major innovative ingredients characterizing human language and distinguishing it from other animal communication systems. In the famous heuristic terminology of Hauser et al. (2002) and Fitch et al. (2005), these two may constitute the representative cases of FLN (the narrow language faculty, as opposed to FLB or the broad language faculty) unique to human language. This is not to say that we can ignore whatever small gap there is between other components of language and their nonhuman counterparts in our explanation of language evolution. For example, the conceptual capacity may be shared by humans and chimpanzees, but it does not follow that we can explain away the evolution of the human conceptual capacity by explaining the chimpanzee homologue/analogue because they are not the same thing after all. Continuity does not guarantee identity, and to argue otherwise would get one trapped into the fallacy of anthropomorphism. In this sense, all ingredients of human language should belong to FLN, to a lesser or greater extent.

With this caveat in mind, this paper outlines two possible avenues of research that may come close to solving the two puzzles mentioned above; these puzzles I take to be the core of what is sometimes called "the logical problem of language evolution" or "Darwin's problem" (Fujita 2002, 2007; Boeckx 2009; Hornstein 2009). In particular, it will be argued that the reduction of syntactic computation to the minimal operation of Merge makes it possible to compare it to other language-independent human and nonhuman capacities, thereby opening a new path towards a better understanding of how language first emerged in the human lineage, and ultimately towards a theory of language which may attain "evolutionary adequacy" (Fujita 2009).

2 Merge and Its Recursiveness

Adopting the minimalist program (Chomsky 1995, et seq.) for the biological studies of language (biolinguistic minimalism) has one major methodological advantage over other approaches (including the pre-minimalist government and binding theory of the early- and mid-1980s) particularly when we address evolutionary issues. This is because, by nullifying all the unexplained parts of UG in favor of general physical/mathematical laws acting upon the evolution and development of every complex system in nature ("the third factor" in Chomsky's 2005 sense), it promises to render the logical problem of language evolution more accessible than was once thought to be. Chomsky (2005: 8) states, "... the more varied and intricate the conditions specific to language, the less hope there is for a reasonable account of the evolutionary origins of UG." From an evolutionary viewpoint, the strong minimalist thesis SMT, to the effect that "language is an optimal solution to interface conditions that FL must satisfy" (Chomsky 2008: 135), can be understood as a grand expectation that there are no such language-specific conditions that are not explained on principled grounds.

Current minimalist syntax succeeds in minimizing the uniquely human syntactic computation by reducing it to the simplest operation of binary Merge and its recursive application. In fact, Chomsky (2010: 52) reformulates SMT as in (1):

(1) SMT: Interfaces + Merge = Language

For our discussion, this is nothing but a succinct description of the two puzzles of language evolution: In order to explain how language evolved, one must explain the evolution of syntax (now reduced to recursive Merge) and the evolution of the interfaces.

By definition, Merge takes two syntactic objects, α and β, and combines them into a set. When this newly formed set, γ, is subject to another application of Merge, we have an instance of recursive Merge. The term *recursion* (in syntax) is sometimes understood in different ways among researchers and that has been the main cause of some misdirected discussion found in the literature. For example, Everett (2005, 2009) has famously claimed that the Amazonian language Pirahã does not have recursion by showing that the language lacks complementation, relativization and other patterns of self-embedding (XP appearing inside another XP of the same category). Bickerton (2009) goes so far as to suggest that recursion is a theoretical artifact, to be replaced by iteration (and lexical properties).

Their observations, though insightful in their own right, do not speak against our well-defined syntactic recursion in terms of recursive Merge and its universal nature. For one thing, even a simplex sentence in Pirahã requires recursive application of Merge for its derivation. What is crucial here is the distinction between what may be called "derivational recursiveness" and "representational recursiveness"; the former refers to the property of an operation applying to its own output, the latter to the (concomitant) property of self-embedding that one may find in the resulting phrase structure. Needless to say, it is only in the derivational interpretation that recursive Merge makes sense, whereas representational self-embedding no longer has its place in minimalist syntax, because the bare phrase structure theory (Chomsky 2001, 2008) excludes such categorial notations as DP and VP from phrase structure. For example, the expression *John's father's house* (an instance of what Roeper 2011 calls Indirect Recursion; see also Roeper 2007) is viewed as recursive not because it has one DP inside another DP but simply because its derivation is made possible by recursive application of Merge. Likewise, the simplex sentence *John likes Mary* is derivationally just as recursive as *John thinks Mary knows Bill believes ... Jane is hungry*. It follows that the total absence of clausal embedding in Pirahã does not justify the exclusion of recursive Merge from UG, whether or not its absence in this language is because of such cultural factors as suggested by Everett's (2009) "Immediacy of Experience Principle." The situation is similar to the lack of overt wh-movement in Japanese and Chinese; no one has taken this observation to mean that the option of moving a wh-phrase overtly should be excluded from UG.

Also, recursion and iteration make a lot of difference both in the process of structure building and in the configurational nature of the resulting structure. Whereas iteration only gives rise to a flat, linear string of symbols, recursion yields the type of hierarchical structure that has long been known to be an essential property of human language. It should be stressed that iterative concatenation alone is not sufficient to create a hierarchical structure. Compare [A-B][C-D] vs. [[[[A-B]-C]-D]; string-wise, they look the same, but only the latter is hierarchically structured and can be generated only by recursive Merge. One might try to apply Merge "iteratively" to derive the latter structure. To do so, one would take the output of a prior application of Merge (say, [A-B]) as a subunit and combine it with another object (C), but this would already be an instance of recursive Merge. In other words, the same string of symbols can be generated or analyzed either recursively or iteratively, and this fact should not be ignored when we assess recent experimental studies with certain species of songbirds which purport to show that they can learn recursive syntactic structures (Gentner et al. 2006; Abe and Watanabe 2011).

Based on these and other observations, let us separate the operation Merge *per se* from its recursive application, and hypothesize that each of them is an independent capacity which needs a distinct evolutionary explanation. In other words, Merge need not be recursive by nature, and nonrecursive Merge is also a logical possibility (call this "Core-Merge"); it is only that human syntax uses Merge recursively (and iteratively, of course).

(2) Recursive Merge = Core-Merge + Recursion

This picture helps further narrow down the genuine component of FLN and make the logical problem of language evolution even more accessible. Along similar lines, several authors, including Fukui (2011), Boeckx (2009) and Hornstein (2009), have already contended that what is truly unique to human language is not Merge but its recursive application. Details aside, these authors share the view that the labeling algorithm plays a crucial role in allowing Merge to apply recursively. Hornstein (2009: 55) proclaims that "labeling ... is the central innovation of UG, the change that enables the peculiar architecture of natural language to emerge." If so, studies of language evolution can be focused on the origin of labeling, while Merge may safely be attributed to FLB as something explicable as a case of descent with modification.

While I fully appreciate the heuristic merit of this kind of reductionist thinking, I would like to point to the fact that to the extent that labeling as the innovative component of human syntax has to remain shrouded in evolutionary mystery, we are in fact restating the central problem of the origin of syntax instead of solving it. My alternative view is that Merge is susceptible of a natural evolutionary account and that the same is true of the labeling operation ("Label") if it can be reduced to Merge. Specifically, I propose:

(3) a. Label is a form of Merge, and needs no separate evolutionary explanation.
 b. Merge (including both Move and Label) is a descendant of the motor control capacity for concrete object combination.

Before expounding these points below, let me first make it clear that finding the origin of syntax in a cognitive domain that is neither linguistic nor uniquely human does not support the view that syntax as we have it today is not an autonomous, domain-specific cognitive module that is unique to the humans. The concept of modularity, together with that of evolvability, assumes a central position in current evo-devo paradigm in biology, and for the most part the situation is the same with today's biolinguistic program. Precluding the classical genetic determinism, our common goal is to attain a better understanding of the dynamic process in which any domain-specific trait develops in the individual, and evolves in the species, from preexisting domain-general ones. The following discussion is only meant to present another possible line of research that seems worth pursuing in order to reach that goal.

3 Label as Merge

As a working hypothesis, let us first assume with Fukui (2011), Boeckx (2009) and Hornstein (2009) that recursive Merge is a complex operation comprising the simple binary combinatorial device (Core-Merge) that concatenates two syntactic atoms (lexical items) into a set and a distinct labeling operation (Label; virtually the same as Fukui's 2011 "Embed").

(4) Recursive Merge:
 a. Core-Merge: $(\alpha, \beta) \Rightarrow \{\alpha, \beta\}$
 b. Label: $\{\alpha, \beta\} \Rightarrow \{\gamma, \{\alpha, \beta\}\}$, where $\gamma = \alpha$ or β.

In general terms, while Core-Merge yields a set of two syntactic atoms, Label identifies this newly formed set with one of its members, turning it to another syntactic atom to which Core-Merge can apply again. Suppose for the sake of discussion that without a label, a set is "invisible" to Core-Merge, basically in accordance with the suggestions made by those authors above. To illustrate:

(5)
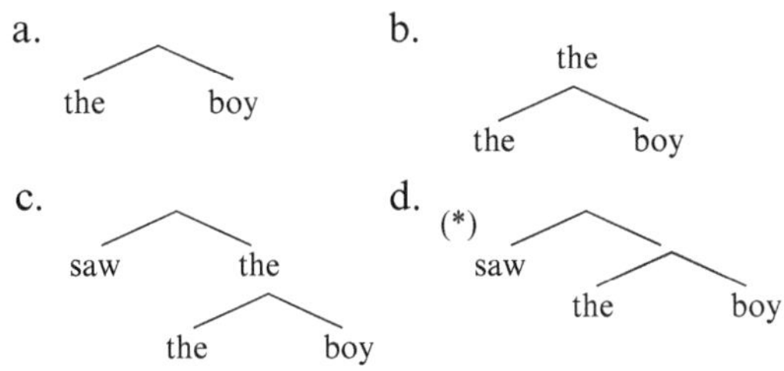

Core-Merge applies to *the* and *boy*, yielding (5a), to which Label applies and converts it to (5b). Since (5b) is now identified as a lexical item (*the*), it can undergo further application of Core-Merge, as in (5c). There is no possibility of applying Core-Merge before Label, as in (d), since here the set {the, boy} remains unidentified. Endocentricity naturally follows from this mode of Label.

There can be different ideas about how the label of a given set can be determined, about why labels are necessary for subsequent application of Core-Merge, or even about whether labels are not eliminable at all. One reasonable speculation will go as follows. Suppose that Core-Merge is always feature-driven. Two syntactic atoms, α and β, can undergo Core-Merge only if a certain formal feature ("edge-feature" of Chomsky 2008 or some possible extension of it) of either α or β detects and agrees with a matching feature of the other. Note that although Core-Merge itself is a symmetric operation, its actual application always takes place in an asymmetric way, with the edge-feature of either α or β "attracting" the other and not vice versa. The choice of the label may be an automatic reflex of this feature reaction; when the edge-feature of α (EF(α)) attracts β, α becomes the label as the active feature holder, and EF(α) is kept active for subsequent operations. Labeling a set amounts to specifying its edge-feature, without which the set remains invisible to syntax. Other rationales for the necessity of labels are easy to imagine (see Narita 2009a for some discussion on the technical details of Label), but I will not bother to discuss them as I will argue directly that Label is not an independent operation distinct from Merge. Nothing below hinges on the correctness of the reasoning above.

To see why Label can be taken to be an instance of Merge, let us first note the striking formal parallelism between Label and Move (Internal Merge). Given $\{\alpha, \beta\}$, Move may apply to it by first copying some syntactic object γ inside α or β, Merging γ and $\{\alpha, \beta\}$, and consequently yielding a new set $\{\gamma, \{\alpha, \beta\}\}$. Label works in almost the same manner; it first copies some syntactic object γ and maps $\{\alpha, \beta\}$ to $\{\gamma, \{\alpha, \beta\}\}$ by combining, or Merging, the two, the sole difference being that γ is itself α or β in the case of Label. (External) Merge applying to γ and $\{\alpha, \beta\}$ also yields the set $\{\gamma, \{\alpha, \beta\}\}$, and this time the difference is that γ is external to α and β.

(6) Merge: $\{\alpha, \beta\} \Rightarrow \{\gamma, \{\alpha, \beta\}\}$
   a. External Merge, where γ is external to α and β.
   b. Internal Merge (Move), where γ is internal to α or β.
   c. Label, where γ is α or β.

In other words, Label is a strictly localized version of Move, and to the extent that Move comes free once Merge is available (Chomsky 2005 and elsewhere), we should expect that Label also comes free. Alternatively (and more appropriately, I think), we may think of Move and Label both as evolutionary exaptations from (Core-)Merge. In either case, we now have a new picture: Label does not make Merge recursive; it is already a minimally recursive Merge.

(7) Core-Merge + Recursion $\Rightarrow$ Merge, Move, Label

Two consequences follow. First and foremost, there is no need to seek for an independent evolutionary scenario for Label. Instead, we can focus on the origin of Merge as the one central issue for studies of human syntax evolution. Secondly, we may regard endocentricity as epiphenomenal, because recursive Merge now does not require prior application of Label. At least this seems to be the case in the domain of compound formation, where exocentricity is ubiquitous (*scarecrow*, *pickpocket*, etc.). It is a minimalist assumption that compounding is also served by Merge, the one and only generative engine of human language. In fact, to accommodate exocentric compounds, we may allow Label to apply externally.

(8) External Label: $\{\alpha, \beta\} \Rightarrow \{\gamma, \{\alpha, \beta\}\}$, where γ is external to α and β.

Needless to say, this raises the question of why External (exocentric) Label is usually limited to compounding, to which I have no clear answer. It is interesting to note, however, that once External Label as in (8) is admitted, there does not seem to be any principled reason to preclude the possibility that the label γ is selected from inside α or β, in a manner analogous to Internal Merge. We then add Internal Label, too.

(9) Internal Label: $\{\alpha, \beta\} \Rightarrow \{\gamma, \{\alpha, \beta\}\}$, where γ is internal to α or β.

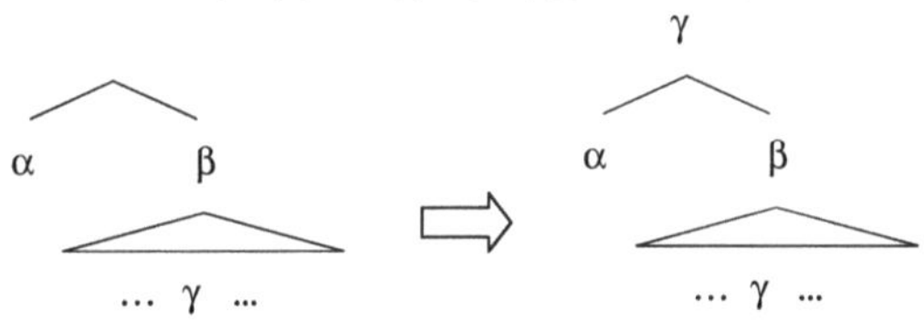

This possibility may be substantiated, for example, by Internally Headed Relative Clauses (IHRCs), the label of which must be an argument residing inside the relative clause. If something like this is correct, then Label has two sporadic extended forms, External and Internal, in addition to its canonical, strictly local Label, which is naturally expected if Label is nothing but a form of Merge.

4 Action Merge

On the assumption that Merge is the fundamental syntactic operation subsuming both Move and Label, we can set about seeking its evolutionary root(s). At this point, comparative ethological methods come into play as a highly instructive tool, as they have always been in many branches of evolutionary and developmental biology. The Merge-based bare phrase structure theory, in sharp contrast to the earlier theory of phrase structure rules and the X-bar theory, renders the comparison of human syntax with other human and nonhuman mental capacities fairly easy to make, and just for this reason it is obviously an optimal theory of syntax for studies of language evolution. Thanks to the advent of biolinguistic minimalism, we are witnessing the long-awaited crossdisciplinary collaboration of generative grammar and cognitive ethology for the first time in the history of linguistics. This is the right time for generative linguists to take into serious consideration many important works researchers in other fields have achieved so far. The acclaimed species-specificity and domain-specificity of the human language faculty, not to mention its innateness, is no excuse for disregarding researches exploring possible connections between language and other traits.

Among many other capacities that seem more or less comparable to Merge (navigation, gestures, numbers, social intelligence, etc.), the kind of hierarchically organized serial object manipulation that can be observed in tool using and tool making deserves special attention for its striking formal similarity to linguistic structure building by Merge. When capuchin monkeys crack nuts with stones, for example, they virtually combine or "Core-Merge" (NUT, STONE), sometimes followed by another Core-Merge of (NUT, STICK), to obtain the meat inside the nut by using a stick. This particular sequential behavior should be seen as an analogue of iteration rather than recursion, however, since the second operation does not apply to the output of the first one. But when chimpanzees crack nuts with a stone anvil and a stone hammer, their behavior is more like recursion as shown in (10) (see Visalberghi et al. 2009 for the recent discovery of wild capuchins acting largely the same way):

(10) a. Merge (NUT, ANVIL) ⇒ {NUT, ANVIL}
b. Merge (HAMMER, {NUT, ANVIL}) ⇒ {HAMMER, {NUT, ANVIL}}

In fact, it has been conventional in studies of animal cognition, including cognitive primatology, to describe this kind of sequential action by means of a phrase-structure-like representation. (10) can be easily mapped onto (11a), and (11b) will correspond to a case in which the animal uses another stone as a meta-tool, as a wedge to stabilize the anvil (Matsuzawa 2001; Biro et al. 2006 suspect, however, that this use of a meta-tool is accidental rather than deliberate):

(11) a.

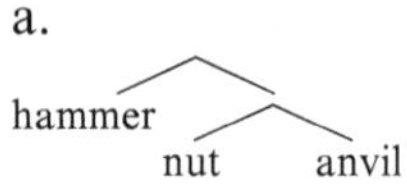

b.

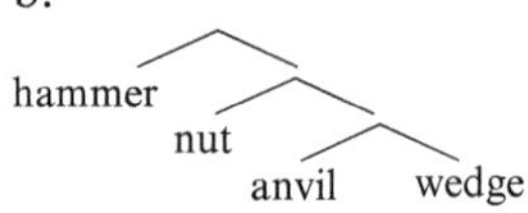

Of course, there is no obvious label in these structures, nor is it clear to what extent they can be assimilated to a genuine phrase structure of human language. What is important to note here, however, is the existence of the basic, Merge-like combinatorial operation in these animal behaviors which can be applied recursively.

The evolutionary link between language and tools has been repeatedly suggested, primarily because of the mental computation they equally require to combine the right elements (concrete objects or abstract symbols) in the right way, to create the right thing for fulfilling the purpose. Stout (2010) identifies three major kinds of capacities that are equally involved in the evolution of language and of tool use. They are shared general capacities like working memory; shared social context, including cognitive capacities for mental state attribution and joint attention; and shared more specific neural substrates, including the inferior frontal gyrus and the mirror-neuron system. I argue that the suggested link between language and tools can be further strengthened by directly connecting tool use to Merge rather than to language at large, with the former serving as the evolutionary precursor to the latter. This causal relation can be supported by the formal parallelism that exists between sequential object manipulation (call this "Action Syntax" by means of "Action Merge") and syntactic derivation by Merge coupled with the developmental fact that object manipulation precedes syntax.

In fact, quite independently of our minimalist theorizing, Greenfield (1991) already noticed and paid serious attention to this kind of parallelism between language and action through her studies of cognitive development in children (see also Greenfield et al. 1972; Greenfield 1998; Lashley 1951). She observed that young children's developing skills in hierarchically organized object manipulation, as typically exhibited in cup nesting and tool using, precede their language development and serve as a preadaptation for it, and argued that a similar situation may hold true of the evolution of language in the species. In connecting the ontogeny and phylogeny of human language via their common precursor, her pioneering cognitive developmental work well anteceded the evo-devo approach to language evolution in today's biolinguistic minimalism.

Greenfield pointed out that there are three distinct developmental stages in children's Action Grammar (our Action Merge), from the simplest Pairing strategy via the Pot strategy to the most complex Subassembly strategy. It is extremely interesting to note that these strategies neatly correspond to the development of linguistic structure building by means of different modes of applying Merge. In terms of nesting cups, these three combinatorial methods can be represented as follows (Greenfield 1991 and elsewhere):

(12) The three strategies of Action Grammar (Action Merge):

a. Pairing Strategy

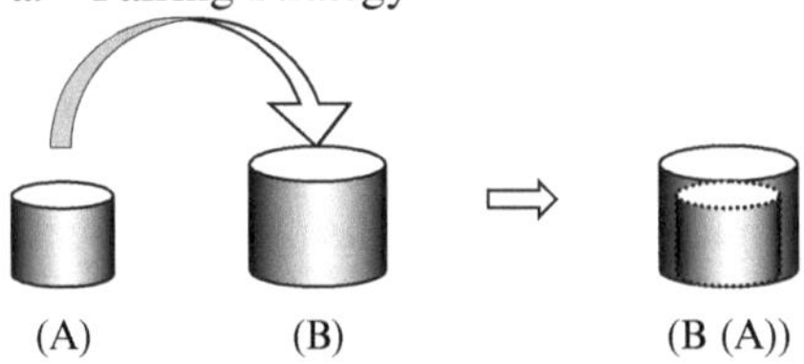

b. Pot Strategy

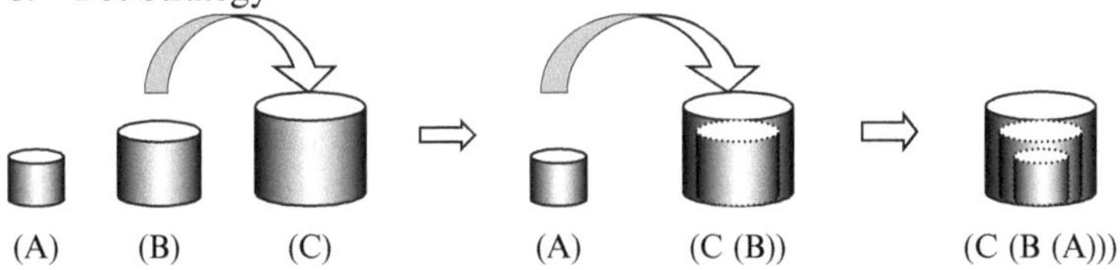

c. Subassembly Strategy

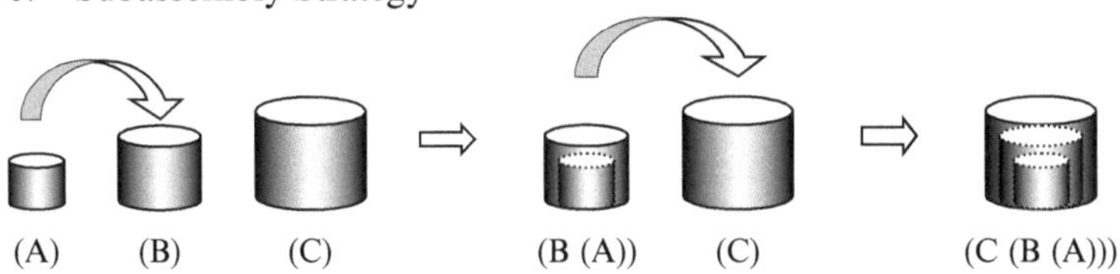

In the first, most elementary Pairing strategy, one just combines the two cups into one object by putting the smaller cup (A) into the larger one (B). This strategy corresponds to the case of Core-Merge. In the next, Pot strategy, the Pairing strategy applies repeatedly and combines three (or more) cups into one object, first by putting the middle-sized cup (B) into the largest (C), then by putting the smallest (A) into (B) + (C). This strategy is similar to applying Core-Merge to the same syntactic atom (here (C)) recursively, as if EF(C) (the edge-feature of (C)) attracts (B) and (A) sequentially.

The third strategy, called the Subassembly strategy, is crucial in understanding the complex nature of human syntax and action. This strategy may at first appear not very different from the second one, but in fact their gap is immense. In this case one first puts (A) into (B), and then takes the complex object (A) + (B) as a subunit for the next operation, which puts it into (C). Here there must be two, instead of one, edge-features acting as the attractors. First, EF(B) attracts (A), and then EF(C) attracts (A) + (B). Thus the Subassembly strategy switches the attractor at each step, and this property seems to be what gives rise to the type of recursion that characterizes human language syntax. What deserves special attention here is that in this Subassembly strategy we can see a rudimentary form of Label at work; one handles (A) + (B) as if this complex object is (B).

We learn from recent studies in comparative cognitive ethology that the Subassembly strategy is almost uniquely human (chimpanzees trained linguistically in captivity are said to be the only exception, and it remains to be seen whether wild chimpanzees also have the same capacity potentially), while the Pairing and Pot strategies are shared among other primates and nonprimates equally. Tokimoto and Okanoya (2004) demonstrate that even degus (*Octodon degu*) can perform the Pot strategy but never the Subassembly strategy. From these observations, it is only natural to suspect that the uniquely human Subassembly strategy in Action Merge is crucially involved in the formation of the also uniquely human linguistic Merge. And even if it finally turns out that Subassembly lies within other animals' capacity (I am rather inclined to believe so—imagine when you fish termites with a twig and bring them together to your mouth), it will not modify the following discussion much; it only strengthens the more favorable view that syntax but not its precursor is uniquely human.

Greenfield's (1991) contention was that Action Grammar corresponds to phoneme combination in word formation, but in response to comments and criticisms by Tomasello (1991), Swann (1998) and others, Greenfield (1998) restored the original position of Greenfield et al. (1972) and agreed that the proper object of combination is not the phoneme but the word. This move makes every sense in the context of Merge-based bare phrase structure theory; as just noted, it is possible to project the three strategies of Action Merge directly onto different strategies of set formation by Merge.

As an illustration, consider the two different ways of building the nominal expression *green tea cup*, which has two different derivations, with corresponding distinct semantic interpretations.

(13) a. Pot-Merge: Merge (green, {tea, cup}) ⇒ {green, {tea, cup}}

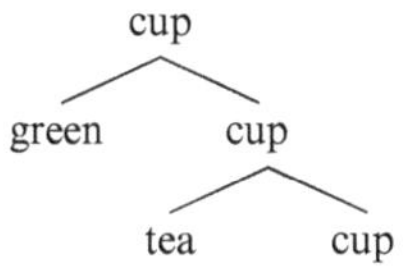

b. Sub-Merge: Merge ({green, tea}, cup) ⇒ {{green, tea}, cup}

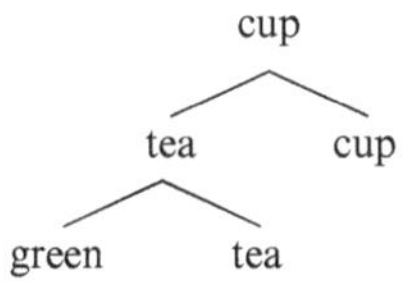

In (13a), one first Core-Merges {tea, cup}, then Merging {green} with this set in a manner parallel with the Pot strategy of Action Merge. Call this mode of Merge application "Pot-Merge." This yields a right-branching structure, and the resulting expression means something like "a tea cup which is green." Here the attractor is uniformly EF(cup), which fact is reflected in the way this structure is Labeled. In (13b), on the other hand, one first Core-Merges {green, tea}, which is then Merged with {cup}, in a manner parallel with the Subassembly strategy. Call this mode of Merge application "Sub-Merge." The resulting structure is left-branching, and the expression as a whole now means "a cup for green tea." In this case, we have two distinct attractors, EF(tea) for the first Merge and EF(cup) for the second Merge. The number of possible derivations with Sub-Merge multiplies as we add objects to combine. Thus *green tea cup cleaner* can be {{{green, tea}, cup}, cleaner}, {{green, {tea, cup}}, cleaner}, {green, {{tea, cup}, cleaner}} or even {{green, tea}, {cup, cleaner}}, though the last one may not be semantically very natural.

The difference between Pot-Merge and Sub-Merge does seem to reflect some important aspect of human syntax, as corroborated by Roeper and Snyder's (2005) observation concerning crosslinguistic variation in possible root compounding patterns. They report that while English allows both right-branching and left-branching compound formation (*child book club* can be either [child [book club]] or [[child book] club]), Swedish allows only the right-branching structure (*barn bok klub* is always [barn [bok klub]]). Without discussing Roeper & Snyder's elaborate account of this discrepancy, here let us just note that the peculiar behavior of Swedish compounding is compatible with the supposition that Sub-Merge is computationally different from, and probably more complex than, Pot-Merge, echoing the putative human monopoly of the Subassembly strategy in Action Merge: After all, it is the last strategy to emerge in child development, at around the age of 20 months (Conway and Christiansen 2001). Needless to say, the fact that Swedish bans the application of Sub-Merge in compound formation does not mean at all that the language lacks it altogether; otherwise, Swedish would have virtually no phrasal syntax. This is so since even the phrase *saw the boy* requires Sub-Merge for its derivation (*saw* gets Merged with the subassembly *the boy*). In fact, Sub-Merge is the nuts and bolts of the recursive syntax of human language, and Pirahã is no exception in this respect.

Throughout the history of generative grammar, certain nodes or phrases have been known to block extraction from within, and have been subject to different forms of formulation under the rubrics of islands, barriers, phases, and so on. The notion of a phase is particularly interesting to us, as it functions as a subassembly in the derivational process. The Phase Impenetrability Condition (PIC; Chomsky 2001, 2008), whatever its precise definition may turn out to be, is very presumably a reflection of the fact that a derivational subassembly, once completed, cannot be probed into by later operations. On the face of it, PIC is a highly language-specific principle that appears to defy any deeper explanation, but when seen this way, it may turn out that PIC can be given a natural place in the evolution of human cognition and language.

5 Solving the Syntax Puzzle

The discussion in the last section amply demonstrated the formal parallelism between Action Merge and Merge. The kind of experimental work made on children's ability to build hierarchical mobiles of plastic straws (Greenfield and Schneider 1977) further supports this position (P. Greenfield, p.c.). This alone, of course, is not evidence for the evolutionary and/or developmental link between action and syntax, nor does it prove that Action Merge is the precursor to Merge, an issue that needs to be explored and settled by multidisciplinary endeavors. As a matter of fact, it is also a logical possibility that Action Merge and Merge are both descendants of an older common precursor or, as pointed out by Tom Roeper (p.c.), that they represent separate evolutionary steps. While I have to leave this issue open for future research, I stress the point again that by minimizing the superficial complexity of human language syntax to recursive Merge, biolinguistic minimalism succeeds in rendering syntax amenable to a comparative ethological analysis, which is itself a great contribution to the progress of evolutionary and developmental studies of language.

Chomsky (2009: 30) remarks: "For both evolution and development, there seems little reason to postulate precursors to unbounded Merge." In his view, Merge is unbounded from the beginning. Everyone understands that young children go through the developmental stages of one- and two-word utterances before they exhibit the full expressive power of unboundedly recursive Merge. This is primarily due to their limitation in Merge-independent cognitive and physical capacities, and does not speak against the innate nature of unbounded Merge.

In evolution, however, it might seem more natural to suppose a transitional process from bounded to unbounded Merge, a transition made possible by various factors including the enhancement of working memory in the enlarged brain. To the extent that Action Merge can be linked to the emergence of Merge in the species, one might be inclined to take it to be the precursor to bounded (rather than unbounded) Merge, given the observation that Action Merge is obviously bounded.

It is worth noting, however, that the distinction between bounded and unbounded Merge is made largely for theoretical purposes. In actual practice, recursive application of Merge is of course bounded for a variety of familiar reasons. But if the unbounded recursiveness of Merge holds only in principle, the same can be said of Action Merge, too. The huge gap between unbounded syntactic structure building and severely bounded cup nesting may in fact be only apparent, and it is possible that the path from Action Merge to Merge was not overwhelmingly distant for H. sapiens fully armed with other unique biological capacities.

Marc Hauser, in his recent discussion of what he claims are the four ingredients of "humaniqueness," points out a crucial difference between animal tools and human tools as a reflection of the mental gap between animals and humans; while animal tools are usually "composed of a single material, designed for a single function," human tools are generally composed of "different materials, each with a particular functions, all wrapped up into a single tool" (Hauser 2009: 48). This is strongly reminiscent of evolutionary psychologists' Swiss Army Knife model of the human mind; we may even conjecture that the army knife itself is a product of the army knife-like massively modular architecture of the human cognition.

I would like to suggest that subassembling is a key factor that gives rise to this uniquely human mental make-up. Each tool part, made separately for serving a different purpose, represents a subassembly to be integrated with others into a single whole unit. Sub-Merge is a linguistic counterpart of this mental computational process involved in uniquely human tool making. Hauser goes on to argue that while recursion in animals may be fragmentally observed by "watching their motor systems in action," our unique recursive thinking evolved by the "release of recursion from its motor prison to other domains of thought" (p. 49). In other words, what was once motor recursion was later extended to cognitive recursion in the human evolution, including but not limited to Merge. This is a nice illustration of descent with modification, and the evolutionary scenario from Action Merge to Merge accords well with this general picture. Empirical validation of this scenario towards the goal of solving the syntax puzzle will be one major topic for the biolinguistic studies of language evolution.

A particularly informative clue to the evolutionary link between tools and syntax may be obtained from recent developments of cognitive archaeology and neuroarchaeology, where the evolution of stone tool making serves as a good indicator of the evolution of human cognitive capacities including language and their neural correlates. Moore (2010), for example, proposes to analyze different modes of stone flaking in terms of Action Grammar and argues that "the most complex action grammars arose after 270 kya" (p. 13). This in turn suggests, to the extent that Action Grammar (Action Merge) is the precursor to syntactic Merge, the latter came into being only more recently, in consistency with the popular view that human language emerged about 100–50 kya, somewhere in the evolutionary history of H. sapiens.

Chomsky (2008: 137) speculates that "... some ancestor, perhaps about 60,000 years ago, underwent a slight mutation rewiring the brain, yielding unbounded Merge." What was this brain rewiring like, and how did it yield Merge? My conjecture is that the brain region originally dedicated to Action Merge later extended to subserve syntactic Merge. Faisal et al. (2010), based on their functional brain imaging studies, observe that activation of left PMv (ventral premotor cortex) is detected both in the simpler "Oldowan" toolmaking (which began with H. habilis about 2.6 mya) and in the more sophisticated "Acheulean" toolmaking (which began with H. erectus about 1.7 mya), which result may be understood as pointing to the importance of this area (approximately BA 6) as the neural correlate of the precursor to Merge. The suggested rewiring then may have taken place from BA 6 to the adjacent BA 44/45 (Broca's area), thereby extending the capacity of physical combination to that of mental/symbolic combination. Yusa (2012) goes further to argue on the basis of neuroimaging data that BA 45 is the locus for domain-specific syntactic Merge while BA 44 implements domain-general Merge for both action and syntax (see also Fujita 2012), which, if correct, helps fine-tune Friederici et al.'s (2006) experimental studies focusing on BA 44's activation in the processing of hierarchical structure (see also Berwick et al. 2013). Other scenarios readily suggest themselves, and in the absence of more conclusive evidence for the moment, I will refrain from discussing them.

6 Solving the C-I Interface Puzzle

The understanding that recursive Merge, although in and of itself a highly domain-specific computational device unique to human language, is nevertheless an evolutionary descendant of a capacity originating in a language-independent motor cognitive domain, provides a valuable research guideline for exploring the evolution of another innovative component of the human language faculty; the interfaces. Recall that how syntax came to be connected to semantics and phonology (the interface puzzle) constitutes one major component of the logical problem of language evolution (Sect. 1). In this section, I will limit my discussion to the syntax-semantics interface (the Conceptual-Intentional or C-I interface) and argue that we can trivialize the C-I interface puzzle to the extent that we view our conceptual capacity largely as an evolutionary descendant of recursive Merge. A similar argument may apply to the syntax-phonology interface (the sensory-motor or S-M interface), too.

Recent studies on the C-I interface in biolinguistics may be properly characterized in terms of the naturalization, or *biologization*, of semantics, as boosted by Uriagereka (2008), Hinzen (2006, 2008), and Pietroski (2005, 2008) (see also Narita 2009b for an overview). The basic idea behind "biosemantics" (Hinzen 2008) may be summarized as follows: Given that the Merge-based recursive syntax is a natural/biological object, semantics is also a natural/biological object insofar as it is determined ("carved out," as Uriagereka puts it) by syntax. In other words, the biosemantic program is an attempt to understand semantics as part of "biosyntax."

Suppose, for example, that syntactic structure building by Merge is directly linked to semantic structure building. This is not to say that syntax is regulated by, or works for, semantics, but rather that an autonomous syntactic computation circumscribes possible conceptual structure formation. We may say that syntax is a blind watchmaker, in the sense that syntax is blind to semantics but still it is able to yield a highly elaborate format for semantic interpretation.

That this is indeed true at least for core cases is easy to confirm if we take note of the virtual isomorphism between the multi-layered VP structure adopted in minimalist syntax and the corresponding (lexical) conceptual structure, a possibility once pursued, though on totally different theoretical grounds, by Generative Semantics in the mid-1960s by means of lexical decomposition. As a well-studied exemplary case, consider the double object verb *give* and its prepositional dative object counterpart with their VP-structures and purported conceptual structures:

(14) a. Mary gives John a book.
b. [$_{vP}$ Mary *v* [$_{VP}$ John V a book]]
c. [Mary CAUSE [John HAVE a book]]

(15) a. Mary gives a book to John.
b. [$_{vP}$ Mary *v* [$_{VP}$ a book V John]]
c. [Mary CAUSE [a book GO-TO John]]

If, as often argued, *v* can be equated with an abstract causative verb and V as a verb of possession (14b) and of movement (15b), the conceptual structures (14c) and (15c) are automatically read off of these syntactic structures, which renders the syntax-to-semantics mapping trivial. The C-I interface puzzle will be solved accordingly; in effect, there will be no C-I interface and therefore no puzzle to be solved (cf. Hinzen 2011). These examples illustrate the thematic, or internal, side of what Chomsky (2008 and elsewhere) calls the duality of semantics (External Merge defines θ-structure), and the other, nonthematic side related to externalization (Internal Merge affects discoursial and scopal interpretation) can be readily accommodated in a similar fashion.

I argued elsewhere (Fujita 1996), based on an even more detailed split VP structure, that certain peculiar syntactic and semantic discrepancies observed between double object and dative object verbs may be explained in purely syntactic terms. Works along that line, in retrospect, were a good precedent for current biosemantic studies. When Merge is extended or exapted to apply to semantic units, syntax becomes the tool for conceptualization, and it makes every sense to say that the uniquely human conceptual capacity is made possible by Merge. Take, for example, the familiar case of Agent-Causer alternation in the thematic interpretation of the subject as in (16a, b).

(16) a. Mary (intentionally) gave John a book.
b. The exam gave John a headache.

This type of thematic alternation is observed with causative verbs in general (*John {intentionally/inadvertently} broke the vase*). Fujita (1996) proposed to treat this property by projecting Agent and Causer onto different subject positions in a three-layered VP structure as in (17) (ignoring the then-extant AgrOP projection):

(17) $[_{VP1}$ Agent V1 $[_{VP2}$ Causer V2 $[_{VP3}$ V3 ...]]]

Here V2 is the verbalizer and the abstract causative head, while V1 licenses the Agentive external argument. This allowed me to explain the observation (Pesetsky 1995) that backward binding is possible when the subject has the Causer reading but not when it has the Agent reading, on the assumption that the binding object has to move to a position between V1 and V2 (a position from which it c-commands the Causer but not the Agent) for Case-theoretic reasons.

(18) a. $[_{VP1}$ Agent V1 [Obj $[_{VP2}$ Causer V2 $[_{VP3}$ V3 (Obj)]]]]
b. * Each other's friends hit John and Mary.
c. Each other's pictures amused John and Mary.

It also allowed me to derive the fact that prepositional dative object verb *give*, in contrast to the double object version, lacks the Causer subject reading (**The exam gave a headache to John*; Oehrle 1976), from the assumption that minimality blocks the Case-driven movement of the object crossing the Causer subject. (See Bruening (2010) for a recent detailed discussion on this restriction and an explanation of apparent counterexamples to it.)

It is interesting to note that current morphosyntactic studies centering on VP and Voice suggest that something like (17) was indeed on the right track. Harley (2013), for example, tries to show based on an analysis of Hiaki verbal morphology that the external argument is not introduced by the causative *v* but by the higher Voice head:

(19) $[_{VoiceP}$ Subj Voice $[_{ApplP}$ Appl $[_{vP} v$ $[_{VoiceP}$ Voice $[_{VP}$ ...]]]]] (adapting Harley's (32))

Details aside, the proposals in (17) and (19) share the view that Agent subject is truly external to the canonical argument structure realized by the core VP structure. Correspondingly, one can build a conceptual structure like (20) for (14a), by applying Merge recursively, with Internal Merge creating the higher occurrence of *Mary*.

(20) [Mary DO [Mary CAUSE [John HAVE a book]]]

Other, more complex conceptual structures which are hierarchically organized can be equally constructed by Merge. In short, it may be envisaged that syntax is the foundation for human conceptualization and thought (see also Calvin 1996), and that Merge is one major determinant of how our mind works.

It is sometimes objected that the kind of abstract syntax described above is more complex than necessary, and goes against the minimalist spirit. Most notably, Culicover and Jackendoff (2005, 2006) criticize the hierarchical split VP structure in favor of a flat VP structure [V NP1 NP2] adopted in their "simpler syntax" framework. Compare:

(21)
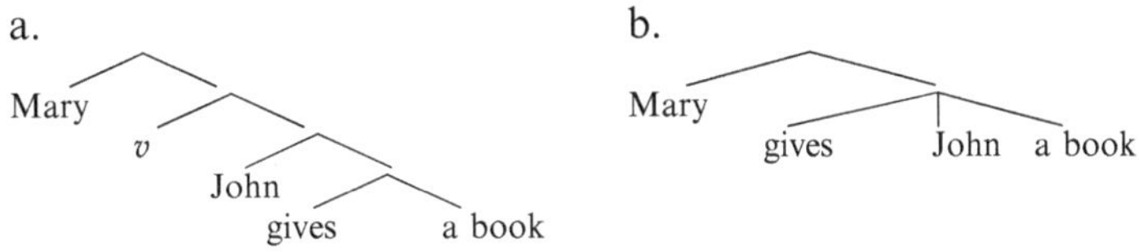

I should point out two things in response to their claim. Firstly, simplicity in this case is a theory-dependent notion and there is no direct comparison between two theories of syntax if they adopt different notions of simplicity. C&J adopt a nonderivational, purely representational theory of syntax, according to which structural simplicity is largely determined by the number of sub-constituents and invisible structures (Culicover and Jackendoff 2006). In light of the Merge-based derivational theory, however, their tripartite VP structure requires something more than the elementary binary combinatorial operation and is certainly computationally more complex. In short, simplicity is in the eye of the beholder, and the real question is not so much which structure is simpler, as which notion of simplicity is more likely to offer a profound insight in approaching one's target of inquiry.

And that leads to my second point. Insofar as one's major concern lies in solving the C-I interface puzzle, it is most likely that the Merge-based derivational theory is the simpler, or more explanatory, candidate, because it allows one to solve it in the most effective way. This view is particularly helpful for evolutionary studies because it encourages us to explore the possibility that at least part of the uniquely human conceptual capacity emerged from Merge combining and recombining preexisting concepts.

(22)
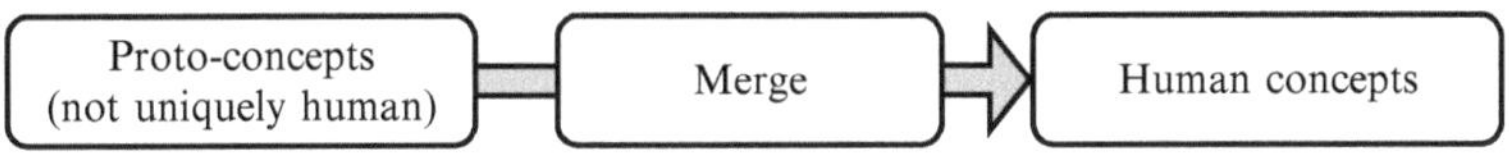

On the other hand, the C&J model of mutually autonomous syntax and semantics needs to elucidate independent evolutionary paths for syntax and semantics, plus the one for their later interlocking. The "simpler syntax" is not simple enough in terms of evolutionary adequacy because it has to explain the evolutionary origins of syntactic structure, conceptual structure and their linking independently of each other.

More importantly, however, on the basis of our hypothesis that syntax made human semantics possible, we can go even further to take seriously the possibility that syntax gave rise to the human lexical capacity, too. Within generative grammar, there has been a long tradition of "lexicalism," according to which word formation takes place not in syntax but in a separate module of the lexicon. We also have a sharply different view which may be called "anti-lexicalism"; it attempts to dispense with the lexicon and handle word formation in syntax. Distributed Morphology (Marantz 1997, et seq.) is one representative approach of this latter position, but many other proposals address the fundamental issue of to what extent the lexicon can be reduced to syntax, including Hale and Keyser's (1993 and elsewhere) "l-syntax" approach to argument structure inter alia.

Indeed, this line of exploration can be traced back to Chomsky's (1970) "lexicalist position," which, quite contrary to the commonly misrepresented view that derived nominals are listed as nouns in the lexicon, argues that they can be syntactically accommodated in the categorial component of the base and that lexical items can be left categorially unspecified within the lexicon (see Marantz 1997). (23a) describes the relevant part of the model of grammar at that time. Compare it with (23b), the present-day Merge-based bare phrase structure model:

(23) a.

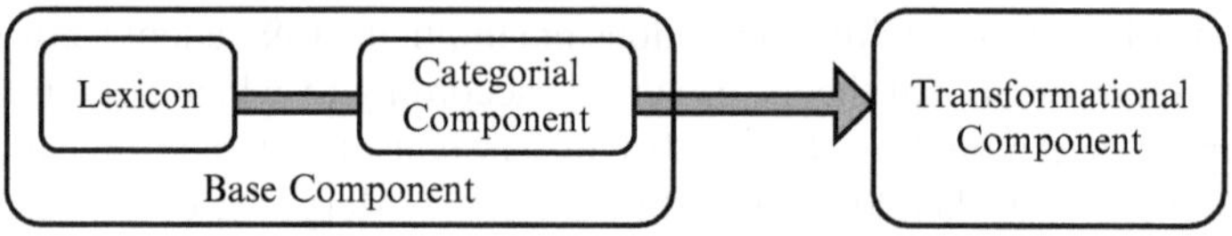

b.

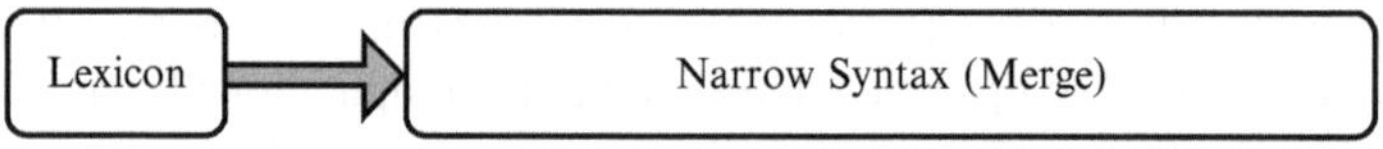

The distinction between the categorial component and the transformational component disappears in (23b) and both are integrated into narrow syntax. It is only natural to assume, then, that what once belonged to the categorial component (such as nominalization) now belongs to narrow syntax, not to the lexicon. Such being the case, it can safely be said that anti-lexicalism, and not lexicalism, constitutes the default hypothesis for minimalist theorizing.

Whether or not the lexicon needs to exist as an independent module in our faculty of language remains to be seen, but my contention here is that anti-lexicalism should also serve as a productive research strategy for studies of language evolution. Suppose the lexicon does exist; then we immediately face the question of how it evolved, and anti-lexicalism offers one possible scenario that deserves careful scrutiny. Specifically, the uniquely human rich and generative lexicon may have evolved from its more restricted predecessor by means of Merge.

(24)

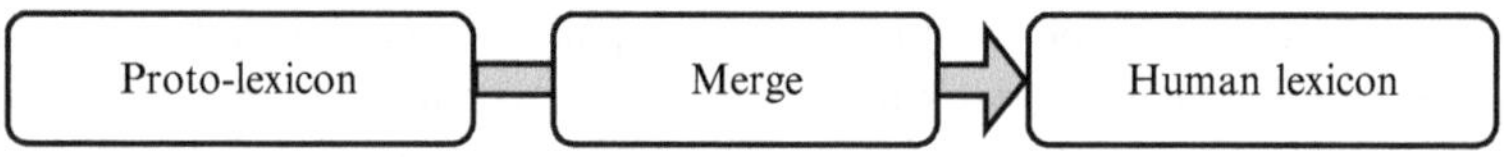

Suppose our ancestors initially had a very limited set of rudimentary word-like symbols, each a direct linking of sound and meaning. How could this small repertoire of symbols be turned into such a system as can generate an infinite array of words which connect sound and meaning only indirectly via our minds? By Merge, I believe. By mentally decomposing those proto-words into atomic units (concepts and sounds) and reorganizing and recombining them into more articulate and elaborate word-like expressions, one can enhance one's lexical power, which would not be too difficult if one had a recursive combinatorial operation like Merge and especially if words are subassemblies for structure building to be constructed by Sub-Merge.

To the extent that something like this conjecture is on the right track, the emerging picture is that words and sentences are equally generated by Merge, which renders the traditional dichotomy between syntax and the lexicon almost meaningless. Merge is what brought human language into existence, in perfect harmony with the Strong Minimalist Thesis. I am in utmost agreement with Piattelli-Palmarini (2010: 161), who, noting the importance of the syntactic nature of words for the purpose of studies of language evolution, states: "The no-escape-from-syntax lesson ... should redirect a more productive inquiry into the evolution of language." As Berwick (2011) puts it, all you need is Merge. And there is no escape from Merge.

All this is not to deny the autonomy of syntax and (lexical) semantics altogether; after all, apart from the suggested core domains (θ-related and discoursial interpretations), there are many properties which are exclusively syntactic or semantic, and conceptual structure should contain much more than can be automatically read off of syntactic computation. But that kind of modular autonomy evolves and develops through the usual process of descent with modification (cf. Marcus 2006) and need not be postulated from the start. To say the least, we can move a step further towards a better understanding of language evolution by not disregarding the syntactic origin of the human conceptual capacity. Here lies the advantage of adopting the biosemantic approach to evolutionary studies.

Last but not least, the above discussion suggests that we need to take a more flexible attitude towards the simple dichotomy of FLN and FLB; it no longer makes sense just to say that only recursion (recursive Merge) belongs to FLN. Instead, we should ask which parts of the C-I system require Merge and therefore belong to FLN, and which other parts have to be sought and found somewhere else in the broader domain of primate (and nonprimate) cognition.

7 Concluding Remarks

In this paper, I have proposed two lines of productive research for studies of language evolution. First, on the assumption that Merge is the single elementary computational device of the human language faculty subsuming both Move and Label, the search for its evolutionary precursor(s) has to be done in the broader context of comparative cognitive ethology. Greenfield's work on Action Grammar provides an excellent testing ground for such a multidisciplinary endeavor. The idea that Merge is an evolutionary (and developmental) descendant of the capacity of sequential object manipulation (Action Merge) deserves serious investigation. Next, the evolution of the C-I interface can be reduced to a trivial issue if we can show, following the biosemantic program, that the core aspects of the human conceptual capacity derive from (bio)syntax, instead of assuming that there was already such a capacity before the emergence of Merge, which somehow had to evolve in order to establish the linking between syntax and semantics. These proposals are at best speculative at the present moment and have to be consolidated by works from every corner of cognitive and biological sciences, which is exactly why biolinguistics has to occupy a central position in modern scientific enterprise.

Acknowledgment The research presented here has been partially supported by the Japan Society for the Promotion of Science Grant-in-Aid for Scientific Research (A) (Grant No. 23242025). At various stages, I have benefited from personal communications with Cedric Boeckx, Anna Maria Di Sciullo, Naoki Fukui, Patricia Greenfield, Takashi Hashimoto, Masayuki Ike-uchi, Roger Martin, Shigeru Miyagawa, Hiroki Narita, Kazuo Okanoya, Masanobu Ueda, Juan Uriagereka and Noriaki Yusa. I would like to express my deep gratitude to Tom Roeper and Peggy Speas as well as to anonymous reviewers for their critical comments on earlier versions of this paper. Usual disclaimers apply.

References

Abe, K., & Watanabe, D. (2011). Songbirds possess the spontaneous ability to discriminate syntactic rules. *Nature Neuroscience, 14*, 1067–1074.

Berwick, R. C. (2011). All you need is Merge: Biology, computation, and language from the bottom up. In A. M. Di Sciullo & C. Boeckx (Eds.), *The biolinguistic enterprise: New perspectives on the evolution and nature of the human language faculty* (pp. 461–491). Oxford: Oxford University Press.

Berwick, R. C., Friederici, A. D., Chomsky, N., & Bolhuis, J. J. (2013). Evolution, brain, and the nature of language. *Trends in Cognitive Sciences, 17*(2), 89–98.

Bickerton, D. (2009). Recursion: Core of complexity or artifact of analysis? In T. Givón & M. Shibatani (Eds.), *Syntactic complexity: Diachrony, acquisition, neuro-cognition, evolution* (pp. 531–543). Amsterdam: John Benjamins.

Biro, D., Sousa, C., & Matsuzawa, T. (2006). Ontogeny and cultural propagation of tool use by wild chimpanzees at Bossou, Guinea: Case studies in nut cracking and leaf folding. In T. Matsuzawa, M. Tomonaga, & M. Tanaka (Eds.), *Cognitive development in chimpanzees* (pp. 476–508). Tokyo: Springer.

Boeckx, C. (2009). The nature of merge: Consequences for language, mind, and biology. In M. Piattelli-Palmarini, J. Uriagereka, & P. Salaburu (Eds.), *Of minds & language: A dialogue with Noam Chomsky in the Basque country* (pp. 44–57). Oxford: Oxford University Press.

Bruening, B. (2010). Double objects constructions disguised as prepositional datives. *Linguistic Inquiry, 41*(2), 287–305.

Calvin, W. H. (1996). *How brains think: Evolving intelligence, then and now*. New York: Basic Books.

Chomsky, N. (1970). Remarks on nominalization. In R. A. Jacobs & P. S. Rosenbaum (Eds.), *Readings in English transformational grammar* (pp. 184–221). Waltham: Ginn.

Chomsky, N. (1995). *The minimalist program*. Cambridge, MA: MIT Press.

Chomsky, N. (2001). Derivation by phase. In M. Kenstowicz (Ed.), *Ken Hale: A life in language* (pp. 1–52). Cambridge, MA: MIT Press.

Chomsky, N. (2005). Three factors in language design. *Linguistic Inquiry, 36*(1), 1–22.

Chomsky, N. (2008). On phases. In R. Freidin, C. P. Otero, & M. L. Zubizarreta (Eds.), *Foundational issues in linguistic theory: Essays in honor of Jean-Roger Vergnaud* (pp. 133–166). Cambridge, MA: MIT Press.

Chomsky, N. (2009). Opening remarks. In M. Piattelli-Palmarini, J. Uriagereka, & P. Salaburu (Eds.), *Of minds & language: A dialogue with Noam Chomsky in the Basque country* (pp. 13–43). Oxford: Oxford University Press.

Chomsky, N. (2010). Some simple evo devo theses: How true might they be for language? In R. K. Larson, V. Déprez, & H. Yamakido (Eds.), *The evolution of language: Biolinguistic perspectives* (pp. 45–62). Cambridge: Cambridge University Press.

Conway, C. M., & Christiansen, M. H. (2001). Sequential learning in non-human primates. *Trends in Cognitive Sciences, 5*(12), 539–546.

Culicover, P. W., & Jackendoff, R. (2005). *Simpler syntax*. Oxford: Oxford University Press.

Culicover, P. W., & Jackendoff, R. (2006). The simpler syntax hypothesis. *Trends in Cognitive Sciences, 10*(9), 413–418.
Everett, D. L. (2005). Cultural constraints on grammar and cognition in Pirahã. *Current Anthropology, 46*(4), 621–646.
Everett, D. L. (2009). Pirahã culture and grammar: A response to some criticisms. *Language, 85*(2), 405–442.
Faisal, A., Stout, D., Apel, J., & Bradley, B. (2010). The manipulative complexity of Lower Paleolithic stone toolmaking. *PLoS One, 5*(11), e13718.
Fitch, W. T., Hauser, M. D., & Chomsky, N. (2005). The evolution of the language faculty: Clarifications and implications. *Cognition, 97*(2), 179–210.
Friederici, A. D., Bahlmann, J., Heim, S., Schubotz, R. I., & Anwander, A. (2006). The brain differentiates human and non-human grammars: Functional localization and structural connectivity. *PNAS, 103*(7), 2458–2463.
Fujita, K. (1996). Double objects, causatives, and derivational economy. *Linguistic Inquiry, 27*(1), 146–173.
Fujita, K. (2002). Review of *Biolinguistics: Exploring the biology of language* by Lyle Jenkins. *Gengo Kenkyu* [Linguistic Research], *121*, 165–178.
Fujita, K. (2007). Facing the logical problem of language evolution. *English Linguistics, 24*(1), 78–108.
Fujita, K. (2009). A prospect for evolutionary adequacy: Merge and the evolution and development of human language. *Biolinguistics, 3*(2–3), 128–153.
Fujita, K. (2012). The evolution of syntactic computation and the language faculty. In K. Fujita & K. Okanoya (Eds.), *Constructing evolutionary linguistics* (pp. 55–75). Tokyo: Hituzi Shobo (in Japanese).
Fukui, N. (2011). Merge and bare phrase structure. In C. Boeckx (Ed.), *The Oxford handbook of linguistic minimalism* (pp. 73–95). Oxford: Oxford University Press.
Gentner, T. Q., Fenn, K. M., Margoliash, D., & Nusbaum, H. C. (2006). Recursive syntactic pattern learning by songbirds. *Nature, 440*, 1204–1207.
Gould, S. J., & Vrba, E. S. (1982). Exaptation: A missing term in the science of form. *Paleobiology, 8*, 4–15.
Greenfield, P. M. (1991). Language, tools, and brain: The ontogeny and phylogeny of hierarchically organized sequential behavior. *Behavioral and Brain Sciences, 14*(4), 531–595.
Greenfield, P. M. (1998). Language, tools, and brain revisited. *Behavioral and Brain Sciences, 21*(1), 159–163.
Greenfield, P. M., & Schneider, L. (1977). Building a tree structure: The development of hierarchical complexity and interrupted strategies in children's construction ability. *Developmental Psychology, 13*(4), 299–313.
Greenfield, P. M., Nelson, K., & Saltzman, E. (1972). The development of rulebound strategies for manipulating seriated cups: A parallel between action and grammar. *Cognitive Psychology, 3*(2), 291–310.
Hale, K., & Keyser, S. J. (1993). On argument structure and the lexical expression of syntactic relations. In K. Hale & S. J. Keyser (Eds.), *The view from building 20: Essays in linguistics in honor of Sylvain Bromberger* (pp. 53–109). Cambridge, MA: MIT Press.
Harley, H. (2013). External arguments and the Mirror Principle: On the distinctness of Voice and *v*. *Lingua, 125*, 34–57.
Hauser, M. (2009, September). Origin of the mind. *Scientific American, 301*, 44–51.
Hauser, M. D., Chomsky, N., & Tecumseh Fitch, W. (2002). The faculty of language: What is it, who has it, and how did it evolve? *Science, 298*, 1569–1579.
Hinzen, W. (2006). *Mind design and minimal syntax*. Oxford: Oxford University Press.
Hinzen, W. (2008). Prospects for an explanatory theory of semantics. *Biolinguistics, 2*(4), 348–363.
Hinzen, W. (2011). Emergence of a systemic semantics through minimal and underspecified codes. In A. M. Di Sciullo & C. Boeckx (Eds.), *The biolinguistic enterprise: New perspectives on the evolution and nature of the human language faculty* (pp. 417–439). Oxford: Oxford University Press.

Hornstein, N. (2009). *A theory of syntax: Minimal operations and universal grammar*. Cambridge: Cambridge University Press.
Lashley, K. S. (1951). The problem of serial order in behavior. In L. A. Jeffress (Ed.), *Cerebral mechanisms in behavior: The Hixon symposium* (pp. 112–136). New York: Wiley.
Marantz, A. (1997). No escape from syntax: Don't try morphological analysis in the privacy of your own lexicon. *UPenn Working Papers in Linguistics, 4*(2), 201–225.
Marcus, G. F. (2006). Cognitive architecture and descent with modification. *Cognition, 101*(2), 443–465.
Matsuzawa, T. (2001). Primate foundations of human intelligence: A view of tool use in nonhuman primates and fossil hominids. In T. Matsuzawa (Ed.), *Primate origins of human cognition and behavior* (pp. 3–25). Tokyo: Springer.
Moore, M. W. (2010). "Grammar of action" and stone flaking design space. In A. Nowell & I. Davidson (Eds.), *Stone tools and the evolution of human cognition* (pp. 13–43). Colorado: University Press of Colorado.
Narita, H. (2009a). Full interpretation of optimal labeling. *Biolinguistics, 3*(2–3), 213–245.
Narita, H. (2009b). The naturalist program for neo-Cartesian biolinguisitcs. In T. Shinya & A. Imaoka (Eds.), *Proceedings of the 24th annual meeting of the Sophia University Linguistic Society* (pp. 55–91). Tokyo: Sophia University Linguistic Society.
Oehrle, R. (1976). *The grammatical status of the English dative alternation*. Doctoral dissertation, MIT, Cambridge, MA.
Pesetsky, D. (1995). *Zero syntax: Experiencers and cascades*. Cambridge, MA: MIT Press.
Piattelli-Palmarini, M. (2010). What is language, that it may have evolved, and what is evolution, that it may apply to language. In R. K. Larson, V. Déprez, & H. Yamakido (Eds.), *The evolution of language: Biolinguistic perspectives* (pp. 148–162). Cambridge: Cambridge University Press.
Pietroski, P. M. (2005). *Events and semantic architecture*. Oxford: Oxford University Press.
Pietroski, P. M. (2008). Minimalist meaning, internalist interpretation. *Biolinguistics, 2*(4), 317–341.
Ridley, M. (2004). *Evolution* (3rd ed.). Oxford: Blackwell.
Roeper, T. (2007). *The prism of grammar: How child language illuminates humanism*. Cambridge, MA: MIT Press.
Roeper, T. (2011). The acquisition of recursion: How formalism articulates the child's path. *Biolinguistics, 5*(1–2), 57–86.
Roeper, T., & Snyder, W. (2005). Language learnability and the forms of recursion. In A. M. Di Sciullo (Ed.), *UG and external systems: Language, brain and computation* (pp. 155–169). Amsterdam: John Benjamins.
Stout, D. (2010). Possible relations between language and technology in human evolution. In A. Nowell & I. Davidson (Eds.), *Stone tools and the evolution of human cognition* (pp. 159–184). Colorado: University Press of Colorado.
Swann, P. (1998). Greenfield on language, tools, and brain. *Behavioral and Brain Sciences, 21*(1), 155–159.
Tokimoto, N., & Okanoya, K. (2004). Spontaneous construction of 'Chinese boxes' by Degus (*Octodon degu*): A rudiment of recursive intelligence? *Japanese Psychological Research, 46*(3), 255–261.
Tomasello, M. (1991). Objects are analogous to words, not phonemes or grammatical categories. *Behavioral and Brain Sciences, 14*(4), 575–576.
Uriagereka, J. (2008). *Syntactic anchors: On semantic structuring*. Cambridge: Cambridge University Press.
Visalberghi, E., Addessi, E., Truppa, V., Spagnoletti, N., Ottoni, E., Izar, P., & Fragaszy, D. (2009). Selection of effective stone tools by wild bearded capchin monkeys. *Current Biology, 19*, 213–217.
Yusa, N. (2012). Hierarchical structure and recursive computation in Broca's area. In K. Fujita & K. Okanoya (Eds.), *Constructing evolutionary linguistics* (pp. 77–94). Tokyo: Hituzi Shobo (in Japanese).

Index

T. Roeper and M. Speas (eds.), *Recursion: Complexity in Cognition*, Studies in Theoretical Psycholinguistics 43, DOI 10.1007/978-3-319-05086-7,

MIX
Papier aus verantwortungsvollen Quellen
Paper from responsible sources
FSC® C105338

If you have any concerns about our products,
you can contact us on
ProductSafety@springernature.com

In case Publisher is established outside the EU,
the EU authorized representative is:
Springer Nature Customer Service Center GmbH
Europaplatz 3, 69115 Heidelberg, Germany

Printed by Libri Plureos GmbH
in Hamburg, Germany